recent advances in

phytochemistry

volume 9

XOCHIPILLI

The statue of the Aztec divinity Xochipilli ("Prince of Flowers") shown above was unearthed more than a century ago from the slopes of Popocatepetl, and is now on exhibit in the Museo Nacional de Antropologia, Mexico City. He appears in an attitude of ecstasy, the result of hallucinogenic intoxication. The statue is adorned with sundry stylized designs of hallucinogenic flowers and the sacred mushroom *Psilocybe aztecorum* Heim. This fungus also appears to have been spoken of as a "flower" (xochitl), a metaphor of enhancement in the Nahuatl language spoken by the Aztecs. Cross sections of the mushroom yield the flower-like boss with 6-fold symmetry on the base of the statue and the similar designs with 3-fold symmetry on the knee caps. They are accompanied by butterflies representing souls of departed spirits. The flower of *Heimia salicifolia* is shown on the right leg; this plant induces an hallucinogenic state in which sounds appear to come from great distances in space, and time, and memory. On the left leg appears the emerging morning glory flower *(Rivea corymbosa).* A necklace of flowers is mentioned in Nahuatl poetry and indeed the necklace of the statue may be of morning glory flowers as they appear when closed at dusk. Tendrils and open flower forms occur elsewhere on the statue, along with the flower of *Nicotiana Tabacum* and *Calea Zacatechichi,* a plant used in divination. (Abstracted from R.G. Wasson. 1973. Botanical Museum Leaflets (Harvard) 23: 8.

recent advances in phytochemistry

volume 9

Edited by

V. C. Runeckles

Department of Plant Science
The University of British Columbia
Vancouver, British Columbia, Canada

PLENUM PRESS • NEW YORK AND LONDON

Library of Congress Catalog Card Number 67-26242
ISBN 0-306-34709-1

Proceedings of the Fourteenth Annual Meeting of the Phytochemical Society of North America held in August 1974 at Western Carolina University in Cullowhee, North Carolina

A Division of Plenum Publishing Corporation
227 West 17th Street, New York, N.Y. 10011

United Kingdom edition published by Plenum Press, London
A Division of Plenum Publishing Company, Ltd.
Davis House (4th Floor), 8 Scrubs Lane, Harlesden, London, NW10 6SE, England

Printed in the United States of America

LIST OF CONTRIBUTORS

K.D. Brunnemann, American Health Foundation, New York, New York.

James A. Duke, Plant Taxonomy Laboratory, Plant Genetics & Germplasm Institute, U.S. Department of Agriculture, Beltsville, Maryland.

G.B. Gori, National Cancer Institute, Bethesda, Maryland.

D. Hoffmann, American Health Foundation, New York, New York.

Joseph Kuć, Department of Plant Pathology, University of Kentucky, Lexington, Kentucky.

S. Morris Kupchan, Department of Chemistry, University of Virginia, Charlottesville, Virginia.

John C. Mitchell, Division of Dermatology, University of British Columbia, Vancouver, British Columbia.

L.A. Mitscher, Division of Natural Products Chemistry, The Ohio State University, Columbus, Ohio.

Koji Nakanishi, Department of Chemistry, Columbia University, New York, New York.

Charlotte Ressler*, Division of Protein Chemistry, Institute of Muscle Disease, and Department of Biochemistry, Cornell University Medical College, New York, New York.

Richard Evans Schultes, Botanical Museum, Harvard University, Cambridge, Massachusetts.

---

*Present address: Department of Pharmacology, Schools of Medicine & Dental Medicine, University of Connecticut Health Center, Farmington, Connecticut.

A. Ian Scott, Sterling Chemical Laboratory, Yale University, New Haven, Connecticut.

Monroe E. Wall, Research Triangle Institute, Research Triangle Park, North Carolina.

E.L. Wynder, American Health Foundation, New York, New York.

PREFACE

For centuries it has been recognized that plants relate to human health and well-being in many ways beyond their fundamental role as primary sources of food and energy. Many of the unique plant constituents have pronounced effects on animal systems or in the human body; some of them are potentially harmful and represent a risk in the use of a particular plant or in the exposure to it, others are useful as medicinal agents in the treatment of diseases. Many of the latter are extracted from plant materials on a large scale for marketing as drugs and even more of them have served as structural prototypes which inspired chemists to synthesize analog drugs with even more desirable properties. Clearly, today's drug therapy had its origins in the exploration and exploitation of pharmacologically active plant constituents. It is therefore appropriate that a symposium of the Phytochemical Society of North America was devoted to this subject.

The present volume consists of eleven papers dealing with various aspects of the topic "Phytochemistry as Related to Disease and Medicine", which were presented at the Fourteenth Annual Meeting of the Phytochemical Society of North America held in August 1974 at Western Carolina University in Cullowhee, N.C. Plant hallucinogens are the subject of the first three chapters. The first, by Schultes, reviews the occurrence of hallucinogenic agents in plants, in tabular form. Emphasis is on plants from Central and South America. The next two chapters deal with marijuana and its constituents; Wall discusses the chemistry and metabolism of the cannabinoids while Hoffmann and his co-workers present the results of a study of the comparative carcinogenicity of the smokes of tobacco and marijuana cigarettes. The fourth chapter, by Duke, speculates on the roles of plant chemistry in folk medicine, and utilizes the 1000 crop matrix system developed by the U.S. Department of Agriculture.

Other harmful or potentially harmful effects of plant constituents are reviewed in the next chapters. Mitchell describes the subject of allergic responses to plants and plant constituents, and Kuć weighs the evidence for and against the occurrence of teratogenic compounds in potatoes infected with *Phytophthora infestans*. Ressler then discusses the plant neurotoxins which cause lathyrism. The search for new medicinal agents in plants is the focus of other chapters; one, by Kupchan, on plant antitumor agents is followed by a discussion by Scott of recent work on the biomimetic synthesis of indole alkaloids, a group of compounds which have come to prominence because of the antitumor activity of vinblastine and vincristine. In turn, this is followed by the extensive review by Mitscher of antimicrobial agents from higher plants. Finally, a contribution by Nakanishi reports the structure of azadirachtin, an insect feeding deterrent of plant origin, thus touching upon an indirect influence which plants could have on human health through control of insect populations.

We would like to thank the authors as well as many others who have been involved in the preparation of this volume. Our thanks go to the following companies whose financial contributions towards the cost of this symposium are greatly appreciated:

Abbott Laboratories, Pharmaceutical Products Division
Hoffman-LaRoche, Inc.
Ortho Pharmaceutical Corp.
Syntex (U.S.A.), Inc.
The Upjohn Company
Wyeth Laboratories, Inc.

A special word of appreciation must go to Mrs. Karen Veilleux who retyped the whole volume and gave birth to a child and the final typescript almost simultaneously.

Victor C. Runeckles
Heinz G. Floss
Kenneth R. Hanson

CONTENTS

Chapter One

PRESENT KNOWLEDGE OF HALLUCINOGENICALLY USED PLANTS: A TABULAR STUDY

RICHARD EVANS SCHULTES

Botanical Museum, Harvard University
Cambridge, Massachusetts

INTRODUCTION

Of the many plants known to have hallucinogenic activity, only a few have ever been purposefully employed in magico-medical, religious or other ceremonies. The reasons for this parsimonious use of psychoactive species is not wholly clear, but it is probable that man has learned by trial and error that some hallucinogenic species are otherwise too toxic for safety. Whatever the reason, it will be clear from the following pages that the chemistry even of those hallucinogenic plants valued highly in primitive societies is often unknown. Furthermore, it will be obvious that many of these hallucinogens have only very recently been botanically identified.

Inasmuch as I have published numerous detailed papers — both technical and popular — and have been a joint author of a book on hallucinogens, it has occurred to me that the most serviceable way of summarizing our present knowledge might be in tabular form. It is my hope that data thus set out may be more easily and more quickly consulted. The four hallucinogenic species mentioned in connection with the fronticepiece to this book are included in the illustrations (indicated in the Tables by an asterisk against the plate number). There is appended a relatively representative bibliography designed to help the specialist easily to find more information than that contained in the summarized tables.

PLATE 1. *Justicia pectoralis* Jacquin var. *stenophylla* Leonard.

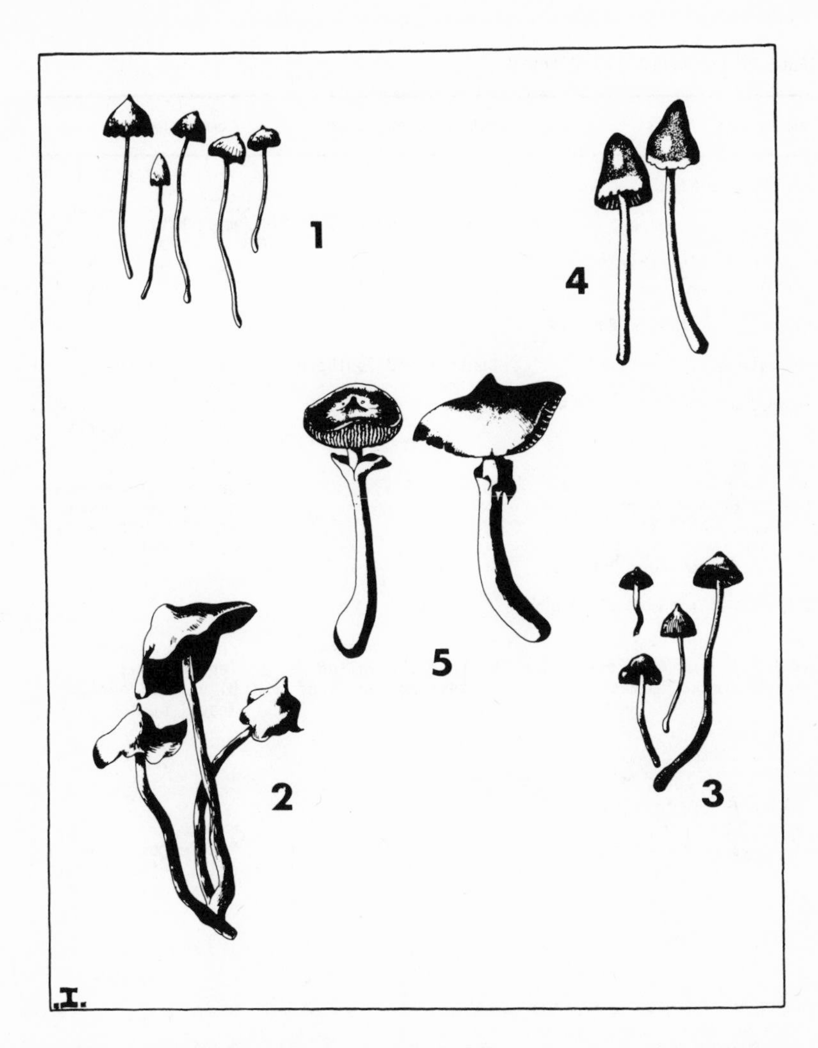

PLATE 2. Several mushrooms reported as hallucinogenic agents in Mexico.

1. *Psilocybe mexicana* Heim
2. *P. zapotecorum* Heim
3. *P. semperviva* Heim & Cailloux
4. *Panaeolus sphinctrinus* Fries
5. *Stropharia cubensis* Earle

(Drawn from Heim: Champignons toxiques et hallucinogènes)

TABLE OF NEW WORLD HALLUCINOGENS

Family and Plant	Region Where Used	Common Names
ACANTHACEAE (Acanthus Family)		
Justicia pectoralis var. *stenophylla* [and possibly other species] PLATE 1	Upper Orinoco-Rio	mashihiri
AGARICACEAE (Agaric Family)		
Conocybe siliginoides *Panaeolus sphinctrinus* PLATE 2 *Psilocybe aztecorum; caerulescens; Hoogshagenii; mexicana; mixaeensis; Wassonii; zapotecorum; et al. spp.* PLATE 2* *Stropharia cubensis* PLATE 2	Central and Southern Mexico	teonanacatl; teyhuintli; tehuintlinanacatl (anciently); siwatsitsintli; cui-ya; nti-ni-se; nti-si-tho-ye-le-nta-ha; ya-nte; nti-ki-so; t-ha-na-sa; to-shka
CACTACEAE (Cactus Family)		
Lophophora Williamsii (Echinocactus Williamsii; Anhalonium Williamsii)	Central and Northern Mexico; the United States; and parts of Canada	peyote; peyotl (anciently); hikuli; hikouri; mescal button
Lophophora diffusa		
Trichocereus Pachanoi	Ecuador; Peru; Bolivia	San Pedro; agua-calla; cimora (preparation)
COMPOSITAE (Composite Family)		
Calea Zacatechichi PLATE 3*	Oaxaca, Mexico	zacatechichi; thlepelakano
CONVOLVULACEAE (Morning Glory Family)		
Ipomoea violacea (Ipomoea tricolor)	Southern Mexico	badoh negro; piule negro

When Identified	How Used	Active Constituents
1968; 1971	Alone as snuff (?)	?
1956-1969	Ingested	Psilocybine; psilocine
1939		
1956-1969		
1887-1888	Fresh or dried tops of plant ingested	Mescaline; pellotine; anhalonidine; anhalamine; hordenine; lophophorine; and 32 other phenylethylamine or isoquinoline alkaloids and related bases
1944; 1967		
1959	Prepared in a drink with other Mescaline plants	Large amounts of pellotine; small amounts of anhalidine, anhalamine and lophophorine; traces of mescaline; anhalonidine; N-methylmescaline; O-methylpellotine
1968	Leaves made into tea and also smoked	?
1960	Seeds ingested	Ergine; isoergine; chanoclavine; elymoclavine; ergometrine; plus several minor alkaloids

TABLE OF NEW WORLD HALLUCINOGENS (Cont.)

Family and Plant	Region Where Used	Common Names
Rivea corymbosa (*Turbina corymbosa)* PLATE 4*	Southern Mexico	ololiuqui (anciently); piule; la señorita; bitoo; yuca-yaha; bador
LABIATAE (Mint Family)		
Coleus Blumei	Oaxaca, Mexico (native of Asia)	el nene; el ahijado
Coleus pumilus	Oaxaca, Mexico (native of Asia)	el macho
Salvia divinorum PLATE 5	Oaxaca, Mexico	pipiltzintzintli (anciently); hierba de la Virgen; hierba de la Pastora; shka-Pastora
LEGUMINOSAE (Pea Family		
Anadenanthera peregrina (*Piptadenia peregrina)*	Orinoco basin, Colombia, Venezuela and adjacent Brazil; possibly disjunctly in Brazilian Amazon; anciently, West Indies	cohoba; yopo; ñiopa; paricá

PLATE 3.

Calea zacatechichi Schlecht.

When Identified	How Used	Active Constituents
1903-1941	Seeds ingested	Ergine; isoergine; chanoclavine; elymoclavine; lysergol; plus several minor alkaloids
1962	Leaves crushed and ingested	?
1962	Leaves crushed and ingested	?
1962	Leaves crushed and ingested	?
1801	Leaves prepared in form of snuff	Same tryptamines and beta-carbolines as found in *Virola* (q.v.) plus 5-hydroxy-N, N-dimethyltryptamine

PLATE 4.

Rivea corymbosa (L) Hall.

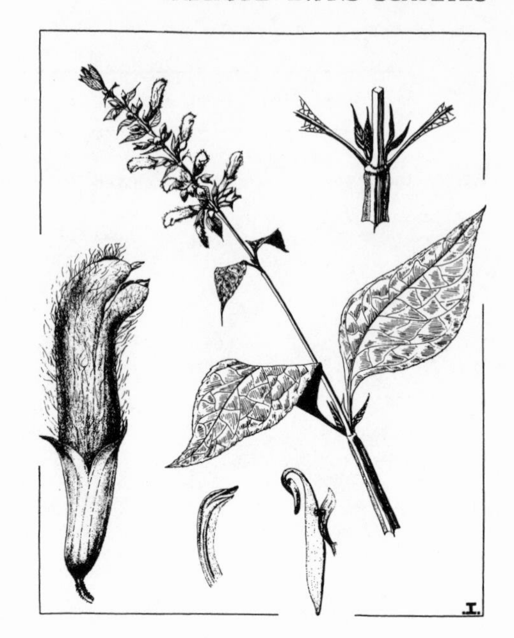

PLATE 5.
Salvia divinorum Epling & Jativa.

TABLE OF NEW WORLD HALLUCINOGENS (Cont.)

Family and Plant	Region Where Used	Common Names
Anadenanthera colubrina	Southern Peru; Bolivia; Northern Argentina	huilca; vilca; cebil; sebil
Mimosa hostilis PLATE 6	Eastern Brazil	vinho de jurema
Sophora secundiflora	Northern Mexico; Southwestern United States	red bean; mescal bean; colorines; frijolillos
LYCOPERDACEAE (Puff Ball Family)		
Lycoperdon marginatum; mixtecorum PLATE 7	Oaxaca, Mexico	gi-i-sa-wa; gi-i-wa
LYTHRACEAE (Loosestrife Family)		
Heimia salicifolia PLATE 8*	Central Mexico	sinicuichi

PLATE 6.
Mimosa hostilis (Mart.) Benth.

When Identified	How Used	Active Constituents
1916	Seeds prepared in form of snuff	N,N-dimethyltryptamine; 5-hydroxy-N,N-dimethyl-tryptamine; 5-hydroxy-N,N-dimethyltryptamine-N-oxide
1881; 1946	Root prepared in drink	N,N-dimethyltryptamine
1965	Seeds ingested	Cytisine
1958	Ingestion of plant	?
1936	Leaves slightly wilted are crushed in water and liquor and set out to ferment	Lythrine; cryogenine; heimine; sinine; vertine; lythridine; nesodine; lyofoline

TABLE OF NEW WORLD HALLUCINOGENS (Cont.)

Family and Plant	Region Where Used	Common Names
MALPIGHIACEAE (Malpighia Family)		
Banisteriopsis Caapi; inebrians	Western Amazon (Bolivia, Brazil, Colombia, Ecuador, Peru); uppermost Orinoco (Colombia, Venezuela); Pacific Coast (Colombia, Ecuador)	ayahuasca; caapi; yajé; pinde; natema; kahi
Tetrapteris methystica	Brazil-Colombian boundary	caapi
MYRISTICACEAE (Nutmeg Family)		
Virola calophylla; calophylloidea; theiodora PLATE 9	Northwest Amazon (Brazil, Colombia) and headwaters of Orinoco (Venezuela)	ebena; nyakwana; yato; yakee; paricá

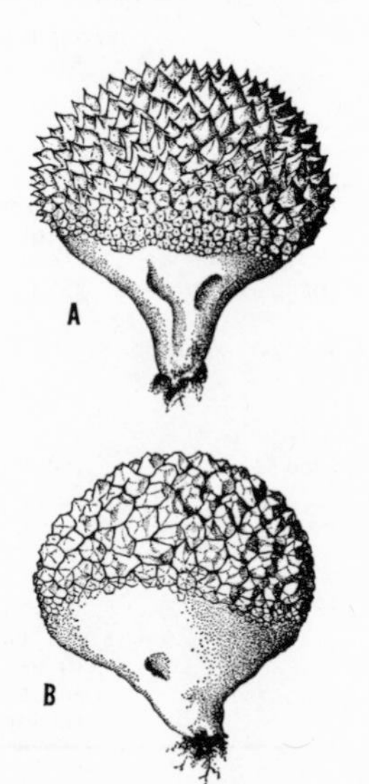

PLATE 7.

A. *Lycoperdon marginatum*
B. *Lycoperdon mixtecorum*

When Identified	How Used	Active Constituents
1851	Bark prepared in a cold-water drink or boiled decoction. Said occasionally to be chewed or prepared for snuffing	Harmine; harmaline; tetrahydroharmine; harmol
1954	Bark prepared in cold water drink	?
[1938]; 1954; 1968 [1966; 1974; 1966]	Resin in bark prepared in snuff. Occasionally licked raw or ingested in form of dried pellets. Some evidence *Virola* may be smoked	Tryptamines and β-carbolines; N,N-dimethyltryptamine; N-methyltryptamine; 5-methoxy-N,N-dimethyltryptamine; 5-methoxy-N-methyltryptamine; 2-methyl-6-methoxy-1,2,3,4-tetrahydro-β-carboline; 1,2-dimethyl-6-methoxy-1,2,3,4-tetrahydro-β-carboline

PLATE 8.

Heimia salicifolia Link & Otto

PLATE 9.
Virola theiodora (spr. ex Bth.) Warburg

TABLE OF NEW WORLD HALLUCINOGENS (Cont.)

Family and Plant	Region Where Used	Common Names
SOLANACEAE (Nightshade Family)		
Brunfelsia grandiflora	Locally and disjunctly in Western Amazonia (Bolivia; Colombia; Ecuador; Peru)	chiric-caspi; chiric-sanango; borrachero
Cestrum laevigatum	Brazil	dama da noite
Datura inoxia (D. meteloides); Stramonium; discolor; Wrightii; ceratocaula	United States, Mexico	jimson weed; thorn apple; toloatzin (anciently); a-neg-la-kya; torna-loco
Datura (Brugmansia) aurea; arborea; candida; dolichocarpa; sanguinea; suaveolens; vulcanicola	Andean highlands (Colombia; Ecuador; Peru; Chile) and (*D. suaveolens*) Westernmost Amazon (Colombia; Ecuador; Peru) and Pacific Coast of Colombia	borrachero; huancacachu; huanto; chamico; sampanilla; floripondio; maicoa; toa; tonga
Iochroma fuchsioides	Colombian Andes	borrachero

PLATE 10.

Latua pubiflora (Gris.) Phil.

When Identified	How Used	Active Constituents
1967	Drink prepared from leaves and bark	?
1973	Leaves smoked	?Possibly saponines; gitogenin; digitogenin
During 1800's	Seeds or leaves usually prepared in drink	A variety of tropane alkaloids, mainly scopolamine, hyoscyamine, atropine
During past century	Seeds or leaves usually prepared in drink	A variety of tropane alkaloids, mainly scopolamine, hyoscyamine, atropine
1970	Leaves prepared in form of tea	?Possibly alkaloids

TABLE OF NEW WORLD HALLUCINOGENS (Cont.)

Family and Plant	Region Where Used	Common Names
Latua pubiflora PLATE 10	Chile	árbol de los brujos; latué
Methysticodendron Amesianum	Southern Andes of Colombia	culebra borrachero; mitskway borrachero

TABLE OF OLD WORLD HALLUCINOGENS

Family and Plant	Region Where Used	Common Names
AGARICACEAE (Agaric Family)		
Amanita muscaria PLATE 11	India (anciently); north-eastern and western Siberia	soma (anciently); fly agaric
APOCYNACEAE (Dogbane Family)		
Tabernanthe Iboga PLATE 12	Gabon, Congo	iboga

PLATE 11.

Amanita muscaria Fries ex Linnaeus

(Drawn from Heim: Champignons toxiques et hallucinogènes)

The "fly agaric," used as an hallucinogenic agent in Siberia.

When Identified	How Used	Active Constituents
1858	Leaves prepared in drink. Fresh fruits reputedly also toxic	Hyoscyamine; scopolamine
1955	Tea of leaves drunk	Mainly scopolamine; atropine; two unidentified alkaloids

When Identified	How Used	Active Constituents
1967 (India); 1730 (Siberia)	Dried mushroom ingested or drunk in fruit juices or reindeer milk	Muscimole; other minor ibotenic acid derivatives
[1864]; 1889	Root ingested	Ibogaine and eleven other indole alkaloids

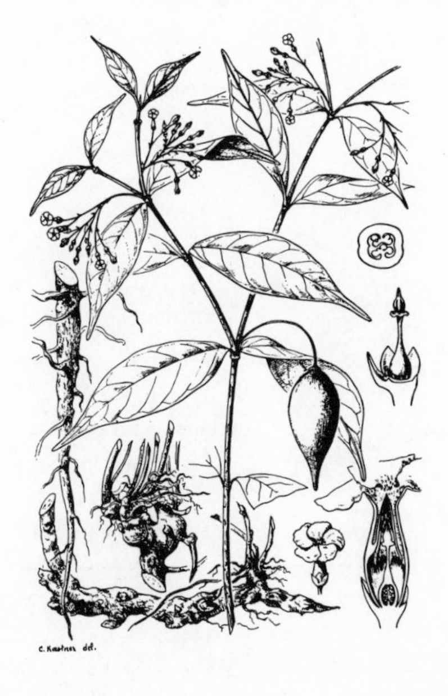

PLATE 12.
Tabernanthe Iboga

TABLE OF OLD WORLD HALLUCINOGENS (Cont.)

Family and Plant	Region Where Used	Common Names
CANNABACEAE (Hemp Family)		
Cannabis sativa, indica, ruderalis	Genus originally Asiatic, now used virtually on all continents	hemp; marihuana; maconha
MYRISTICACEAE (Nutmeg Family)		
Myristica fragrans	Southeast Asia	mace; nutmeg; madashaunda
SOLANACEAE (Nightshade Family)		
Atropa Belladonna	Europe (in Middle Ages)	belladonna
Hyoscyaminus niger	Europe (in Middle Ages)	henbane
Mandragora officinarum	Europe (in Middle Ages)	mandrake
ZINGIBERACEAE (Ginger Family)		
Kaempferia Galanga	New Guinea	galanga, maraba
ZYGOPHYLLACEAE (Caltrop Family)		
Peganum Harmala	Asia	Syrian rue, harmala

When Identified	How Used	Active Constituents
Anciently	Resin eaten or smoked; crushed leaves and inflorescences smoked; dried plant powdered and drunk in water or milk (India: *bhang*); dried plant powdered and made into candies with sugar, spices (India: *majun*); dried pistillate inflorescences smoked with tobacco or eaten or drunk (India: *ganja*); pure resin eaten with spices (India: *charas*); often taken mixed with other narcotics, e.g. seeds of Datura	Mainly tetrahydrocannabinol. Less active or questionably active are the more than 20 known cannabidiolic acids, cannabidiol or related compounds
Anciently	Nutmeg ingested in powdered form or sniffed.	Probably myristicine and other aromatic fractions of oil of nutmeg
Anciently	An ingredient of witches' brews	Hyoscyamine (principally); scopolamine; atropine
Anciently	An ingredient of witches' brews	Hyoscyamine (principally); scopolamine; minor alkaloids; tropine, scopine, apoatropine, cuscohygrine
Anciently	An ingredient of witches' brews	Hyoscyamine (principally); scopolamine; nor-hyoscyamine; cuscohygrine (mandragorine)
1962	?	?Possibly a constituent of the essential oils
Anciently	?	Harmine, harmaline

SELECTED BIBLIOGRAPHY

1. Altschul, S. von R. 1972. "The Genus *Anadenanthera* in Amerindian Cultures." Harvard Botanical Museum, Cambridge, Mass.
2. Anderson, E.F. 1969. The biography, ecology and taxonomy of *Lophophora* (Cactaceae). Brittonia 21: 299-310.
3. Appel, H., A. Rother & A.E. Schwarting. 1965. Alkaloids of *Heimia salicifolia*. II. Isolation of nesodine and lyfoline and their correlation with other lythraceous alkaloids. Lloydia 28: 84-89.
4. Barrau, J. 1958. Nouvelles observations au sujet des plantes hallucinogènes d'usage autochtone en Nouvelle-Guinée. Journ. Agric. Trop. Bot. Appl. 5: 377-378.
5. _____. 1962. Observations et travaux récents sur les végétaux hallucinogènes de la Nouvelle-Guinée. Journ. Agric. Trop. Bot. Appl. 9: 245-249.
6. Beverly, R. 1705. "History and Present State of Virginia." R. Parker, London.
7. Biocca, E. 1966. Viaggi tra gli indi alto Rio Negro -- alto Orinoco. Consiglio Nazionale delle Richerche, Rome 2: 235-252.
8. Blomster, R.N., A.E. Schwarting & J.M. Bobbit. 1964. Alkaloids of *Heimia salicifolia*. Lloydia 27: 15-24.
9. Bodendorf, K. & H. Kummer. 1962. Uber die Alkaloide in *Latua venenosa*. Pharm. Zentralhalle 101: 620-622.
10. Boke, N.H. & E.F. Anderson. 1970. Structure, development and taxonomy in the genus *Lophophora*. Am. Jour. Bot. 57: 569-578.
11. Bravo, H.H. 1967. Una revisión del género *Lophophora*. Cact. Succ. Mex. 12: 8-17.
12. Bristol, M.L., W.C. Evans & J.F. Lampard. 1969. The alkaloids of the genus *Datura*, section Brugmansia. Part VI. Tree Datura drugs. (*Datura candida* cvs.) of the Colombian Sibundoy. Lloydia 32: 123-130.
13. Bruhn, J.G. & C. Bruhn. 1973. Alkaloids and ethnobotany of Mexican peyote cacti and related species. Econ. Bot. 27: 241-251.
14. _____ & B. Holmstedt. 1974. Early peyote research: an interdisciplinary study. Econ. Bot. (in press).
15. Bunge, A. 1847. Beiträge zur Kenntniss der Flora Russlands und der Steppen Central-Asiens. Mem. Sav. Etr. Petersb. 7: 438.
16. Campbell, T.N. 1958. Origin of the Mescal Bean Cult. Am. Anthrop. 60: 156-160.

17. Chagnon, N.A., P. Le Quesne & J.M. Cook. 1971. Yanomamó hallucinogens: anthropological, botanical and chemifindings. Curr. Anthrop. 12: 72-74.
18. Chen, A.L. & K.K. Chen. 1939. Harmine, the alkaloid of caapi. Quart. Journ. Pharm. Pharmacol. 12: 30-38.
19. Cooper, J.M. 1949. Stimulants and narcotics. In "Handbook of South American Indians" (J.H. Steward, Ed.) Bur. Am. Ethnol. Bull. No. 143, U.S. Gov't. Printing Office, Washington, D.C. pp. 525-558.
20. Crosby, D.M. & J.L. McLaughlin. 1973. Cactus alkaloids XIX. Crystallization of mescaline HCl and 3-methoxytyramine HCl from *Trichocereus Pachanoi*. Lloydia 36: 416-418.
21. Der Marderosian, A.H. 1965. Nomenclatural history of the Morning Glory, *Ipomoea violacea*. Taxon 14: 234-240.
22. _____, H.V. Pinkley & M.F. Dobbins IV. 1968. Native use and occurence of N,N-dimethyltryptamine in the leaves of *Banisteriopsis Rusbyana*. Am. Journ. Pharm. 140: 137-147.
23. _____ & H.W. Youngken, Jr. 1966. The distribution of indole alkaloids among certain species and varieties of *Ipomoea, Rivea* and *Convolvulus* (Convolvulaceae). Lloydia 29: 35-42.
24. Deulofeu, V. 1967. Chemical compounds isolated from *Banisteriopsis* and related species. In "Ethnopharmacologic Search for Psychoactive Drugs" (D. Efron, Ed.) Public Health Serv. Publ. No. 1645, U.S. Gov't. Printing Office, Washington, D.C. 393-402.
25. Diguet, L. 1928. "Les cactacées utiles du Mexique". Arch. Hist. Nat., Soc. Nat. Acclim. France, Paris.
26. Douglas, B., J.L. Kirkpatrick, R.F. Raffauf, R.F. Ribeiro & J.A. Weisbach. 1964. Problems in Chemotaxonomy. II. The major alkaloids of the genus *Heimia*. Lloydia 27: 25-31.
27. Downing, D.F. 1962. The chemistry of the psychotomimetic substances. Quart. Rev. 16: 133-162.
28. Ducke, A. 1938. Plantes nouvelles. Arch. Inst. Biol. Veg. 4: 3.
29. Elger, F. 1928. Uber das Vorkommen von Harmin in einer südamerikanischen Liane (yagé). Helv. Chim. Acta II: 162-166.
30. Emboden, W.A. Jr. 1972. "Narcotic Plants". Macmillan Co., New York.
31. Eugster, C.H. 1968. Wirkstoffe aus dem Fliegenpilz. Naturwissenschaften 55: 305-313.

32. Evans, W.C., V.A. Major & M. Pethan. 1965. The alkaloids of the genus *Datura*, section Brugmansia. Part III. *Datura sanguinea* R. and P. Planta Medica 13: 353-358.
33. ____ & W. Wellendorf. 1959. The alkaloids of the roots of *Datura*. J. Chem. Soc. [Org.] 1959: 1406.
34. Fadiman, J. 1965. *Genista canariensis*: a minor psychedelic. Econ. Bot. 19: 383-384.
35. Farnsworth, N.R. 1968. Hallucinogenic plants. Science 162: 1086-1092.
36. ____ 1969. Some hallucinogenic plants and related plants. In "Current Topics in Plant Science" Academic Press, New York. pp. 367-399.
37. ____ 1972. Psychotomimetic plants and related higher plants. J. Psyched. Drugs 5: 67-74.
38. ____ 1974. Psychotomimetic plants II. Psyched. Drugs 6: 83-84.
39. Ferris, J.P., C.B. Boyce & R.C. Briner. 1966. Lythraceae alkaloids. Structure and stereochemistry of the major alkaloids of *Decodon* and *Heimia*. Tetrahedron Lett. 3641-3649.
40. Fish, M.S., N.M. Johnson & E.C. Horning. 1955. *Piptadenia* alkaloids. Indole bases of *Peperegrina* (L.) Benth. and related species. J. Am. Chem. Soc. 77: 5892-5895.
41. Fodor, G. 1967. The tropane alkaloids. In "The Alkaloids". Vol. 9. pp. 269-303. Academic Press, New York.
42. Friedberg, C. 1956. Rapport sommaire sur une mission au Pérou. J. Agric. Trop. Bot. Appl. 6: Nos. 8-9.
43. ____ 1965. Des Banisteriopsis utilesés comme drogue en Amérique du Sud. J. Agric. Trop. Bot. Appl. 12: 403-437; 550-594; 729-780.
44. Furst, P.T. 1971. *Ariocarpus retusus*, the 'false peyote' of Huichol tradition. Econ. Bot. 25: 182-187.
45. García-Barriga, H. 1958. El yajé, caapi o ayahuasca -- un alucinógeno amazónico. Univ. Nac. Colombia, No. 23, 59-76.
46. Gonçalves de Lima, O. 1946. Óbservações sôbre o vinho de jurema utilizado pelos indios Pancarú de Tacaratú (Pernambuco). Arqu. Inst. Pesquisas Agron. 4: 45-80.
47. Guzmán, H.G. 1959. Sinopsis de los conocimientos sobre los hongos alucinógenos mexicanos. Bol. Soc. Bot. Mex., 24: 14-34.
48. Haller, A. & E. Heckel. 1901. Sur l'ibogaine, principe actif d'une plante du genre *Tabernaemontana* originaire de Congo. Comptes Rend. 133 : 850-853.

49. Hartwich, C. 1911. "Die menschlichen Genussmittel." Chr. Herm., Tauchnitz, Leipzig.
50. Heim, R. 1956. Les champignons divinatoires utilisés dans les rites des indiens Mazatèques. Comptes Rend. 242: 965-968.
51. _____ 1956. Les champignons divinatoires . . . dans les pays Mije, Mazatèque, Zapoteque et Nahua. Comptes Rend. 242: 1389-1395.
52. _____ 1957. Les agarics hallucinogènes du genre Psilocybe. Comptes Rend. 244: 659-700.
53. _____ 1963. "Les Champignons Toxiques et Hallucinogènes". N. Boubée et Cie., Paris.
54. _____ 1965. Les substances indoliques produites par les champignons toxiques et hallucinogènes. Bull Méd. Leg. 8: 122-141.
55. _____ 1967. "Nouvelles Investigations sur les Champignons Hallucinogènes." Edit. Mus. Nat. Hist. Nat., Paris.
56. _____ & R.G. Wasson. 1958. "Les Champignons Hallucinogènes du Mexique." Edit. Mus. Nat. Hist. Nat., Paris.
57. _____ & R.G. Wasson. 1962. Une investigation sur les champignons sacrés des Mistèques. Comptes Rend. 254: 788-791.
58. Hennings, P. 1888. Eine giftige Kaktee, *Anhalonium Lewinii* n. sp. Gartenfl. 37: 410-412.
59. Hoffer, A. & H. Osmond. 1967. "The Hallucinogens." Academic Press, New York.
60. Hofmann, A. 1961. Chemical, pharmacological and medical aspects of psychotomimetics. J. Exp. Med. Sci. 5: 31-51.
61. _____ 1961. Die Wirkstoffe der mexikanischen Zauberdroge 'Ololiuqui'. Planta Medica 9: 354-367.
62. _____ 1963. The active principles of the seeds of *Rivea corymbosa* and *Ipomoea violacea*. Bot. Mus. Leafl., Harvard Univ. 20: 194-212.
63. _____ 1968. Psychotomimetic agents. In "Chemical Constitution and Pharmacodynamic Action" (A. Burger, ed.) Vol. 2, pp. 169-235. M. Dekker, New York.
64. Holmstedt, B. 1965. Tryptamine derivatives in epené, an intoxicating snuff used by some South American Indian tribes. Arch. Int. Pharmacodyn. Ther. 156: 285-305.
65. _____ & J.-E. Lindgren. 1967. Chemical constituents and pharmacology of South American snuffs. In

"Ethnopharmacologic Search for Psychoactive Drugs". (D. Effron, Ed.) pp. 339-373. Public Health Serv. Publ. No. 1645, U.S. Gov't. Printing Office, Washington, D.C.

66. von Humboldt, A. & A. Bonpland. 1852-53. "Personal Narrative of Travels to the Equinoctial Regions of America." (T. Ross, ed. and trans.) Henry C. Bohn, London.
67. Kapadia, G.J. 1970. Peyote constituents: chemistry, biogenesis and biological effects. J. Pharm. Sci. 59: 1699-1727.
68. _____ & H.M. Fales. 1968. Krebs cycle conjugates of mescaline. Identification of fourteen new peyote alkaloid-amides. Chem. Commun. 1968: 1688-1689.
69. _____ & H.M. Fales. 1968. Peyote alkaloids VI. Peyophorine, a tetrahydroisoquinoline cactus alkaloid containing an N-ethyl group. J. Pharm. Sci. 57: 2017-2018.
70. _____ & M.B.E. Fayez. 1970. Peyote constituents: chemistry, biogenesis and biological effects. J. Pharm. Sci. 59: 1699-1727.
71. _____ & R.J. Highet. 1968. Peyote alkaloids III. Structure of peyonine, novel β-phenethylpyrrole from *Lophophora Williamsii*. J. Pharm. Sci. 57: 254-262.
72. _____, N.J. Shah & T.B. Zalucky. 1968. Peyote alkaloids II. Anhalotine, lophotine and peyotine, the quaternary alkaloids of *Lophophora Williamsii*. J. Pharm. Sci. 57: 254-262.
73. Karsten, R. 1920. Berauschende und narkotische Getränke unter den Indianern Südamerikas. Beiträge zur Sittengeschichte der sudamerikanischen Indianern. Act. Acad. Aboensis Hum. 1: 28-72.
74. Lai, A., M. Tin-Wa, E.S. Mika, G.J. Persinos & N.R. Farnsworth. 1973. Phytochemical investigation of *Virola peruviana*, a new hallucinogenic plant. J. Pharm. Sci. 62: 1561-1563.
75. Leary, J.D. 1970. Alkaloids of the seeds of *Datura sanguinea*. Lloydia 33: 264-266.
76. Legler, G. & R. Tschesche. 1963. Die Isolierung von N-Methyltryptamin, 5-Methoxy-N-methyltryptamin und 5-Methoxy-N,N-dimethyltryptamin aus der Rinde von *Piptadenia peregrina* Benth. Naturwissenschaften 50: 94-95.
77. Lewin, L. 1927. "Phantastika -- die Betaubenden und erregenden Genussmittel." Verlag G. Stilke, Berlin.

78. Lumholtz, C. 1902. "Unknown Mexico." pp. 356-379. Scribner, New York.
79. Lundström, J. 1970. Biosynthesis of mescaline and 3,4-dimethoxyphenethylamine in *Trichocereus Pachanoi*, Br. and R. Acta Pharm. Suec. 7: 651-666.
80. _____ 1971. Biosynthesis of tetrahydroisoquinoline alkaloids in *Lophophora Williamsii* (Lem.) Coult. Acta Pharm. Suec. 8: 485-496.
81. 1971. Biosynthetic studies on mescaline and related cactus alkaloids. Acta Pharm. Suec. 8: 275-302.
82. _____. 1972. Identification of new peyote alkaloids: isomers of the main phenolic tetrahydroisoquinolines. Acta. Chem. Scand. 26: 1295-1297.
83. _____ & S. Agurell. 1967. Thin-layer chromatography of the peyote alkaloids. J. Chromatog. 30: 271-272.
84. _____ & S. Agurell. 1968. Gas chromatography of peyote alkaloids. A new peyote alkaloid. J. Chromatog. 36: 105-108.
85. _____ & S. Agurell. 1971. Biosynthesis of mescaline and tetrahydroisoquinoline alkaloids in *Lophophora Williamsii* (Lem.) Coult. Acta Pharm. Suec. 8: 261-274.
86. MacDougall, T. 1960. *Ipomoea tricolor*, a hallucinogenic plant of the Zapotecs. Bol. Centro Inv. Antrop. Mex. No. 6: 6-8.
87. _____. 1968. A composite with psychic properties? Gard. Journ. 18: 105. 1965.
88. Mariani Ramírez. C. 1965. "Témas de Hipnosis." Edit. Andrés Bello, Santiago.
89. McLaughlin, J.L. & A.G. Paul. 1966. The cactus alkaloids. I. Identification of N-methylated tryamine derivatives in *Lophophora Williamsii*. Lloydia 29: 315-327.
90. Mechoulam, R. (Ed.) 1973. "Marijuana." Academic Press, New York.
91. Morton, C.V. 1931. Notes on yajé, a drug plant of Southeastern Colombia. J. Wash. Acad. Sci. 21: 485-488.
92. Naranjo, P. 1969. Etnofarmacología de las plantas psicotrópicas de America. Terapía 24: 5-63.
93. _____. 1970. "Ayahuasca: Religion y Medicina." Edit. Universitaria, Quito.
94. _____ & E. Naranjo. 1961. Estudio farmacodinámico de una planta psicotomimetica: *Coriaria thymifolia* (shanshi). Arch. Criminol. Neuro-Psiquiatr. Discipl. Conexas 9: 600-616.
95. Neumayer, J. & R.A. Shagoury. 1971. Chemistry and

pharmacology of marijuana. J. Pharm. Sci. 60: 1433-1457.
96. O'Connel, F.D. & E.V. Lynn. 1953. The alkaloids of *Banisteriopsis inebrians* Morton. J. Am. Pharm. Assoc. 42: 753-754.
97. Ola'h, G.-M. 1969. "Le Genre *Panaeolus*." Imprimerie Monnager, Le Mans.
98. Pachter, I.J. & A.F. Hopkinson. 1960. Note on the alkaloids of *Methysticodendron Amesianum*. J. Am. Pharm. Assoc. 49: 621-622.
99. Pennington, C.W. 1963. "The Tarahumar of Mexico -- their Environment and Material Culture." University of Utah Press, Salt Lake City.
100. Perrot, E. & Raymond-Hamet. 1927. Le yagé, plante sensorielle des Indiens de la région amazionienne de l'Equateur et de la Colombie. Comptes Rend. 184: 1266-1268.
101. _____ & Raymond-Hamet. 1927. Yagé, ayahuasca, caapi et leur alcoloïde: telepathine ou yageine. Bull. Sci. Pharmacol. 34: 337-347, 417-426, 500-514.
102. Philippi, R.A. 1858. *Latua* Ph., eine neues Genus aus Solanaceen. Bot. Zeit. 16: (33) 241-242.
103. Phokas, G.K. 1959. "Contribution to the Definition of the Drastic Components of Mandrake Root. Thesis. Athens.
104. Pinkley, H.V. 1969. Plant additives to ayahuasca, the South American hallucinogenic drink. Lloydia 32: 305-314.
105. Plowman, T., L.O. Gyllenhaal & J.E. Lindgren. 1971. *Latua pubiflora:* magic plant from southern Chile. Bot. Mus. Leafl., Harvard Univ. 23: 61-92.
106. Plugge, P. C. & A. Rauwerde. 1896. Fortgesetzte Untersuchungen über das Vorkommen von Cystisin in Vershiedenen Papilionaceae. Arch. Pharm. 234: 685-697.
107. Poisson, J. 1965. Note sur le natem, boisson toxique peruvienne et ses alkaloides. Ann. Pharm. Fr. 23: 241-244.
108. Poisson, M.J. 1960. Présence de mescaline dans une Cactacée péruvienne. Ann. Pharm. Fr. 18: 764-765.
109. Pope, H.G., Jr. 1969. *Tabernanthe Iboga* -- an African narcotic plant of social importance. Econ. Bot. 23: 174-184.
110. Popelak, A. & G. Lettenbauer. 1967. The mesembrine alkaloids. In "The Alkaloids" (R.H.F. Manske, ed.). Vol. 9, pp. 467-482. Academic Press, New York.

111. Prance, G.T. 1970. Notes on the use of plant hallucinogens in Amazonian Brazil. Econ. Bot. 24: 62-68.
112. Raffauf, R.F. 1970. "A Handbook of Alkaloids and Alkaloid-containing plants." Wiley-Interscience, New York.
113. Ravicz, R. 1961. La Mixteca en el estudio comparitivo del hongo alucinante. An. Inst. Nac. Antrop. Hist. 13: 73-92.
114. Reko, B.P. 1934. Das mexikanische Rauschgifte Ololiuqui. El Mex. Ant. 3: 1-7.
115. Reko, V.A. 1936. "Magische Gifte -- Rausch- und Betäubungsmittel der Neuen Welt." Ferdinand Enke Verlag, Stuttgart.
116. ____. 1936. Was ist Peyote? Zeitschr. Parapsychol. 4: 7.
117. Reti, L & J.A. Castrillon. 1951. Cactus alkaloids. I. *Trichocereus Terscheckii* (Parmentier) Britton and Rose. J. Am. Chem. Soc. 73: 1767-1769.
118. Ristic, S. & A. Thomas. 1962. Zur Kenntniss von *Rhynchosia pyramidalis* (Pega-Palo). Arch. Pharmaz. 295: 510.
119. Ritchie, E. & W.C. Taylor. 1967. The galbulemima alkaloids. In "The Alkaloids" (R.H.F. Manske, Ed.). Vol. 9, pp. 529-543. Academic Press, New York.
120. Rivier, L. & J.-E. Lindgren. 1972. 'Ayahuasca', the South American hallucinogenic drink: an ethnobotanical and chemical investigation. Econ. Bot. 26: 101-129.
121. Rouhier, A. 1927. "La Plante Qui Fait les Yeux Emerveillés -- le Peyotl." Gaston Doin et Cie., Paris.
122. Safford, W.E. 1915. An Aztec narcotic. Journ. Hered. 6: 291-311.
123. _____ 1916. Identity of cohoba, the narcotic snuff of ancient Haiti. Journ. Wash. Acad. Sci. 6: 547-562.
124. _____ 1917. Narcotics plants and stimulants of the ancient Americas. Ann. Rept. Smithson. Inst. 1916: 387-424.
125. _____. 1920. Daturas of the Old World and New. Ann. Rept. Smithson. Inst. 1920: 537-567.
126. _____. 1922. Daturas of the Old World and New: an account of their narcotic properties and their use in oracular and initiatory ceremonies. Ann. Rept. Smithson. Inst. 1920: 537-567.
127. Santesson, C.G. 1937. Notiz über piule, eine mexikanische Rauschdroge. Ethnolog. Stud. 4: 1-11.
128. _____. 1938. Noch eine mexikanische Piule-Droge, Semina *Rhynchosiae phaseoloidis* DC. Ethnolog. Stud. 6: 179-183.

129. Schultes, R.E. 1937. Peyote (*Lophophora Williamsii*) and plants confused with it. Bot. Mus. Leafl., Harvard Univ. 5: 61-88.
130. _____. 1939. Plantae Mexicanae II. The identification of teonanacatl, a narcotic Basidiomycete of the Aztecs. Bot. Mus. Leafl., Harvard Univ. 7: 37-54.
131. _____. 1941. "A Contribution to our Knowledge of Rivea corymbosa, the Narcotic Ololiuqui of the Aztecs." Harvard Botanical Museum, Cambridge, Mass.
132. _____. 1954. Plantae Austro-Americanae IX. Bot. Mus. Leafl. 16: 202-205.
133. _____. 1954. A new narcotic snuff from the northwest Amazon. Bot. Mus. Leafl., Harvard Univ. 16: 241-260.
134. _____. 1955. A new narcotic genus from the Amazon slope of the Colombian Andes. Bot. Mus. Leafl., Harvard Univ. 17: 1-11.
135. _____. 1957. The identity of the malpighiaceous narcotics of South America. Bot. Mus. Leafl., Harvard Univ. 18: 1-56.
136. _____. 1965. Ein Halbes Jahrhundert Ethnobotanik amerikanischer Halluzinogene. Planta Medica 13: 126-157.
137. _____. 1966. The search for new natural hallucinogens. Lloydia 29: 293-308.
138. _____. 1967. The botanical origin of South American snuffs. In "Ethnopharmacologic Search for Psychoactive Drugs (D. Effron, ed.). pp. 291-306. Public Health Serv. Publ. No. 1645, U.S. Gov't. Printing Office, Washington, D.C.
139. _____. 1969. Hallucinogens of plant origin. Science 163: 245-254.
140. _____. 1969-70. The Plant Kingdom and hallucinogens. Bull. Narcotics 21: (3) 3-16; (4) 15-27; 22 (1) 25-53.
141. _____. 1970. The botanical and chemical distribution of hallucinoges. Ann. Rev. Pl. Physiol. 21: 571-594.
142. _____. 1970. The New World Indians and their hallucinogenic plants. Bull. Morris Arb. 21: 3-14.
143. _____ & A. Hofmann. 1973. "The Botany and Chemistry of Hallucinogens." Charles C. Thomas, Springfield, Ill.
144. _____ & B. Holmstedt. 1968. De plantis toxicariis e Mundo Novo tropicale commentationes II. The vegetal ingredients of the myristicaceous snuffs of the northwest Amazon. Rhodora 70: 113-160.
145. _____ & R.F. Raffauf. 1960. *Prestonía*: an Amazon narcotic or not? Bot. Mus. Leafl., Harvard Univ. 19: 109-122.

146. Schulz, B. 1959. *Lagochilua inebrians* Bge., eine interessante neue Arzneipflanze. Dtsch. Apotheker-Zeit. 99: 1111.
147. Shulgin, A.T., T. Sargent & C. Naranjo. 1967. The chemistry and psychopharmacology of nutmeg and several related phenylisopropylamines. In "Ethnopharmacologic Search for Psychoactive Drugs. (D. Effron, ed.). pp. 202-214. Public Health Serv. Publ. No. 1645, U.S. Gov't. Printing Office, Washington, D.C.
148. Singer, R. 1958. Mycological investigations on teonanacatl, the Mexican hallucinogenic mushroom. Part I. Mycologia 50: 239-261.
149. _____ & A.H. Smith. 1958. Mycological investigations on tionanacatl, the Mexican hallucinogenic mushroom. Part II. Mycologia 50: 262-303.
150. Spruce, R. 1908. "Notes of a Botanist on the Amazon and Andes." (A.R. Wallace, Ed.) Macmillan, London. 2 vol. (1908). Reprinted edition, Johnson Reprint Co., New York (1970).
151. Staub, H. 1943. Uber die chemischen Bestandteile der Mandragorawurzel. 2. Die Alkaloide. Helv. Chem. Acta 26: 944-965.
152. Taber, W.A. & R.A. Heacock. 1962. Location of ergot alkaloids and fungi in the seeds of *Rivea corymbosa* (L.) Hall. f. 'Ololiuqui'. Can. J. Microbiol. 8: 137-143.
153. _____, L.C. Vining & R.A. Heacock. 1963. Clavine and lysergic acid alkaloids in varieties of Morning Glory. Phytochemistry 2: 65-70.
154. Todd, J.S. 1969. Thin-layer chromatography analysis of Mexican populations of *Lophophora* (Cactaceae). Lloydia 32: 395-398.
155. Truitt, E.B., Jr. 1967. The pharmacology of myristicin and nutmeg. In "Ethnopharmacological Search for Psychoactive Drugs. (D. Effron, ed.) pp. 215-222. Public Health Serv. Publ. No. 1645, U.S. Gov't. Printing Office, Washington, D.C.
156. Turner, W.J. & J.J. Heyman. 1960. The presence of mescaline in *Opuntia cylindrica*. J. Org. Chem. 25: 2250.
157. Tyler, V.E. 1966. The physiological properties and chemical constituents of some habit-forming plants. Lloydia 29: 275-291.
158. Urbina, M. 1903. El peyote y el ololiuqui. An. Mus. Nac. Mexico 7: 25-38.
159. Villavicencio, M. 1858. "Geografía de la República del Ecuador." R. Craigshead, New York.

160. Vrkoc, J., V. Herout & F. Sorm. 1961. On terpenes. CXXXIII. Structure of acorenone, a sesquiterpene ketone from sweet flag oil (*Acorus Calamus* L.) Coll. Czechoslov. Chem. Commun. 26: 3183-3185.
161. Warburg, W. 1897. "Die Muskatnuss." Verlag von Wilhelm Engelmann, Leipzig.
162. Wassén, S.H. & B. Holmstedt. 1963. The use of paricá, an ethnological and pharmacological review. Ethnos 1: 5-45.
163. Wasson, R.G. 1962. The hallucinogenic mushrooms of Mexico and psilocybin: a bibliography. Bot. Mus. Leafl., Harvard Univ. 20: 25-73.
164. _____. 1962. A new Mexican psychotropic drug from the Mint Family. Bot. Mus. Leafl., Harvard Univ. 20: 77-84.
165. _____. 1963. Notes on the present status of ololiuhqui and other hallucinogens of Mexico. Bot. Mus. Leafl., Harvard Univ. 20: 161-193.
166. _____. 1967. "Soma: Divine Mushroom of Immortality." Harcourt, New York.
167. _____. 1970. Some of the Aryans: an ancient hallucinogen. Bull. Narcotics 22: 25-30.
168. Wasson, V.P. & R.G. Wasson. 1957. "Mushrooms, Russia and History." Pantheon, New York.
169. Weil, A.T. 1965. Nutmeg as a narcotic. Econ. Bot. 19: 194-217.
170. Willaman, J.J. & H.L. Li. 1970. Alkaloid-bearing plants and their contained alkaloids, 1957-1968. Lloydia, Suppl. 33: (3A).

Chapter Two

RECENT ADVANCES IN THE CHEMISTRY AND METABOLISM OF THE CANNABINOIDS

MONROE E. WALL

Research Triangle Institute
Research Triangle Park, North Carolina

INTRODUCTION

The psychotomimetic activity of the variety of hemp known as *Cannabis sativa* has been known since antiquity. Various preparations of this plant under the names of marijuana, hashish, charas, dagga, and bhang are smoked or chewed by possibly two to three hundred million people. Accordingly, these materials undoubtedly constitute the most widely used group of illicit drugs. The chemistry of these *Cannabis* constituents studied extensively during the 1940's by Todd and his colleagues and the late Roger Adams and his students has been reviewed[1,2,3,4]. However, real progress in the isolation, structural elucidation and synthesis of the various cannabinoids has come only in the last decade, aided by the numerous modern separation methods, analytical techniques, and, most notably, by the development of NMR and mass spectrometry.

Our knowledge of the metabolism of the cannabinoids is even more recent[1-5], and took place only after the synthetic methodology was placed on a firm basis. In turn, as the knowledge of cannabinoid metabolites has increased, new targets have been given to the synthetic chemist, many of which have already been successfully attained. As a result, pharmacologists, clinicians, and biochemists now have at their disposal a large number of pure cannabinoids, unlabeled, radiolabeled or heavy isotope-labeled, and the number is increasing rapidly. This review will attempt to present the somewhat synergistic manner in which chemistry and metabolism have been interwoven to contribute to recent progress. Of necessity, many of the studies cited will be from the author's point of view and are taken from the work of his laboratory.

NOMENCLATURE OF CANNABINOIDS

There are at least four different numbering systems which have been used in publications relating to cannabinoids. Two of the most common are shown in Figure 1. According to these, the major psychotomimetically active constituent of marijuana would be called Δ^9- or Δ^1-tetrahydrocannabinol (Δ^9- or Δ^1-THC). The Δ^9- nomenclature is utilized in the system for numbering dibenzopyrane compounds and will be used throughout this review. This system is utilized by "Chemical Abstracts" and was adopted by the U.S. National Institute of Mental Health (now National Institute on Drug Abuse). The Δ^1-system was introduced by Mechoulam and has certain general advantages when applied to a large variety of cannabinoids, some of which resemble monoterpenoids

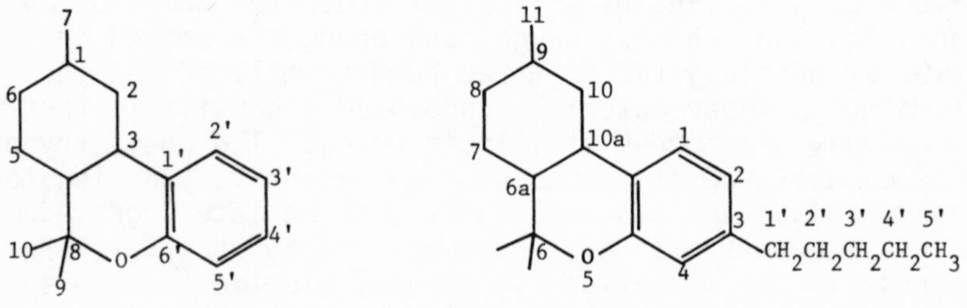

Monoterpenoid Dibenzopyran

Figure 1

rather than dibenzopyranes. The dual nomenclature for other compounds which will be discussed extensively in this review are 11-hydroxy-Δ^9-THC (7-hydroxy-Δ^1-THC) and 11-nor-Δ^9-THC-9-carboxylic acid (7-nor-Δ^1-THC-1-carboxylic acid). For the latter compound we will use the incorrect but convenient term 11-carboxy-Δ^9-THC.

The structure of Δ^9-THC was established by spectroscopic measurements and chemical correlations[6,7]. Its detailed conformational analysis has been reported[8], and indicates that the cyclohexene ring A in the formula shown in Figure 2 exists in the half chair conformation. As a consequence, substituents at positions 7 and 8 can be regarded as axial or equatorial. In this review hydroxylated constituents are named according to the steroid system. Thus, 7α- and 8β-hydroxy are axial; 7β- or 8α-hydroxy are equatorial constituents.

GLC ANALYSIS OF NATURALLY OCCURRING CONSTITUENTS OF MARIJUANA

Figure 3 shows a number of the naturally occurring constituents which are found in marijuana. Of these, Δ^9-THC, cannabidiol and cannabinol are the most important. Δ^8-THC is probably not found *per se* in plant materials but is an artifact which is produced, in part, upon

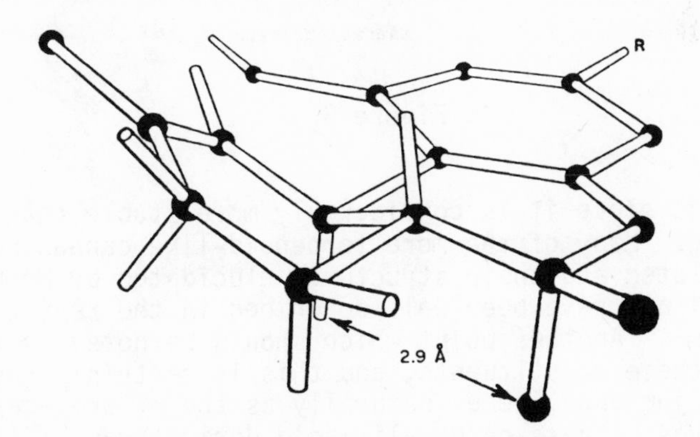

Figure 2. Conformation of Δ^9-THC. (Reference 8. Reprinted with permission of the authors and the Journal of the American Chemical Society.)

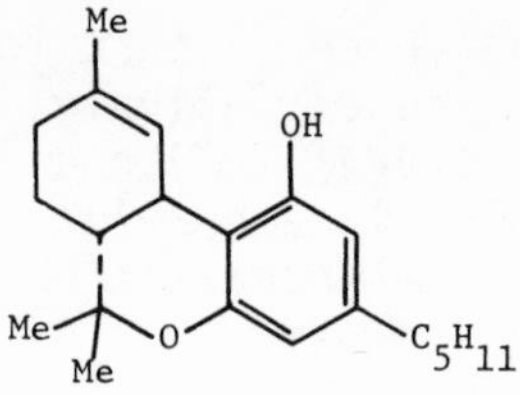

Δ^9-THC

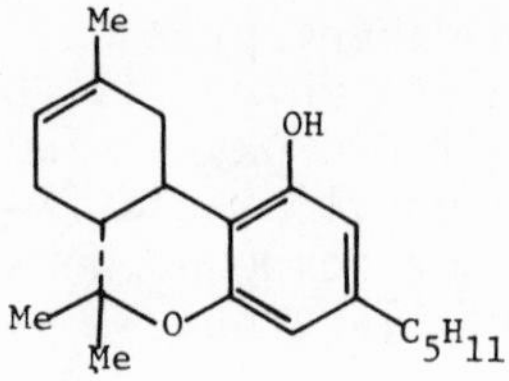

Δ^8-THC

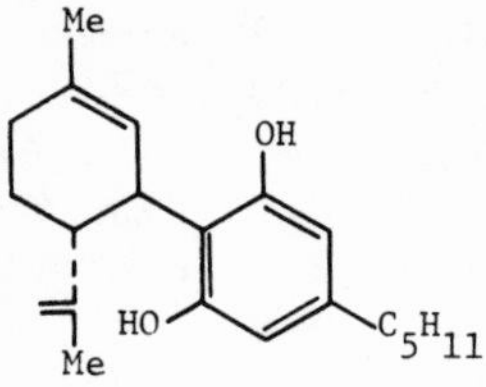

Cannabidiol

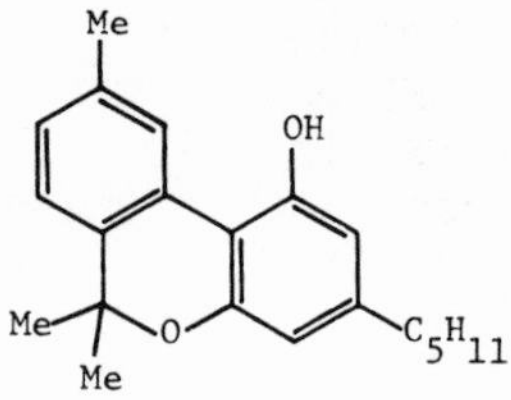

Cannabinol

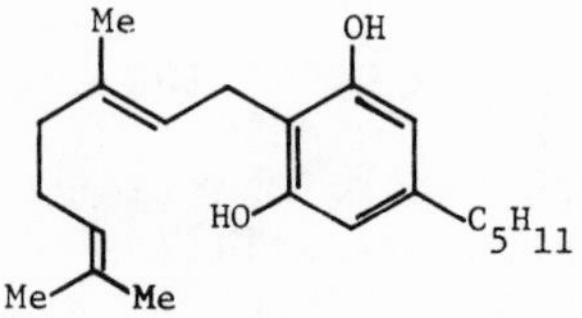

Cannabigerol

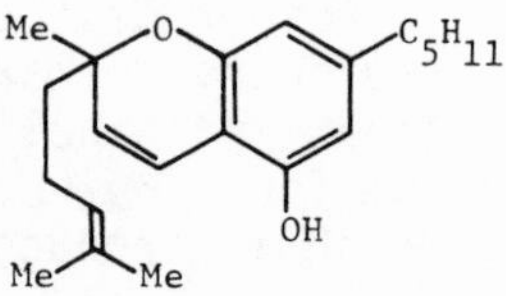

Cannabichromene

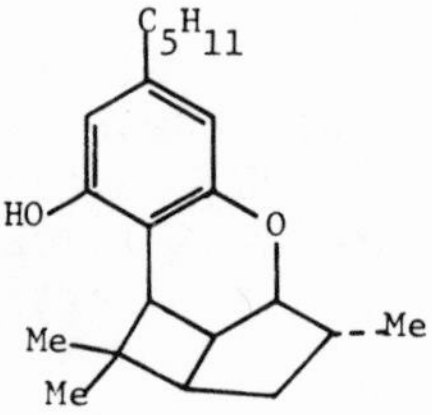

Cannabicyclol

Figure 3

pyrolysis, since it is considerably more stable than the Δ^9-analog. Many of the more terpenoid-like cannabinoids were isolated and their structures elucidated by Mechoulam and Gaoni and have been well described in the review by Mechoulam[1]. Another point which should be noted is that many of these constituents, and this is certainly the case of the major ones, occur naturally as the *o*- or *p*-carboxylic derivatives. These carboxylic acid derivatives are quite unstable to heat, and will slowly decarboxylate at room temperature. However, they can be differentiated by gas-liquid chromatography[9]. (The methodology is illustrated in

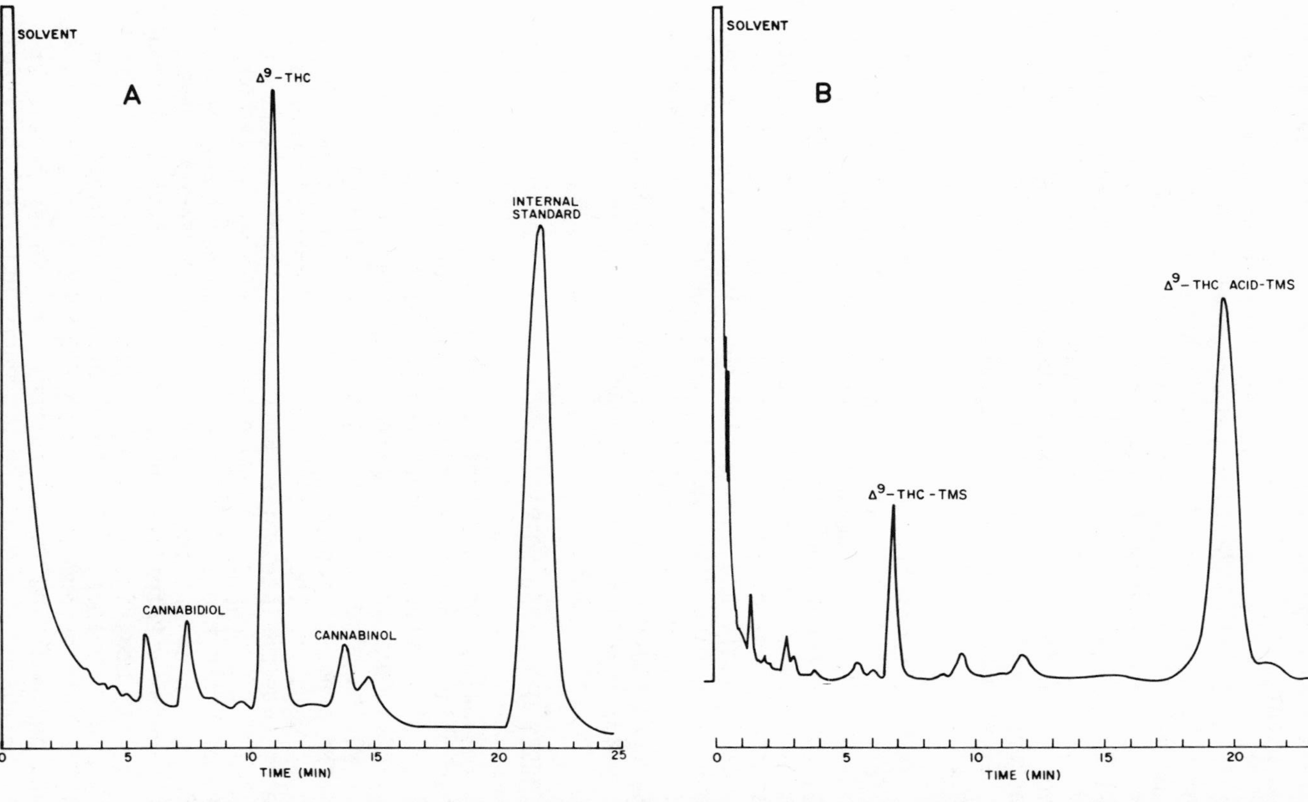

Figure 4. A. Typical GLC of Mexican marijuana extract. B. GLC of derivatized Mexican marijuana extract showing differentiation of Δ^9-THC and Δ^9-THC acid.

Figure 4.) As can be seen, glc of the underivatized marijuana plant extracts shows no signs of any carboxylic acid constituent. However, if the total cannabinoids are converted to the corresponding trimethylsilyl (tms) derivatives, the corresponding carboxylic acid is now stable and can be readily analyzed by glc as a major constituent. From a practical view, however, the underivatized carboxylic THC analogs are so unstable to heat that, when smoked in a normal way by man, the compounds are quantitatively converted to the decarboxylated compounds. The development of accurate glc procedures has permitted facile assessment of the composition of various types of cannabinoid preparations.

There seem to be two major types of *Cannabis* varieties. One form which is exemplified by the type found in Mexico has Δ^9-THC as the predominant constituent. On the other hand, *Cannabis* obtained from Turkey and the near East in general usually has cannabidiol as the major constituent. As another point of interest, compounds analogous to Δ^9-THC, cannabinol, and cannabidiol but with a propyl instead of amyl sidechain have recently been found in samples of Pakistani and Nepalese hashish[10]. In the Nepalese hashish this was a major constituent. Propyl-Δ^9-THC is also biologically active.

SYNTHESIS OF CANNABINOIDS

Following the excellent review on synthetic methods by Mechoulam[1], in this section we will review only methodology of particular importance to the theme of the review. Although various syntheses had been reported previously, the major breakthroughs in this area came in 1967, with the condensation of a pinene derivative, (-)-verbenol, with olivetol in the presence of *p*-toluenesulfonic acid to give an intermediate which could then be further converted to Δ^8-THC as shown in Figure 5.[4] By addition of the elements of hydrochloric acid across the double bond, followed by dehydrochlorination, Δ^9-THC could be produced. Almost simultaneously Petrzilka and coworkers published a method[11-12] which, with modifications, has become the basis for most of the current synthesis of both unlabeled and radiolabeled Δ^9-THC. This method is shown in Figure 6 in a slightly improved and modified form[13] suitable for the synthesis of tritium-labeled Δ^9-THC. The procedure utilizes the readily

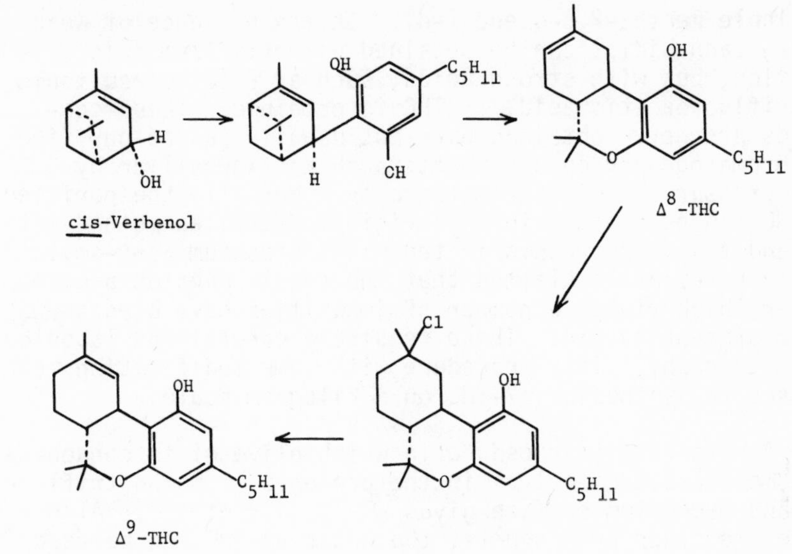

cis-Verbenol

Δ^8-THC

Δ^9-THC

Figure 5

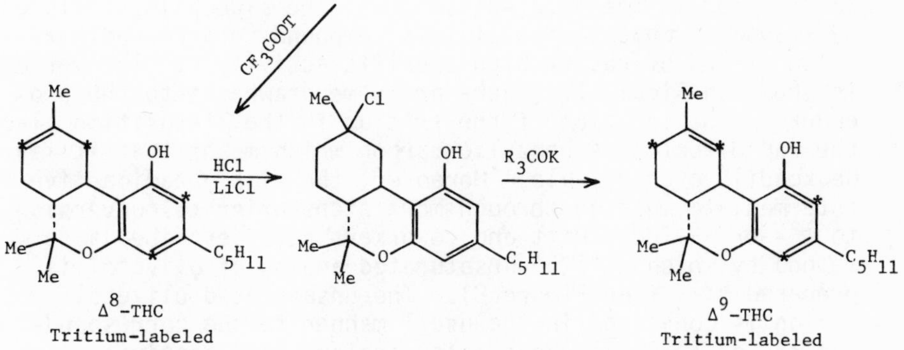

CF_3COOT

HCl
LiCl

R_3COK

Δ^8-THC
Tritium-labeled

Δ^9-THC
Tritium-labeled

Figure 6

available mentha-2,8-diene-1-ol. In the presence of weak acids, cannabidiol can be obtained directly from this reaction, but with strong acids, such as *p*-toluenesulfonic or trifluoroacetic acid, Δ^8-THC is obtained. These compounds are never obtained pure but usually can be purified by chromatography on adsorbents such as florosil or by high pressure liquid chromatography. Normally the purified Δ^8-THC can be treated in the original Petrzilka method with HCl and then dehydrochlorinated with potassium *tert*-amylate. Although Petrzilka claimed that the final reaction proceeds in very high yield, a number of impurities have been shown to be present by glc. These require a careful and laborious chromatography. This procedure with some modifications can be used to synthesize Δ^9-THC on a kilogram scale.

A more recent procedure in which olivetol is condensed with mentha-2,8-dien-1-ol in the presence of boron trifluoride and magnesium sulfate gives Δ^9-THC directly[14]. Although there are other by-products, the crude Δ^9-THC can be readily purified on a small scale. If the procedure is applicable to a large scale, it may well replace the previously published methodology.

These procedures are not applicable to the preparation of tritium-labeled cannabinoids of high specific activity. For certain purposes, particularly studies involving possible receptor sites, it is absolutely required that the radiolabeled compound be *carrier free*. Under these circumstances, methods of a different nature are required, in which the label is in the sidechain. Gill and Jones published one of the earliest of these procedures in which the tritium label was located in the 1',2'-position of the sidechain, followed by standard condensation of this compound to give radiolabeled Δ^8-THC of rather high specific activity[15]. The method is shown in Figure 7. There are some drawbacks to the procedure. The location of the tritium in the 1'-position places the radiolabel on a benzylic carbon which means that it may be readily exchangeable. Moreover, the highly radioactive intermediate must go through more steps prior to conversion to Δ^8- or Δ^9-THC. Pitt and coworkers have described a method by which a 4',5'-unsaturated analog of olivetol is prepared[16-17] (see Figure 8). The unsaturated olivetol analog is converted in the usual manner to the corresponding Δ^8- or Δ^9-4',5'-unsaturated analog. Radioactivity is introduced by tritiation over a homogenous rhodium catalyst. The procedure produces carrier free Δ^8- or Δ^9-THC. The

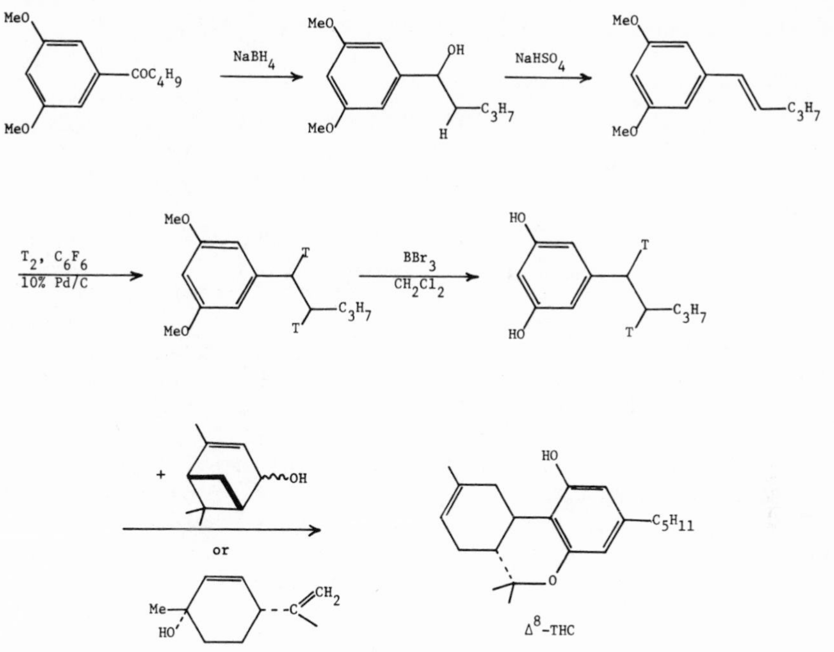

Figure 7

latter compound is unstable and must be used soon after preparation.

In some cases it may be desirable or necessary to use ^{14}C-labeled THC for certain studies, particularly if a metabolic site is involved. We have developed a procedure, outlined in Figure 9, for the synthesis of ^{14}C-Δ^9-THC, labeled at carbon-11. Other methods utilize the synthesis of ^{14}C-labeled olivetol, followed by the conversion of this compound to Δ^8- or Δ^9-^{14}C-labeled THC via condensation with the menthadienol in the standard manner. Our procedure involves conversion of Δ^8-THC to the corresponding $\Delta^{9(11)}$-THC benzyl ether, which is oxidized to give the 9-ketone, and converted to the ^{14}C-labeled $\Delta^{9(11)}$-THC benzyl ether by reaction with the labeled Wittig reagent. Halogenation followed by dehydrohalogenation in the standard manner gives ^{14}C-labeled-Δ^9-THC.

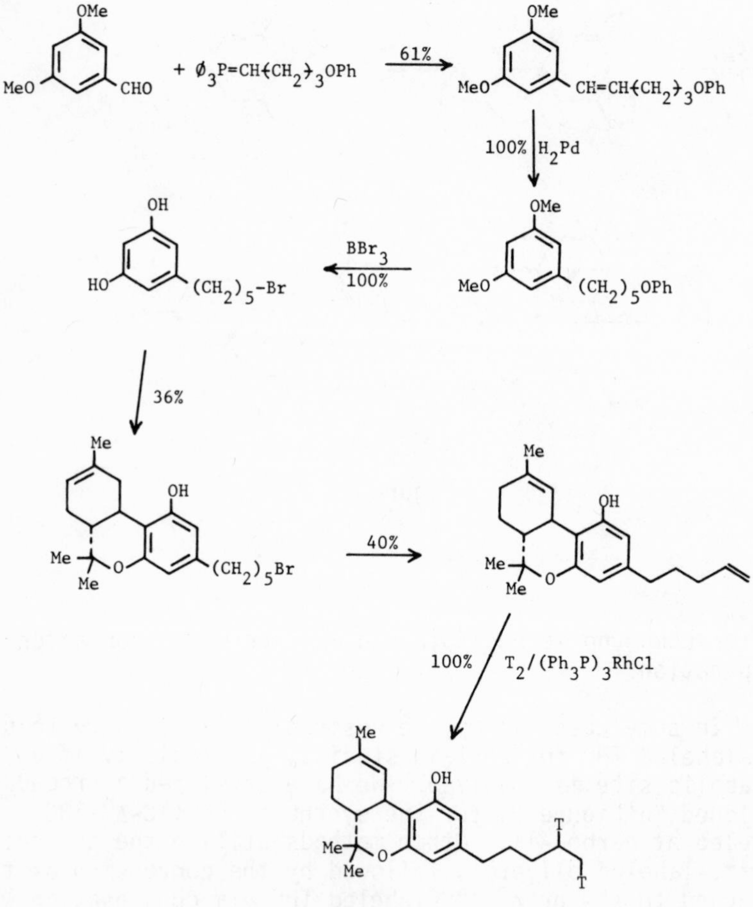

OMe
MeO CHO + $\phi_3P{=}CH(CH_2)_3OPh$ → 61% → MeO CH=CH(CH_2)_3OPh

100% H_2Pd

OMe
MeO $(CH_2)_5OPh$ → BBr_3 100% → OH HO $(CH_2)_5$-Br

36%

Me OH Me Me O $(CH_2)_5Br$ → 40% → Me OH Me Me O

100% $T_2/(Ph_3P)_3RhCl$

Me OH Me Me O T T

Figure 8

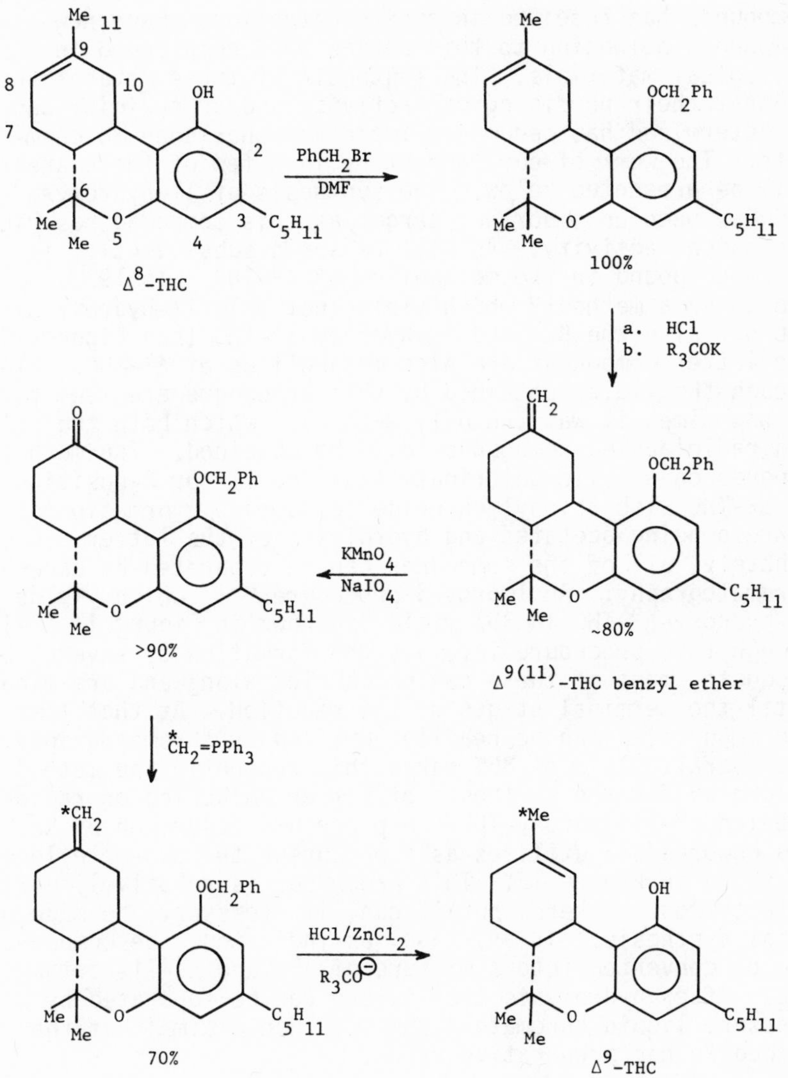

Me 11
9
8
10
OH
7
2
PhCH$_2$Br
DMF
6
Me
O
3
C$_5$H$_{11}$
Me 5
4
Δ^8-THC
Me
OCH$_2$Ph
Me
O
C$_5$H$_{11}$
Me
100%
a. HCl
b. R$_3$COK
CH$_2$
OCH$_2$Ph
O
OCH$_2$Ph
KMnO$_4$
NaIO$_4$
Me
O
C$_5$H$_{11}$
Me
>90%
Me
O
C$_5$H$_{11}$
Me
~80%
$\Delta^{9(11)}$-THC benzyl ether
$\overset{*}{C}H_2$=PPh$_3$
*CH$_2$
OCH$_2$Ph
HCl/ZnCl$_2$
R$_3$CO$^{\ominus}$
Me
O
C$_5$H$_{11}$
Me
70%
*Me
OH
Me
O
C$_5$H$_{11}$
Me
Δ^9-THC

Figure 9

SYNTHESIS OF METABOLITES

In recent years the metabolism of Δ^9-THC and related compounds has received intensive attention. Many new compounds belonging to this series have been found in biological materials. The synthesis of these metabolites so that their physiological activity and/or toxicity can be determined has become an important challenge to chemists. The work of our laboratory in a few of these areas will be presented below. The synthesis of 11-hydroxy-Δ^9-THC has been an important target, as this compound has high biological activity. As will be shown subsequently, it is a key compound in the metabolism of Δ^9-THC. In 1972, we presented a method[19] which yields not only 11-hydroxy-Δ^9-THC but also the 8α- and 8β-hydroxy-Δ^9-THC (see Figure 10). The latter compounds are also metabolites of Δ^9-THC. Although the yields obtained by this procedure are low, for a long time, it was the only method by which both the "cold" and radiolabeled compounds could be obtained. The method depends on allylic chlorination at the 11- or 8-positions of Δ^9-THC with thionyl chloride followed by formation of corresponding acetates and hydrolysis of the latter. Fortunately, all of the compounds can be separated by careful chromatography. An improved procedure[16,17] which yields 11-hydroxy-Δ^9-THC in 35% yield is shown in Figure 11. Although this procedure involves the formation of several byproducts, most of these can be carried along and are minor until the terminal stages of the reaction. At that time the impurities can be readily purified by chromatography. The overall yield of 35% makes this currently the method of choice for the synthesis of either unlabeled or radiolabeled Δ^9-11-hydroxy-THC. A procedure described by Razdan and coworkers[19] utilizes as a precursor the exo-methylene, as shown in Figure 12. This precursor is relatively unavailable to most workers, but it can, if necessary, be made by total synthesis. In any case, as indicated, the compound can be converted into a mixture of Δ^8- and Δ^9-11-acetoxy-THC. After hydrolysis the mixture can be separated by high pressure liquid chromatography. Razdan claims that the procedure has preparative value.

There is also considerable interest in 11-nor-Δ^8- or Δ^9-THC-9-carboxylic acid (11-carboxy-Δ^9-THC). A synthesis for the Δ^8- analog is shown in Figure 13. This method[18] can also be used for the facile synthesis of 11-hydroxy-Δ^8-THC. In the case of the Δ^9- series, the synthesis of

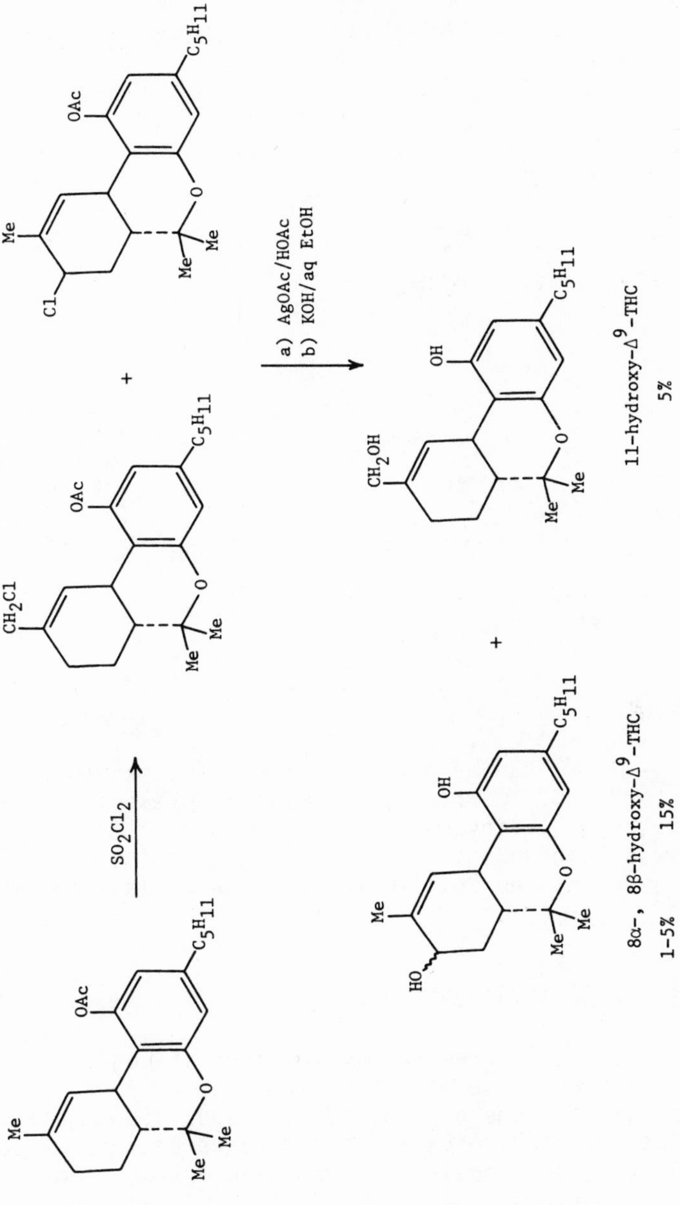

Me
OAc
C_5H_{11}
Me
Me
SO_2Cl_2
CH_2Cl
OAc
C_5H_{11}
Me
Me
+
Cl
Me
OAc
C_5H_{11}
Me
Me
a) AgOAc/HOAc
b) KOH/aq EtOH
CH_2OH
OH
C_5H_{11}
Me
Me
11-hydroxy-Δ^9-THC
5%
+
HO
Me
OH
C_5H_{11}
Me
Me
8α-, 8β-hydroxy-Δ^9-THC
1-5% 15%

Figure 10

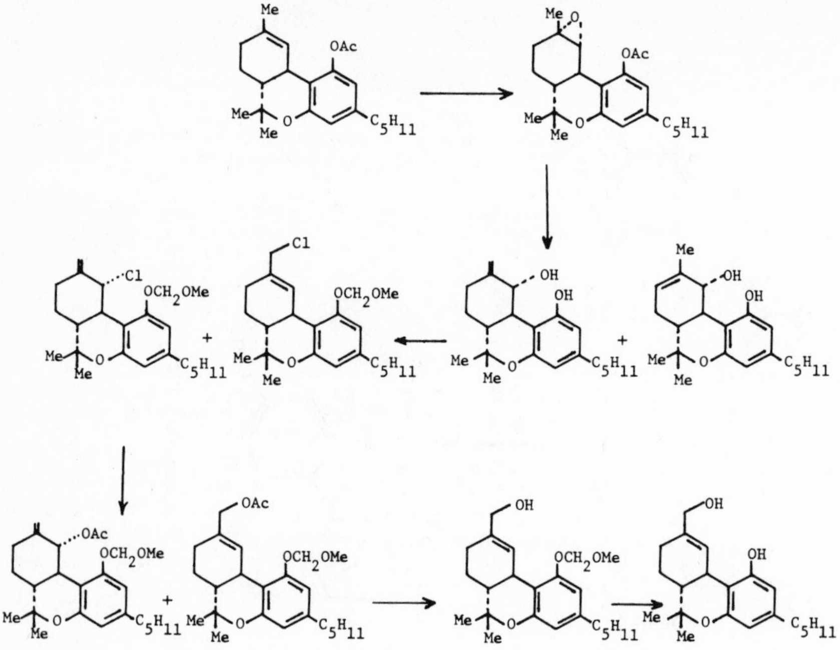

Figure 11

11-carboxy-Δ^9-THC can be prepared by careful oxidation of 11-hydroxy-Δ^9-THC. Currently there is interest in the synthesis of the 8,11-dihydroxy-Δ^9-THC because of the possible use of these compounds in the treatment of glaucoma. The synthesis[16,17] of the 8α,11-dihydroxy analog is shown in Figure 14.

METABOLISM OF Δ^9-TETRAHYDROCANNABINOL

The advances in chemical methodology discussed in the preceding sections have permitted remarkable progress to be made in studies of the physiological disposition and metabolism of the cannabinoids in recent years because of the availability of the requisite pure unlabeled and radiolabeled substrates. Because of the large volume of material to be discussed (see reviews[1-20] for more complete details) only

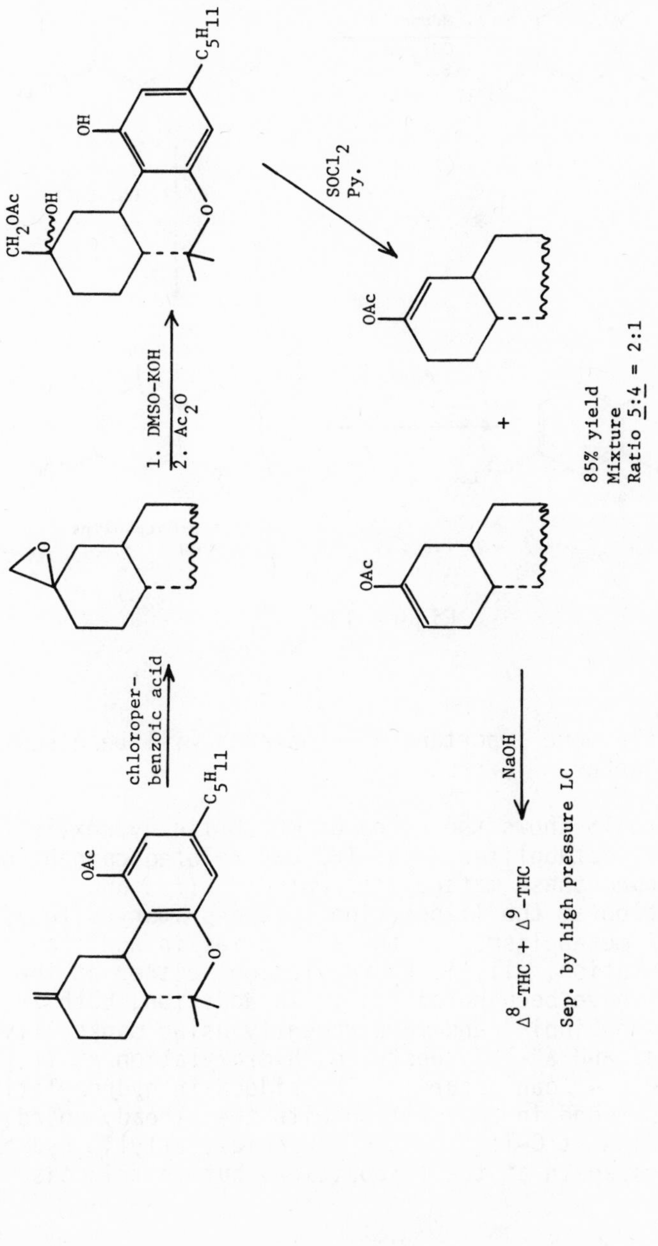

OAc

C_5H_{11}

chloroper-
benzoic acid

O

1. DMSO-KOH
2. Ac_2O

CH_2OAc

OH

OH

C_5H_{11}

$SOCl_2$
Py.

OAc

+

OAc

85% yield
Mixture
Ratio 5:4 = 2:1

NaOH

Δ^8-THC + Δ^9-THC

Sep. by high pressure LC

Figure 12

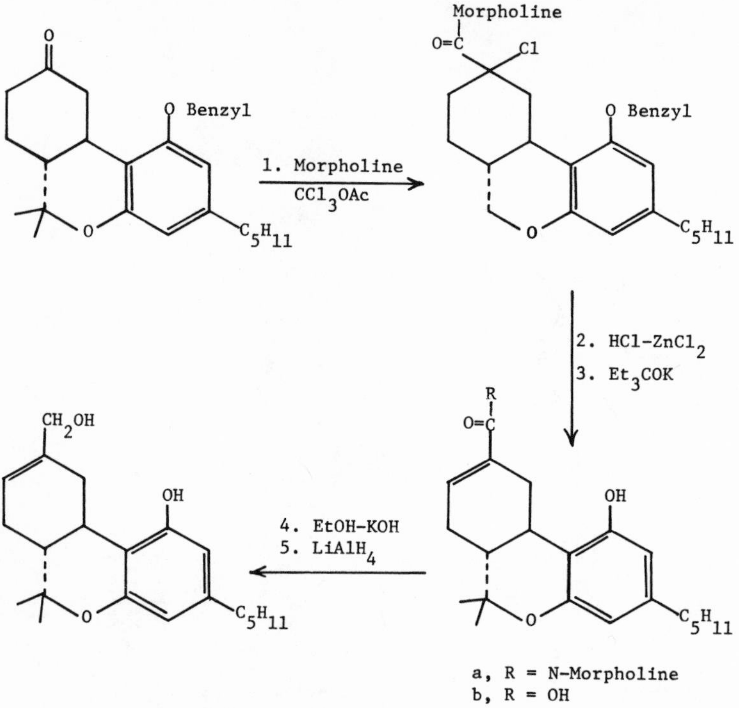

Figure 13

a few of the more important developments will be discussed in this chapter.

Figure 15 shows the sites of metabolic hydroxylation for neutral metabolites of Δ^9-THC and related cannabinoids. We have found these metabolites both *in vitro* and *in vivo*. Hydroxylation at the 11-position is the primary site of microsomal metabolism. In the Δ^9- series, in addition to 11-hydroxylation, allylic hydroxylations either at the 8α- or the 8β- have been noted[21,22]. In addition, both in the case of cannabinol[22] and more recently using monkey liver homogenates and Δ^9-THC substrate, hydroxylation at 1',2',3' and possibly 4' can occur[21]. The sidechain hydroxylation was always found in conjunction with the already noted hydroxylation at C-11. In the Δ^8-series, allylic hydroxylation occurs again at the 11-position, but in this case

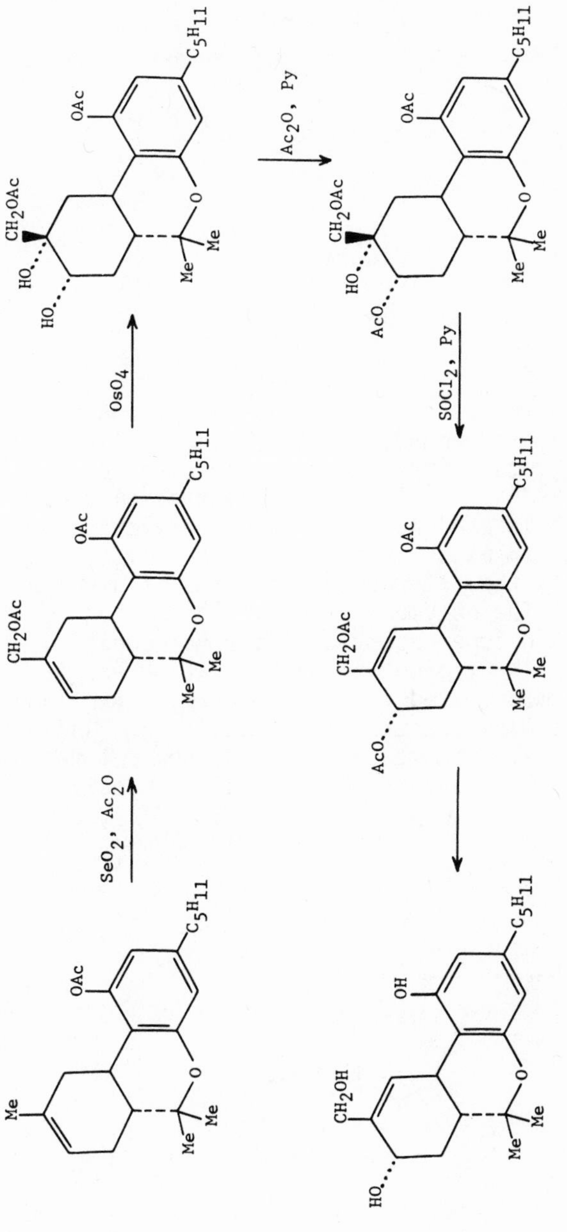

Figure 14

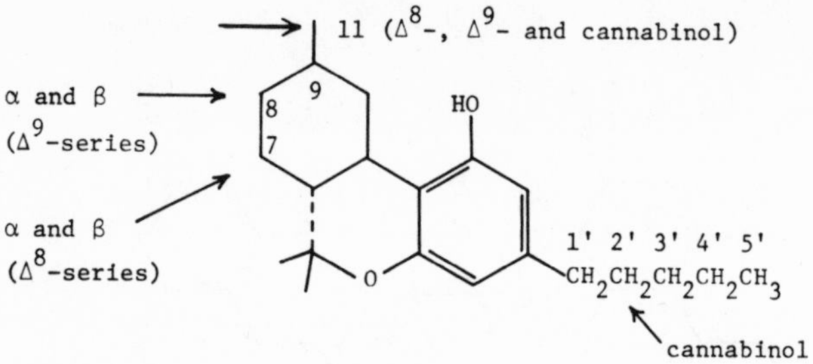

11 (Δ^8-, Δ^9- and cannabinol)
α and β (Δ^9-series)
α and β (Δ^8-series)
HO
1' 2' 3' 4' 5'
$CH_2CH_2CH_2CH_2CH_3$
cannabinol

Figure 15

additional hydroxylation at the 7α- or 7β-position can occur, usually secondarily to the primary 11-hydroxylation [21,22]. Acid metabolites have usually been found after *in vivo* studies[23,24,25]. 11-Carboxy cannabinoids have been found as such in conjunction with hydroxylation at C-8 or in the sidechain as shown in Figure 16. 11-Hydroxy-Δ^9-THC is undoubtedly the intermediate for the acidic 11-carboxy metabolites. The enzyme system for this oxidation is found in the 9000 × g supernatant of liver homogenates, which also contains the microsomal hydroxylation enzymes. Recently we have demonstrated this with monkey liver homogenates, utilizing glc-mass spectrometry techniques. Under these conditions, Δ^9-THC formed a number of neutral metabolites. In addition, 11-carboxy-Δ^9-THC was also unequivocally identified[25,26], as major component *in vitro*, along with

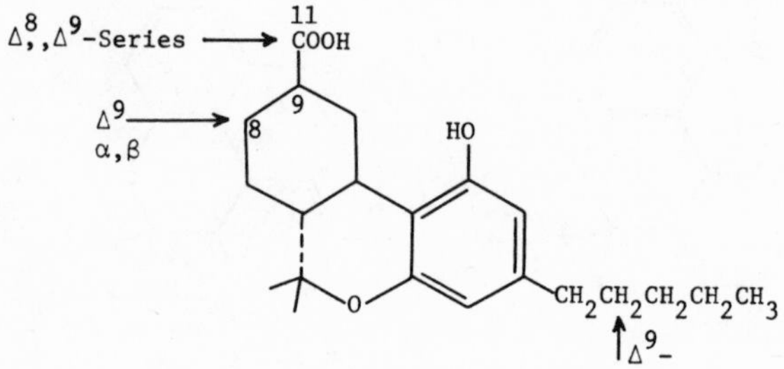

Δ^8,Δ^9-Series
11 COOH
Δ^9 α,β
HO
$CH_2CH_2CH_2CH_2CH_3$
Δ^9-

Figure 16

11-hydroxy-Δ^9-THC.

At the Research Triangle Institute we have been studying the metabolism of Δ^9-THC and related compounds in man for the last five years. During this time we have perfected techniques for the extraction and analysis of radiolabeled substrates and for the qualitative analysis of unlabeled substrates using gas/liquid chromatography-mass spectrometry (glc-ms). Charts 1, 2 and 3 summarize the procedures for the extraction and analysis of human blood, urine and feces following the administration of radiolabeled samples. Many of our studies have been done with radiolabeled Δ^9-THC administered parenterally by the intravenous route. In order to conduct these studies, a special procedure for emulsifying the water-insoluble THC had to be developed[27]. The general methodology is presented in Table 1. The neutral and acidic metabolites can be well separated by thin layer chromatography, and the zones in which these metabolites are found are scraped and then counted in scintillation counters. Figure 17 shows typical thin layer separations obtained in our laboratory. Many of these compounds have been verified by utilizing large samples of plasma or feces; after purification, glc-ms techniques have been utilized for the final identification. Table 2 shows the type of glc separation that can be obtained with these metabolites. It can be seen that Δ^9-THC is well separated from the 8-hydroxy-Δ^9-THC epimers which in turn are well separated from 11-hydroxy-Δ^9-THC. We have demonstrated that many of the THC metabolites can be identified by their glc-ms patterns[25]. When the mass spectrometric data is subjected to computerized techniques, even compounds such as the 8α-hydroxy- and 8β-Δ^9-THC, which have virtually the same retention time, can be distinguished by noting in sequence the mass spectra obtained in the various areas under the total ionization curve. In this case it became evident that 8α-hydroxy-Δ^9-THC, which has a slightly shorter retention time than the 8β-analog, is found in the early portion of the TIC plot. Since the mass spectrum of the former is significantly different from that of the 8β-analog, the two epimers can be conveniently differentiated in this manner. Using the techniques of combined glc-ms, all of the Δ^9-THC metabolites initially identified by radiolabeling techniques were unequivocally shown to be present [25-26].

With this background, we are now in a position to

Chart 1

Standard Analysis of Plasma Sample
Example of 3 ml Plasma

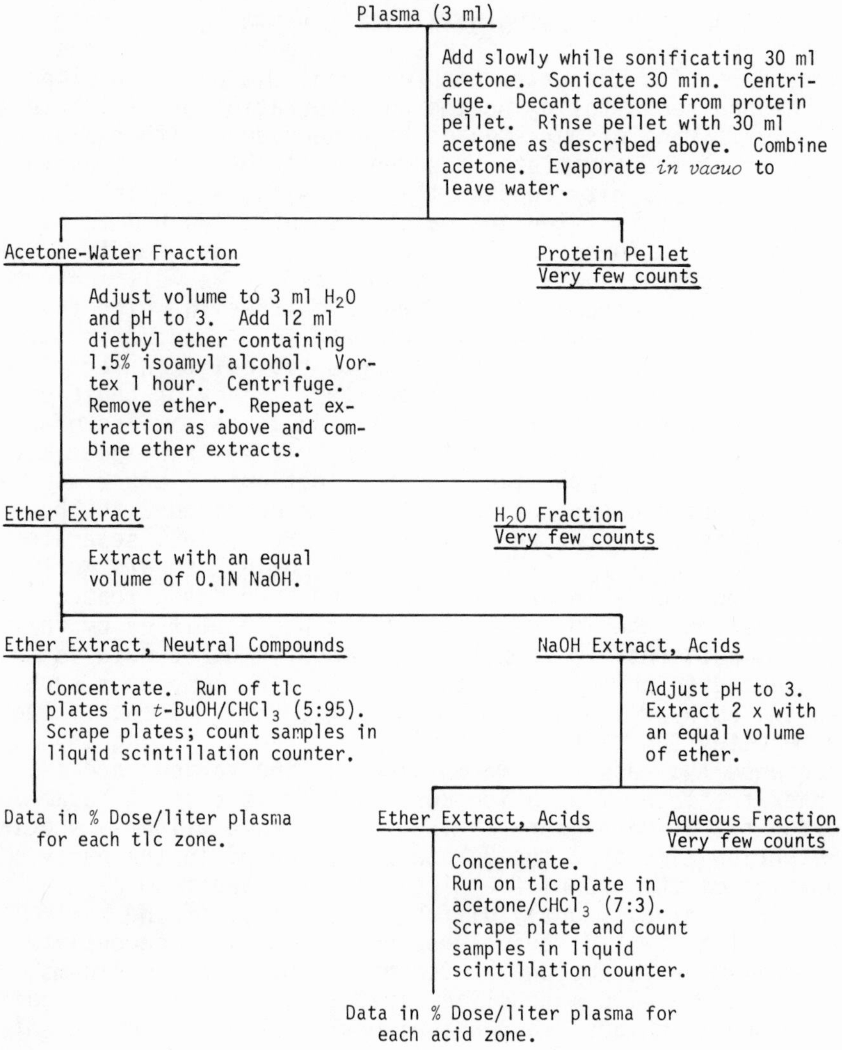

Plasma (3 ml)

Add slowly while sonificating 30 ml acetone. Sonicate 30 min. Centrifuge. Decant acetone from protein pellet. Rinse pellet with 30 ml acetone as described above. Combine acetone. Evaporate *in vacuo* to leave water.

Acetone-Water Fraction

Adjust volume to 3 ml H_2O and pH to 3. Add 12 ml diethyl ether containing 1.5% isoamyl alcohol. Vortex 1 hour. Centrifuge. Remove ether. Repeat extraction as above and combine ether extracts.

Protein Pellet
Very few counts

Ether Extract

Extract with an equal volume of 0.1N NaOH.

H_2O Fraction
Very few counts

Ether Extract, Neutral Compounds

Concentrate. Run of tlc plates in *t*-BuOH/$CHCl_3$ (5:95). Scrape plates; count samples in liquid scintillation counter.

Data in % Dose/liter plasma for each tlc zone.

NaOH Extract, Acids

Adjust pH to 3. Extract 2 x with an equal volume of ether.

Ether Extract, Acids

Concentrate. Run on tlc plate in acetone/$CHCl_3$ (7:3). Scrape plate and count samples in liquid scintillation counter.

Data in % Dose/liter plasma for each acid zone.

Aqueous Fraction
Very few counts

Chart 2

Standard Analysis of Urine Samples

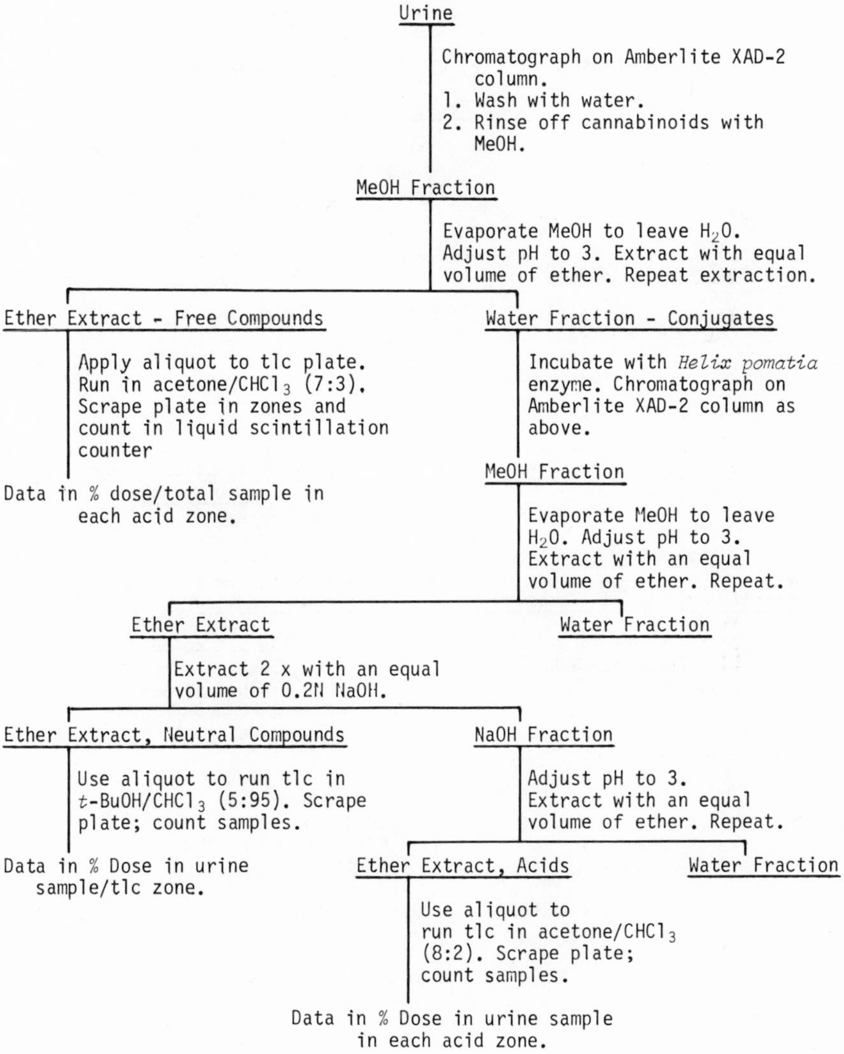

Urine

Chromatograph on Amberlite XAD-2 column.
1. Wash with water.
2. Rinse off cannabinoids with MeOH.

MeOH Fraction

Evaporate MeOH to leave H_2O. Adjust pH to 3. Extract with equal volume of ether. Repeat extraction.

Ether Extract - Free Compounds

Apply aliquot to tlc plate. Run in acetone/$CHCl_3$ (7:3). Scrape plate in zones and count in liquid scintillation counter

Data in % dose/total sample in each acid zone.

Water Fraction - Conjugates

Incubate with *Helix pomatia* enzyme. Chromatograph on Amberlite XAD-2 column as above.

MeOH Fraction

Evaporate MeOH to leave H_2O. Adjust pH to 3. Extract with an equal volume of ether. Repeat.

Ether Extract

Extract 2 x with an equal volume of 0.2N NaOH.

Water Fraction

Ether Extract, Neutral Compounds

Use aliquot to run tlc in *t*-BuOH/$CHCl_3$ (5:95). Scrape plate; count samples.

Data in % Dose in urine sample/tlc zone.

NaOH Fraction

Adjust pH to 3. Extract with an equal volume of ether. Repeat.

Ether Extract, Acids

Use aliquot to run tlc in acetone/$CHCl_3$ (8:2). Scrape plate; count samples.

Data in % Dose in urine sample in each acid zone.

Water Fraction

Chart 3

Standard Analysis of Feces Samples

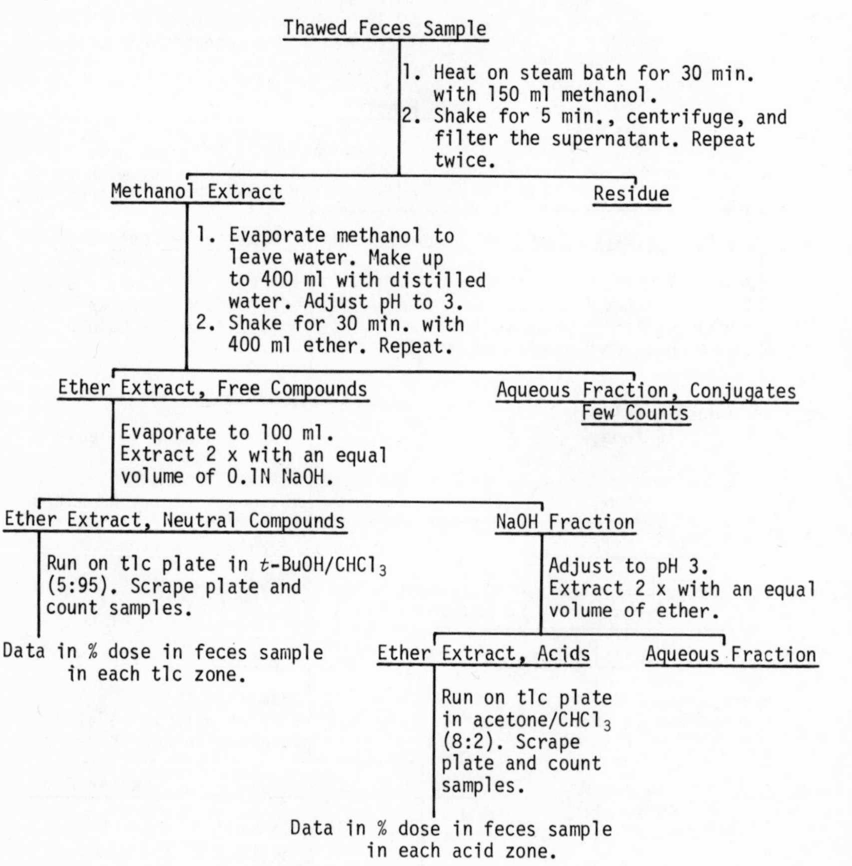
Thawed Feces Sample
1. Heat on steam bath for 30 min. with 150 ml methanol.
2. Shake for 5 min., centrifuge, and filter the supernatant. Repeat twice.
Methanol Extract
Residue
1. Evaporate methanol to leave water. Make up to 400 ml with distilled water. Adjust pH to 3.
2. Shake for 30 min. with 400 ml ether. Repeat.
Ether Extract, Free Compounds
Aqueous Fraction, Conjugates Few Counts
Evaporate to 100 ml. Extract 2 x with an equal volume of 0.1N NaOH.
Ether Extract, Neutral Compounds
NaOH Fraction
Run on tlc plate in t-BuOH/$CHCl_3$ (5:95). Scrape plate and count samples.
Adjust to pH 3. Extract 2 x with an equal volume of ether.
Data in % dose in feces sample in each tlc zone.
Ether Extract, Acids
Aqueous Fraction
Run on tlc plate in acetone/$CHCl_3$ (8:2). Scrape plate and count samples.
Data in % dose in feces sample in each acid zone.

Table 1. Injectable Form of Δ^9-THC and Related Compounds

1. 10 mg THC dissolved in 0.5 ml EtOH
2. Mix with 50 ml HSA, with vigorous stirring
3. Filter through 0.22 μm Millipore filter
4. Administer dose via Harvard Constant Infusion pump at rate of 0.92 ml/min

Table 2. Glc Retention Times of TMS Ethers of Δ^9-THC and Metabolites. Column: 1% OV-17, 6 ft.

Neutrals		
TMS Ether of	220° C (min)	210 → 230° C (min)
Δ^9-THC	2.7	8.0
1'-OH-Δ^8-THC	3.3	9.3
8 -OH-Δ^9-THC	5.3	13.4
8 -OH-Δ^9-THC	5.4	13.5
3'-OH-Δ^8-THC	5.6	13.8
11-OH-Δ^9-THC	7.0	16.1
8α-11-diOH-Δ^9-THC	7.8	17.8
8β-11-diOH-Δ^9-THC	10.0	20.7
Acids		
TMS Ether of	247° C (min)	240° C (min)
11-COOH-Δ^9-THC	4.5	6.0
11-COOH, ξ-OH-Δ^9-THC	8.5	10.8
11-COOH, 8β-OH-Δ^9-THC	6.4	8.6

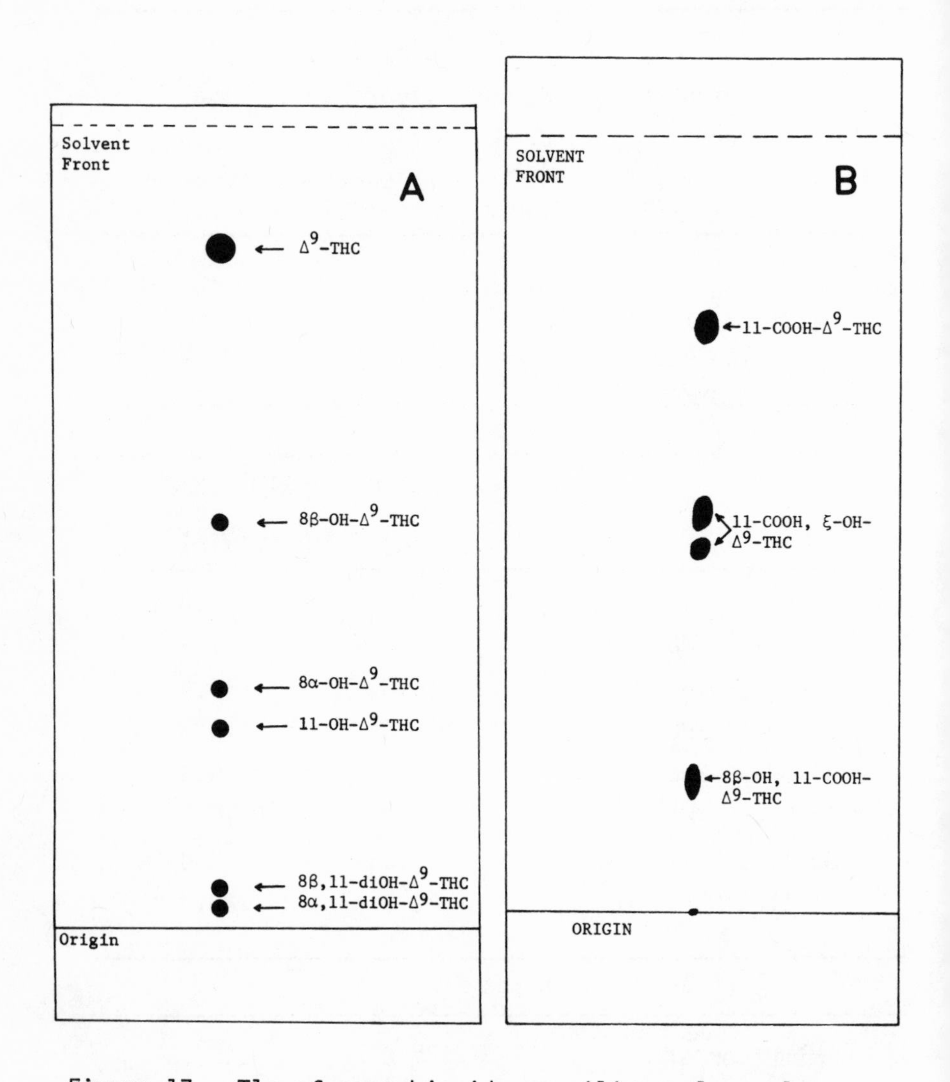

Figure 17. Tlc of cannabinoids on silica gel on glass.
A: Neutral cannabinoids; solvent: *t*-BuOH/$CHCL_3$ (5:95).
B: Acidic cannabinoids; solvent: acetone/ $CHCL_3$ (8:2).

discuss the metabolism of Δ^9-THC and its 11-hydroxy analog in man.

The data for total free and conjugated cannabinoids found in the plasm after IV infusion of approximately 4.6 mg of Δ^9-THC are shown in Figure 18, along with the record of the subjective psychological effects. It is apparent that the peak cannabinoid content and the psychological "high" occur at about the same time, i.e., approximately 30 minutes after administration commences. The cannabinoid content falls off rather slowly during the first 90 minutes then continues this decline over 72 hours. It will be noted that most of the cannabinoids were in the free form in the plasma, the conjugated fraction being about 5 times smaller. It can be seen that the total cannabinoid content was approximately 2% of the dose/liter plasma. This means that about 6% of the total dose may be considered to be in the blood at the time of the highest level, approximately 30 minutes after administration commenced. By 90 minutes this value had fallen rather slowly to about 0.75% dose/liter. Figure 19 shows the major constituents found in the plasma when Δ^9-THC and its 11-hydroxy metabolite are administered intravenously and also when Δ^9-THC is administered by smoking.

Turning first to the Δ^9-THC shown in Figure 19A, we see that immediately after administration of the parent substance a number of metabolites are formed. These increase, reach a maximum within 30 minutes and then slowly fall at a rate which is not greatly different from that of the parent substance. The predominant metabolite is 11-carboxy-Δ^9-THC, which has been shown by glc-ms techniques to be present in plasma, feces and urine, along with a much less well defined mixture which we call polar acids or polyhydroxy acids, which are found in equal or greater quantity. The polar acids are extractable with ether from plasma or urine and can be removed from ether with sodium hydroxide. They receive their designation "polar acids" because they do not move from the origin under conditions which readily move monohydroxylated carboxy acids. In consequence, we believe they are polyhydroxylated or possibly di- or trimeric. It is attractive to believe that they are formed by further metabolism of the 11-carboxy-Δ^9-THC, as they are formed almost simultaneously. However, this remains to be proven. Of particular interest is the low level of 11-hydroxy-Δ^9-THC. It can be seen that it is

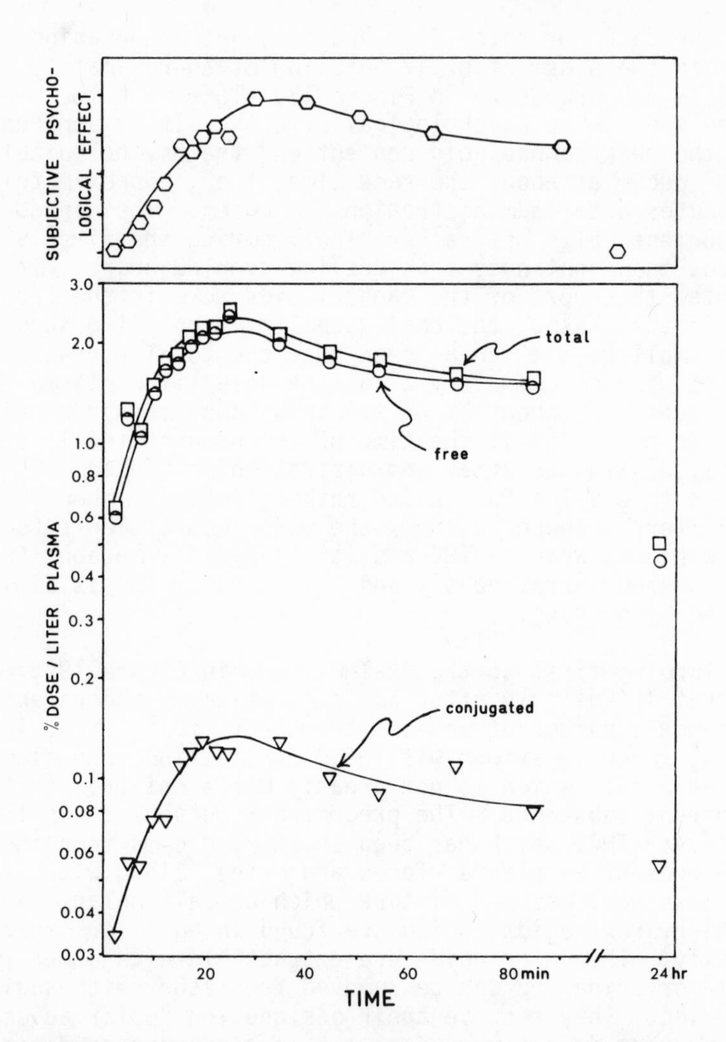

Figure 18. Subjective Psychological Effects and Total, Free, and Conjugated Cannabinoids Found in the Plasma after Administration of 4 mg Δ^9-THC via Smoking. (Average of 5 Subjects

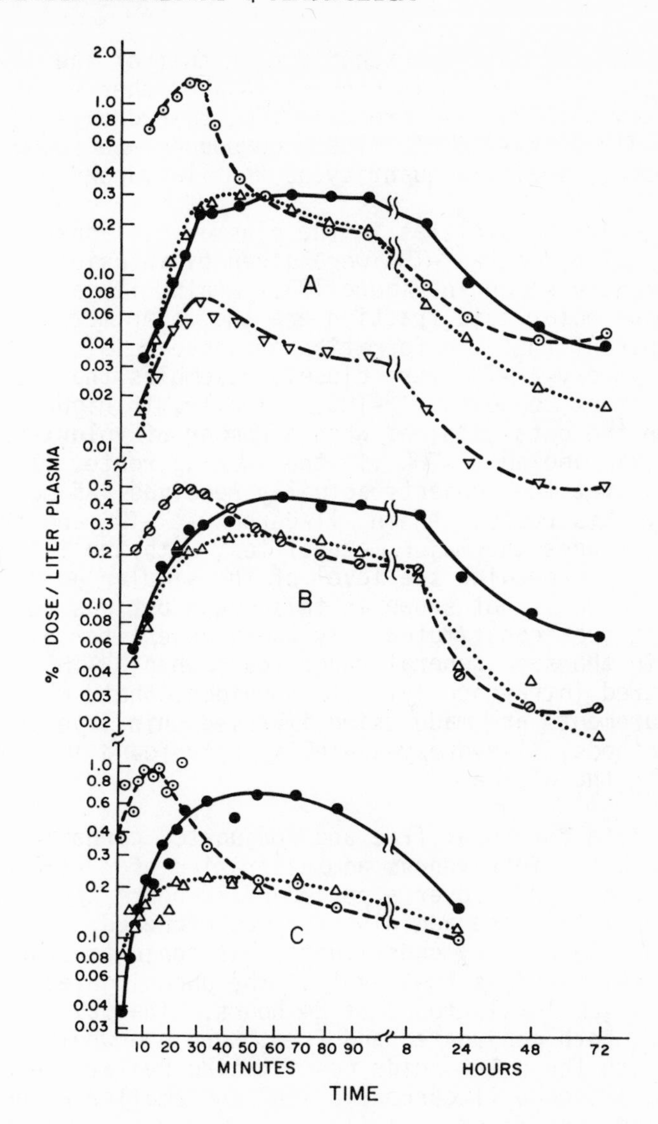

Figure 19. Metabolism of Δ^9-THC (A) and 11-OH-Δ^9-THC (B) administered intravenously and of Δ^9-THC administered by smoking (C).

A: ⊙, Δ^9-THC; △, 11-COOH-Δ^9-THC; ●, polar acids; ▽, 11-OH-Δ^9-THC.

B: ⊘, 11-OH-Δ^9-THC; △, 11-COOH-Δ^9-THC; ●, polar acids.

C: ⊙, Δ^9-THC; △, 11-COOH-Δ^9-THC; ●, polar acids.

found in amounts only one-twentieth of that of the Δ^9-THC at its peak. Not shown in Figure 19A are other minor metabolites such as 8β-hydroxy-Δ^9-THC, 8α-hydroxy-Δ^9-THC and 8,11-dihydroxy-Δ^9-THC. These compounds are found in approximately the same quantity as the 11-hydroxy metabolite.

The major metabolites in the plasma when approximately 4.8 mg of 11-hydroxy-Δ^9-THC were given by infusion to 7 volunteers are shown in Figure 19B. Again, 11-carboxy-Δ^9-THC and the polar acid fraction are the major metabolites. It is apparent that the formation of these acidic substances from 11-hydroxy-Δ^9-THC very closely resembles that found for the parent compound, Δ^9-THC. Finally, in Figure 19C are shown the data obtained when a number of volunteers received radiolabeled Δ^9-THC via the smoking route. It was estimated that the subjects actually received 4-5 mg of Δ^9-THC by this route. Again, 11-carboxy-Δ^9-THC and the polar acids were the major metabolites, with the latter now considerably exceeding the level of the simpler acid. 11-Hydroxy-Δ^9-THC is not shown in this graph but was determined and at its peak constituted only 0.03% dose/liter plasma, a level in the same general range found when Δ^9-THC was administered intravenously. It is evident that, when careful measurements are made using improved thin layer chromatography methods, 11-hydroxy-Δ^9-THC is only found in minor amounts in the plasma.

The data for total free and conjugated cannabinoids in the urine after intravenous administration of Δ^9-THC are shown in Figure 20. Over a period of 72 hours approximately 12% of the total dose administered was excreted. The major portion of the urinary constituents was conjugated and was approximately 3 times the level of the unconjugated fraction at the maximum levels found at 24 hours. The urinary constituents, both conjugated and free, were overwhelmingly acidic, with the polar acids constituting by far the major fraction, although 11-carboxy-Δ^9-THC and smaller amounts of other acids were found. Similar results were found for 11-hydroxy-Δ^9-THC and hence will not be presented.

Turning next to the data found for the analysis of the feces after Δ^9-THC and 11-hydroxy-Δ^9-THC were administered, we find that over a 72 hour period approximately 30% of the total dose was excreted in both cases. In terms of the individual cannabinoids, it was found that there was only a minor proportion of conjugated cannabinoids in the feces.

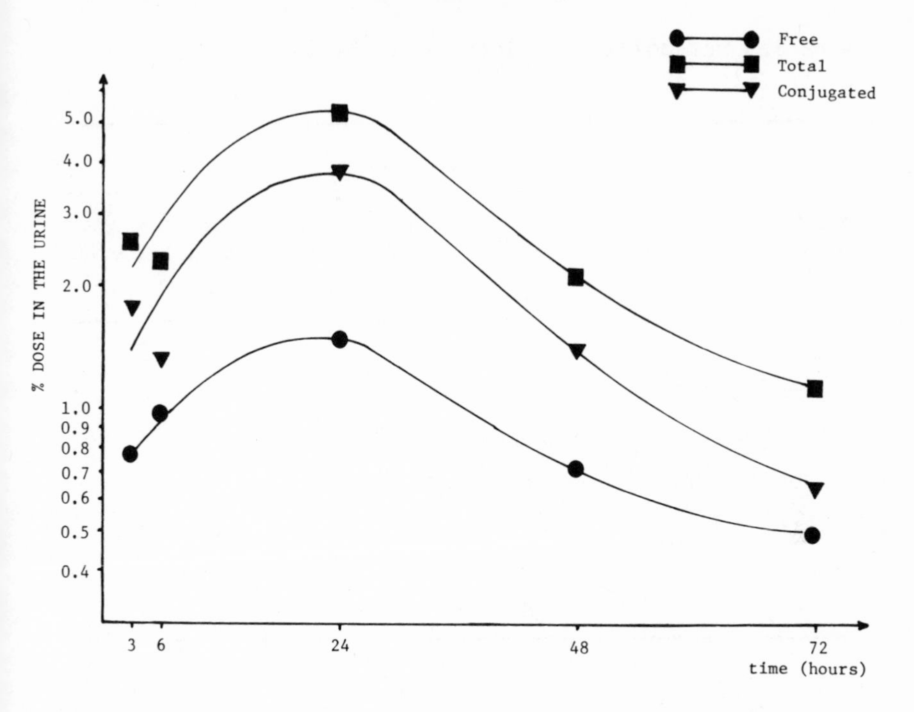

Figure 20. Total, free, and conjugated cannabinoids in urine. (Averages of 6 subjects following intravenous infusion of 4.6 mg Δ^9-THC)

The composition of the metabolites in the 24-hour feces sample is shown in Table 3. In addition to Δ^9-THC, data are also presented for 11-hydroxy- and 8β-hydroxy metabolites. The feces data show that when Δ^9-THC is administered, in marked contrast to the situation in plasma or urine, 11-hydroxy-Δ^9-THC is one of the major metabolites, along with the 11-carboxy and polar acid fraction. 11-Hydroxy-Δ^9-THC shows a rather large amount of unmetabolized starting material, i.e., about 22%, in contrast to the Δ^9- or the 8β-hydroxy compound. 11-Carboxy-Δ^9-THC and the polar acids are the major metabolites as in the case of Δ^9-THC. Again, with the 8β-hydroxy-Δ^9-THC, the polar acids are by far the major metabolites, with a compound shown to be 8,11-dihydroxy-Δ^9-THC as the next major compound found. The 11-hydroxy-Δ^9-THC found after administration of the

Table 3. Metabolites of Intravenously Infused Δ^9-THC, 11-OH-Δ^9-THC and 8β-OH-Δ^9-THC.

Free Cannabinoids in 24 hr feces	Substrate		
	Δ^9-THC	11-OH-Δ^9-THC	8β-OH-Δ^9-THC
	(% of total cannabinoids)		
Polar acids	29	41	72
Intermediate Acids	9	10	-
8,11-diOH-Δ^9-THC	Minor	Minor	24
11-COOH-Δ^9-THC	29	28	-
11-OH-Δ^9-THC	21	22	-
8β-OH-Δ^9-THC	5	-	4
Δ^9-THC	8	-	-

Δ^9- parent substance is most likely formed by metabolism in the liver and not by microorganisms in the gut. This is indicated by the fact that administration of Δ^9-THC rectally to several volunteers and then analysis after 24 hours showed that virtually no change had occurred in the starting Δ^9-THC. It is conceivable, of course, that the 11-hydroxy-Δ^9-THC is conjugated in the liver and is excreted in the bile in this form and is then hydrolyzed during passage through the small and large intestines. Our point is that 11-hydroxylation occurs via the liver microsomal enzyme system.

We interpret the data previously presented to show that the main course of metabolism of Δ^9-THC proceeds via its 11-hydroxy analog as the key compound in the metabolic series. This allylic hydroxylation proceeds rapidly and readily. Subsequently, additional hydroxylation at 8α- or 8β- positions or in the sidechain may occur. All of these metabolic hydroxylations are brought about by the liver microsomal enzyme system. In addition, another enzyme system can convert the 11-hydroxy-Δ^9-THC to the corresponding 11-carboxy-Δ^9-THC, presumably via the common alcohol dehydrogenase mechanism. Although 11-hydroxy-Δ^9-THC is readily formed both *in vitro* and *in vivo*, it is evident from our data that little is found in the plasma. Apparently, this metabolite is preferentially excreted via the bile and, as noted earlier, constitutes a major metabolite found in the feces. Because 11-hydroxy-Δ^9-THC is found in only relatively small quantities in the plasma, we deduce that large amounts of this compound cannot get into the brain. From our studies[28] in which it was found that Δ^9-THC and 11-hydroxy-Δ^9-THC have very similar biological activities in man, we believe that 11-hydroxy-Δ^9-THC cannot be the only active form. We deduce this since, as stated earlier, it is found in such very low quantities in the plasma, and if this indeed were the only active form then one would assume it would have to be biologically very potent. Accordingly, it is our belief that a number of cannabinoids, including Δ^9-THC, 11-hydroxy-Δ^9-THC and several others, including even cannabinol, can be active in the proportions in which they are received at an appropriate receptor site, which is presumably in the brain.

ACKNOWLEDGEMENT

The studies reported in this review were carried out under contracts with the NIDA. Metabolism work was supported under contract HSM-42-71-95 and preceding contracts. The syntheses studies were conducted under contract HSM-42-71-108 and preceding contracts. We wish to express appreciation to a number of my colleagues for both their collaboration and permission to quote some of their unpublished work. Most of the chemical syntheses reported were carried out under the supervision of Dr. Colin G. Pitt. Dr. Mario Perez-Reyes, M.D., School of Medicine, University of North Carolina at Chapel Hill, was responsible for the clinical protocol described in this paper. Ms. Dolores R. Brine was of major assistance in conducting the metabolic investigations.

REFERENCES

1. Mechoulam, R. 1973. "Marijuana: Chemistry, Pharmacology, Metabolism and Clinical Effects" pp. 1-99. Academic Press, New York.
2. Neumeyer, J.L. & R.A. Shagoury. 1971. J. Pharm. Sci., 60: 1433.

3. Mechoulam, R. & Y. Gaoni. 1967. Tetrahedron Lett., 1109.
4. _____ & Y. Gaoni. 1967. Fortsch. Chem. Org. Naturst., 25: 175.
5. Lemberger, L. 1972. In "Advances in Pharmacology and Chemotherapy" (S. Garratini, F. Hawking, A. Golden & I. Kopin, eds.), pp. 221-251. Academic Press, New York.
6. Gaoni, Y. & R. Mechoulam. 1964. J. Am. Chem. Soc., 86: 1646.
7. _____ & R. Mechoulam. 1971. J. Am. Chem. Soc., 93: 217.
8. Archer, R.A., D.B. Boyd, P.V. Demarco, I.J. Tyminski & N.L. Allinger. 1970. J. Am. Chem. Soc., 92: 5200.
9. Davis, K.H., N.H. Martin, C.G. Pitt, J.W. Wildes & M.E. Wall. 1971. Lloydia, 33(4): 453.
10. Merkus, F.W.H.M. 1971. Nature, 232: 579.
11. Petrzilka, T., W. Haefliger, C. Sikemeier, G. Ohloff & A. Eschenimoser. 1967. Helv. Chim. Acta, 50: 719.
12. _____, W. Haefliger, C. Sikemeier. 1969. Helv. Chim. Acta, 52: 1102.
13. Timmons, M.C., C.G. Pitt & M.E. Wall. 1969. Tetrahedron Lett., 36: 3129.
14. Razdan, R.K., H.C. Dalzell & G.R. Handrick. 1974. J. Am. Chem. Soc., 96: 5860.
15. Gill, E.W. & G. Jones. 1972. J. Label. Compounds, 8: 237.
16. Pitt, C.G. & M.E. Wall. 1974. Ann. Rept., Synthesis of Radiolabeled Cannabinoids and Metabolites, Contract PH-42-71-108, NIDA.
17. _____, M. Fowler, S. Sathe, S.C. Srivastava & D. Williams 1975. J. Am. Chem. Soc. (in press).
18. _____, F. Hauser, R.L. Hawks, S. Sathe & M.E. Wall. 1972. J. Am. Chem. Soc., 94: 8578.
19. Razdan, R.K., D.B. Uliss & H.C. Dalzell. 1973. J. Am. Chem. Soc., 95: 2361.
20. Paton, W.D.M. & J. Crown. 1972. "Cannabis and Its Derivatives: Pharmacology and Experimental Psychology" Oxford Univ. Press, London.
21. Wall, M.E., D.R. Brine, G.A. Brine, C.G. Pitt, R.I. Freudenthal & H.D. Christensen. 1970. J. Am. Chem. Soc., 92: 3466.
22. _____. 1971. Ann. N.Y. Acad. Sci., 191: 23.
23. Burstein, S., J. Rosenfeld & T. Wittstruck. 1972. Science 176: 422.
24. Nilsson, I.M., S. Agurell, J.L.G. Nilsson, A. Ohlsson, J.E. Lindren & R. Mechoulam. 1973. Acta Pharmaceutica Suecica, 19: 97.
25. Wall, M.E. & D.R. Brine. 1973. Internat. Symp. Mass

Spectrom. in Biochem. and Med., May 7-8, Milan, Italy.
26. ____, D.R. Brine, C.G. Pitt & M. Perez-Reyes. 1972. J. Am. Chem. Soc., 94: 8579.
27. Perez-Reyes, M., M.C. Timmons, M.A. Lipton, K.H. Davis & M.E. Wall. 1972. Science, 177: 633.
28. ____, M.C. Timmons, M.H. Lipton, H.D. Christensen, K.H. Davis & M.E. Wall. 1973. Experientia 29: 1009.

Chapter Three

ON THE CARCINOGENICITY OF MARIJUANA SMOKE

D. HOFFMANN, K.D. BRUNNEMANN, G.B. GORI AND
E.L. WYNDER

*American Health Foundation, New York, and
National Cancer Institute, Bethesda, Maryland*

INTRODUCTION

Marijuana *(Cannabis sativa* L.*)* has become the second most widely used smoke product in the Western World[1,2,3,4]. Marijuana smoke, however, is inhaled in significantly lower doses than tobacco smoke and its effect is predominantly psychotomimetic with some acute toxic side effects. Nevertheless, information as to the carcinogenicity of this inhalant is needed. Since most marijuana smokers are also cigarette smokers, it needs furthermore to be determined whether marijuana smoke can potentiate the carcinogenic effect of tobacco smoke.

Until now, no data have been published as to the presence of tumorigenic agents in and the carcinogenic effects of marijuana smoke[4,5,6]. Recently, a short-term test on mouse skin showed that the particulate matter of marijuana smoke ("tar") induced dose-related sebaceous gland destruction and epidermal hyperplasia with acenthosis[7,8]. Marijuana smoke has been reported to increase abnormal cell proliferation in epithelial cells of lung explants from mice[9].

The present study was designed to determine the carcinogenic potential of marijuana smoke. In the first phase we smoked marijuana cigarettes under conditions established for tobacco cigarettes[11]; we analyzed the mainstream smoke for toxic and tumorigenic agents and bioassayed its particulate matter for carcinogenicity and tumor-promoting activity on mouse skin. For comparison and as a positive control, we analyzed the smoke of a blended tobacco cigarette and tested its particulate matter for tumorigenicity and tumor-promoting activity.

EXPERIMENTAL METHODS

Marijuana Cigarettes

The marijuana cigarettes were made available by the Division of Cancer Cause and Prevention of the National Cancer Institute and were prepared from marijuana bricks which originated from confiscations. The leaves were separated from the stems and twigs and cut to about 30 cuts per inch. No humectants or other additives were used for the marijuana blend. The paper used for the cigarettes had comparable porosity to the paper used for the tobacco control cigarettes.

Tobacco Cigarettes

Both the tobacco and the marijuana cigarettes were 85 mm long and without filter tips. The tobacco cigarettes were the standard blended cigarettes SEB-I of the Tobacco Working Group of the National Cancer Institute.

Smoking of Cigarettes

The cigarettes for the chemical studies were smoked with 20-channel[12] and single-channel (H. Borgwalt, Hamburg) smoking machines. The cigarettes for the bioassays were smoked with a multiple unit automatic machine.

Analytical Methods

In Table 1 we have summarized the methods used in this study. For each individual determination in which we needed 10 or less cigarettes, we selected the cigarettes by weight and draw resistance. The experimental deviations of the analytical methods averaged between 4 and 6%, and, in the case of polynuclear aromatic hydrocarbons (PAH), within ±8%.

Determination of Cannabinols

For each analysis, 10 cigarettes were smoked on a 20-port Phipps & Bird smoking machine through Cambridge filters. The loaded filters were extracted with 10 ml absolute ethanol, and the extracts were analyzed by glc. The separation was achieved under the following conditions: 2 m x 3 mm stainless steel column filled with 1.4% OV-17 on Gas Chrom Q (100/120 mesh). The temperatures were: injector block, 250°C; column 200°C; and FID, 290°C. The carrier gas was helium, at a flow rate of 40 ml/min. Under these conditions the retention times were: cannabidiol (CBD), 8.5 min.; Δ'-tetrahydrocannabinol (Δ'-THC), 11.5 min.; and cannabinol (CBN), 16.5 min. The detection limit for Δ'-THC was about 10 ng.

Bioassays

For the bioassays we used Swiss albino female mice (Ha/ICR/Mil). The application of the test suspension was started during the second telogen phase of the hair cycle[23].

a. Complete Carcinogenicity

The experimental groups for the complete tumorigenicity assay were started with 100 mice each. Three times weekly we painted a 50% "tar"-suspension on the shaved backs of the mice. On the average 75 mg of "tar" were given per application. All moribund animals were killed as were the remaining mice, 74 weeks after the

Table 1

SMOKE OF MARIJUANA CIGARETTES AND TOBACCO CIGARETTES
METHODS FOR ANALYSIS AND BIOASSAYS

PARAMETER TESTED	NO. OF CIGARETTES	INTERNAL STANDARD	REFERENCE
I. ANALYSIS			
MOISTURE, CIGARETTE FILLER	2 x 10	-	13
PRESSURE DROP	10 x 1	-	14
STATIC BURNING RATE	10 x 1	-	15
CARBON MONOXIDE, CARBON DIOXIDE	4 x 1	-	16
AMMONIA	2 x 10	$^{14}C-CH_3NH_2$	17
HYDROGEN CYANIDE	3 x 1		
DIMETHYL- AND METHYLETHYL-NITROSAMINE	300	$^{14}C-(CH_3)_2N-NO$	19
ORGANIC GAS PHASE CONSTITUENTS	3 x 1	-	14
ISOPRENE	3 x 1	-	UNPUBLISHED
pH	3 x 1	-	20
TOTAL PARTICULATE MATTER, DRY	4 x 4	-	21
NICOTINE	2 x 4	^{14}C-NICOTINE	14
CANNABINOLS	3 x 10	$^{14}C-\Delta^9$-CANNABINOL	THIS STUDY
VOLATILE PHENOLS	2 x 10	^{14}C-PHENOL	14
POLYNUCLEAR AROMATIC HYDROCARBONS	300	^{14}C-NAPHTHALENE ^{14}C PHENANTHRENE ^{14}C BENZO[a]PYRENE ^{14}C-BENZ[a]ANTHRACENE	22
II. BIOASSAYS			
COMPLETE TUMORIGENICITY (100 MICE)	50,000	-	23
TUMOR PROMOTING ACTIVITY (60 MICE)	25,00	-	23

first application. A skin lesion of a diameter of at least 1 mm was regarded as a papilloma if the tumor did not regress during the first 3 weeks. The lateral invasion of a tumor into adjacent skin was macroscopically considered as transformation into a carcinoma.

Continued growth of such a lesion, however, was required along with a histologic confirmation of epitheliomas. During autopsy all other suspicious lesions were examined.

b. Tumor-Promoting Activity

Initiation was carried out by treatment of the backs of the mice with a single 75 μg dose of 7,12-dimethylbenz *a* anthracene (DMBA) in 0.05 ml of acetone during the second telogen phase of the hair cycle and between hours 5 and 7 of the 11-hour light cycle. Ten days after initiation we applied, thrice weekly, 0.1 ml of a 50% acetone suspension of marijuana or tobacco "tar", respectively, on the initiated area (average of 75 mg per application) and, in the case of a control group, 0.1 ml of acetone alone. The experiments were terminated after 12 months of promoter application. Tumors were identified as described in the preceding section.

RESULTS AND DISCUSSION

Smoke Analysis

The smoking of marijuana leaves in the form of cigarettes produces a dense aerosol similar to that from tobacco cigarettes. Both aerosols are generated within and in front of a burning cone. The formation of smoke is a chain of oxidations, hydrogenations, crackings, distillations, and sublimations. The group of toxic agents generated as gas phase constituents defined as those components of which 50% or more pass through a glass fiber filter[11] includes carbon monoxide, nitrogen oxides, ammonia, hydrogen cyanide, and several other volatile organic agents.

Figure 1 compares the gas chromatograms of the mainstream smoke of marijuana and tobacco cigarettes. Qualitatively the organic gas phases are similar; quantitatively only one major difference emerges, namely, the higher concentration of isoprene in tobacco smoke. We found 83 and 310 μg, respectively, isoprene in the mainstream of marijuana and tobacco cigarettes (Table 2). We attribute this difference to the high percentage of terpenoids in the

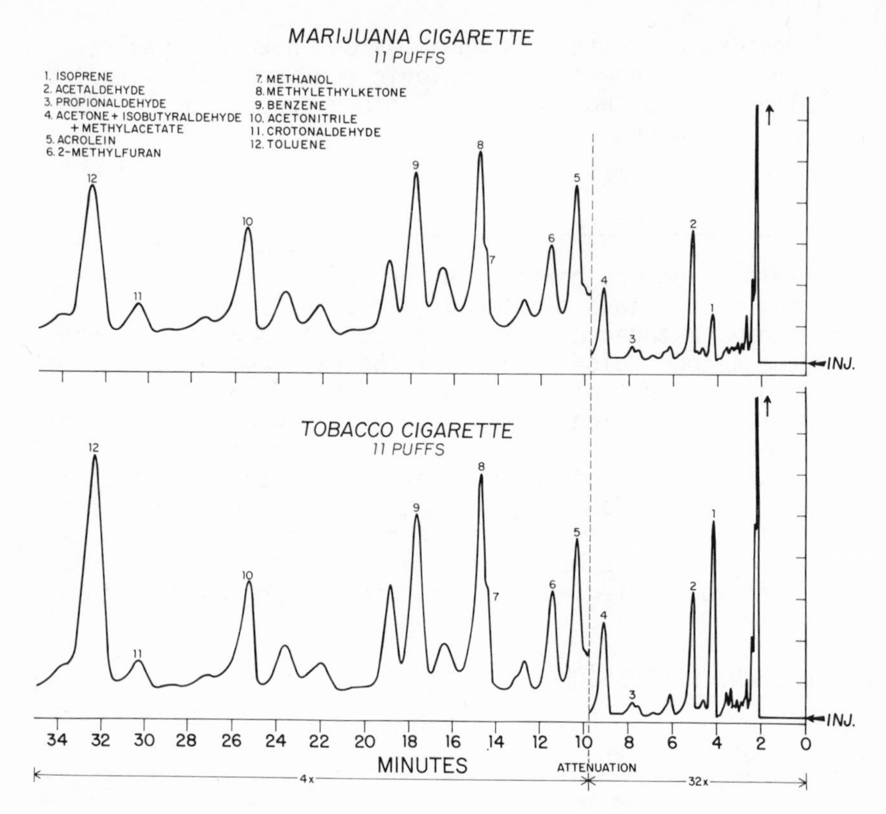

Figure 1. Gas chromatograms of organic compounds in the gas phase of marijuana cigarettes and tobacco cigarettes.

wax-rich layer of tobacco leaves which yield isoprene upon combustion. The burning of a leaf is also influenced by the amount of waxes as indicated by the higher CO concentration in tobacco smoke. In general, the porosity of the cigarette paper is a major factor for the CO concentration in the smoke; however, in this study both cigarettes were wrapped in the same paper.

The amounts of ammonia, hydrogen cyanide, acrolein, acetonitrile, benzene and toluene, and the volatile *N*-nitrosamines are not significantly different between the

Table 2

MARIJUANA AND TOBACCO REFERENCE CIGARETTE
ANALYSIS OF MAINSTREAM SMOKE

	MARIJUANA CIGARETTE 85 mm	TOBACCO CIGARETTE 85 mm
A. ANALYTICAL DATA		
AVERAGE WEIGHT, mg	1115	1100
MOISTURE, %	10.3	11.1
PRESSURE DROP, cm	14.7	7.2
STATIC BURNING RATE, mg/sec.	0.88	
PUFF NUMBER	10.7	11.1
B. MAINSTREAM SMOKE		
I. GAS PHASE		
CARBON MONOXIDE, Vol.%	3.99	4.58
mg	17.6	20.2
CARBON DIOXIDE, Vol.%	8.27	9.38
mg	57.3	65.0
AMMONIA, μg	228	198
HCN, μg	532	498
ISOPRENE μg	83	310
ACETALDEHYDE, μg	1200	980
ACETONE, μg	443	578
ACROLEIN, μg	92	85
ACETONITRILE, μg	132	123
BENZENE, μg	76	67
TOLUENE, μg	112	108
DIMETHYLNITROSAMINE, ng	75	84
METHYLETHYLNITROSAMINE, ng	27	30
PH, 3rd	6.56	6.14
5th	6.57	6.15
7th	6.58	6.14
9th	6.56	6.10
10th	6.58	6.02

two types of smoke. Although the analysis of dimethyl- and methylethylnitrosamine is time-consuming and cumbersome, we determined these agents because they are strong animal carcinogens, though not on mouse skin[24,25]. We detected diethylnitrosamine in both aerosols; however, the amounts were below 5 ng per cigarette. In addition, tobacco smoke contains about 140 ng of nitrosonornicotine as a possible carcinogen[19].

A major toxic factor in tobacco smoke is nicotine, especially if the smoke is alkaline. With increasing pH above 6.0, increasing amounts of nicotine are present in unprotonated form, about 50% at pH 7.8[20]. The unprotonated form of nicotine is the most toxic form of this alkaloid. Although we expected comparable pH values for marijuana smoke and for the smoke of a blended cigarette, this fact had to be established. For this determination the mainstream smoke of a cigarette was passed over a highly sensitive combination electrode. The recorded pH values for marijuana and tobacco smoke are compared in Figure 2. Since the major toxic agents in marijuana smoke, Δ^1- and Δ^6-tetrahydrocannabinol are phenols (Fig. 3), at pH 6.6 they will be present in unionized form.

Since tobacco alkaloids serve as precursors for volatile pyridines, we found these volatile bases in significantly higher concentrations in tobacco smoke. The values for one marijuana cigarette and one tobacco cigarette were: pyridine, 14.4 and 24.8 μg; α-picoline, 7.4 and 13.1 μg;

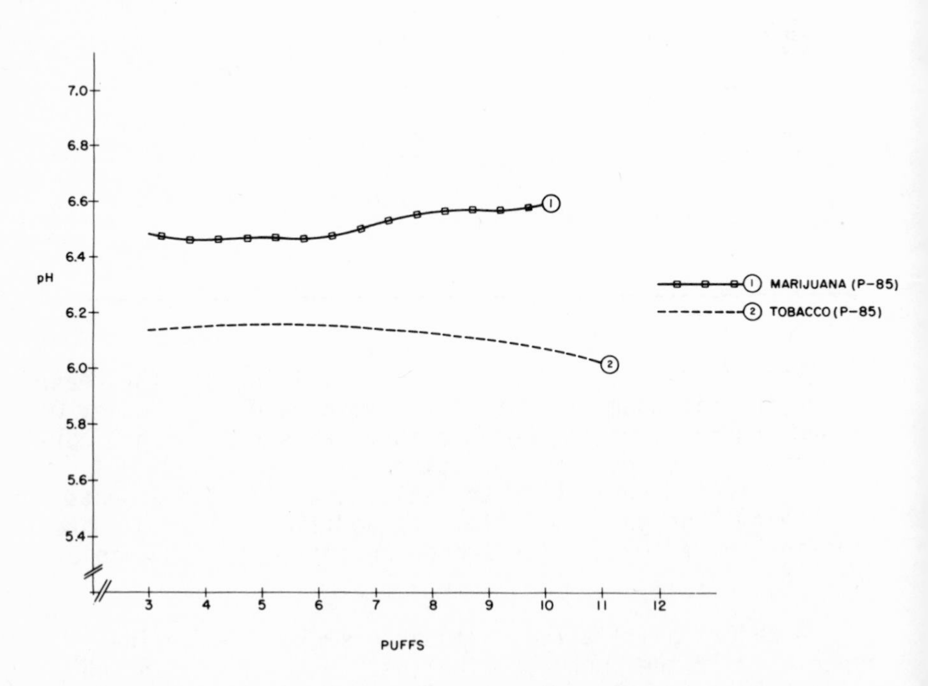

Figure 2. pH of total mainstream smoke of experimental cigarettes: 1) marijuana (P-85), 2) tobacco (P-85).

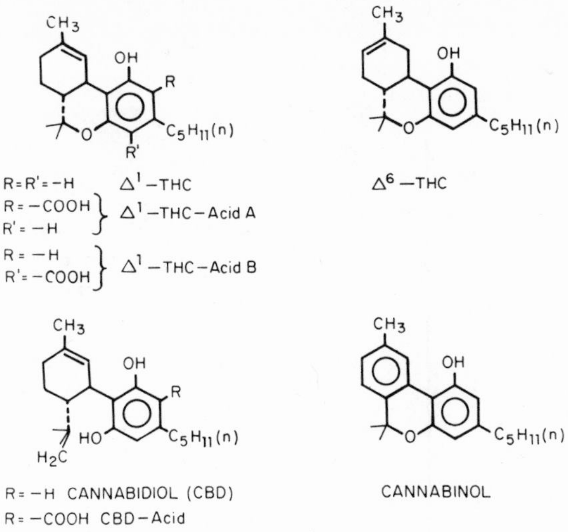

$R = R' = -H$ Δ^1-THC

$R = -COOH$, $R' = -H$ } Δ^1-THC-Acid A

$R = -H$, $R' = -COOH$ } Δ^1-THC-Acid B

Δ^6-THC

$R = -H$ CANNABIDIOL (CBD)

$R = -COOH$ CBD-Acid

CANNABINOL

Figure 3. Structures of cannabinoids.

total volatile pyridines measured, 29.0 and 82.8 μg, i.e. about one third the concentration in marijuana smoke compared to tobacco smoke. In contrast to cigar smoke, it is not expected that the pyridines play an important role in toxicity and/or flavor of marijuana smoke.

The particulate matter of tobacco smoke contains about 50% more volatile phenols than marijuana smoke (Table 3). In the experimental animal, volatile phenols have been shown to act as weak tumor promoters[11].

In general, more than half of Δ^1-THC and (CBD) is present in marijuana as the respective acids, Δ^1-THC acids A and B and cannabidiolic acid. During gas chromatographic analysis, these acids are decarboxylated to the free phenolic cannabinols. Average values for the filler of the experimental marijuana cigarettes were found to be 0.61% of Δ^1-THC, 0.32% of CBD, 0.23% of cannabinol (CBN) and about 0.01% of Δ^6-THC (Fig. 3). These values are rather low compared to Mexican marijuana with 3-4% of Δ^1-THC[2] and suggest that the confiscated marijuana used in this study was probably

Table 3

MARIJUANA AND TOBACCO REFERENCE CIGARETTE
ANALYSIS OF MAINSTREAM SMOKE

	MARIJUANA CIGARETTE 85 mm	TOBACCO CIGARETTE 85 mm
B. MAINSTREAM SMOKE		
<u>II. PARTICULATE PHASE</u>		
TOTAL PARTICULATE MATTER, DRY, mg	22.7*	39.0*
PHENOL, μg	76.8	138.5
o-CRESOL, μg	17.9	24
m- and *p*-CRESOL, μg	54.4	65
2,4- and 2,5-DIMETHYLPHENOL, μg	6.8	14.4
CANNABIDIOL, μg	190	
Δ^9-TETRAHYDROCANNABINOL, μg	820	
CANNABINOL, μg	400	
NICOTINE, μg	-	2850
NAPHTHALENE, ng	3000 (136)	1200 (53)
1-METHYLNAPHTHALENE, ng	6100 (270)	3650 (162)
2-METHYLNAPHTHALENE, ng	3600 (160)	1400 (63)
BENZ[*a*]ANTHRACENE, ng	75 (3.3)	43 (1.9)
BENZO[*a*]PYRENE, ng	31 (1.38)	22.1 (1.0)

*SIDESTREAM SMOKE, DRY TPM, mg 40.7 57.3

diluted with domestic marijuana (>0.1% THC). With a 62 mm cigarette column (23 mm butt length) we smoked about 660 mg of marijuana (10.3% moisture) and with it about 4.0 mg of Δ^1-THC, 2.01 mg of CBD and 1.5 mg of CBN. Since we found o.82 mg of Δ^1-THC, 0.19 mg of CBD and 0.4 mg of CBN in the mainstream smoke of these cigarettes, we observed transfer rates of about 20%, 9%, and 26%, respectively. The low transfer rate of cannabidiol indicates its relatively low stability when compared to Δ^1-THC and CBN. The Δ^1-THC concentration of 3.5% in the particulates of the smoke compared to 0.61% in the cigarette filler demonstrates that the smoking of marijuana represents an enrichment process for the most powerful cannabinol in form of a fine, inhalable aerosol and supports the observation that marijuana intake is preferred in the form of smoke.

The nicotine value of 2.85 mg per cigarette is rather high compared with present day commercial nonfilter cigarettes (<2.0 mg; ref. 26); however, the blend for the standard cigarette was formulated about 8 years ago.

Polynuclear aromatic hydrocarbons (PAH) are formed during incomplete combustions of organic matter. Some of the four- to six-ring condensed aromatic hydrocarbons are known carcinogens to animals including mouse skin and serve as tumor initiators when the particulate matter of inhalants is bioassayed[10,27]. We found significantly higher amounts of PAH in marijuana smoke, not only of the naphthalenes, but also of the weak carcinogen, benz[*a*]anthracene and the strong carcinogen, benzo[*a*]pyrene. These results do not come as a surprise, since tobacco, especially in a commercial cigarette, burns better and contains less cellulose and cellulose-like constituents than marijuana. Qualitatively, both inhalants contain the same type of PAH, although they differ significantly in their concentration. This fact is best demonstrated by establishing PAH profiles[22]. In order to arrive at reliable quantitative data, we add ^{14}C-labeled naphthalene, phenanthrene, benz[*a*]anthracene, and benzo[*a*]pyrene to the "tar" at the beginning of the analysis (Fig. 4). After 2 distribution steps the PAH concentrate is chromatographed on alumina, resulting in 8 PAH fractions. These fractions are analyzed individually by gas chromatography and high speed liquid chromatography. Figure 5 compares the glc profiles for the phenanthrene fractions of the "tars" of four smoking products, and shows the higher concentrations of PAH in the phenanthrene and fluorene fraction of marijuana smoke.

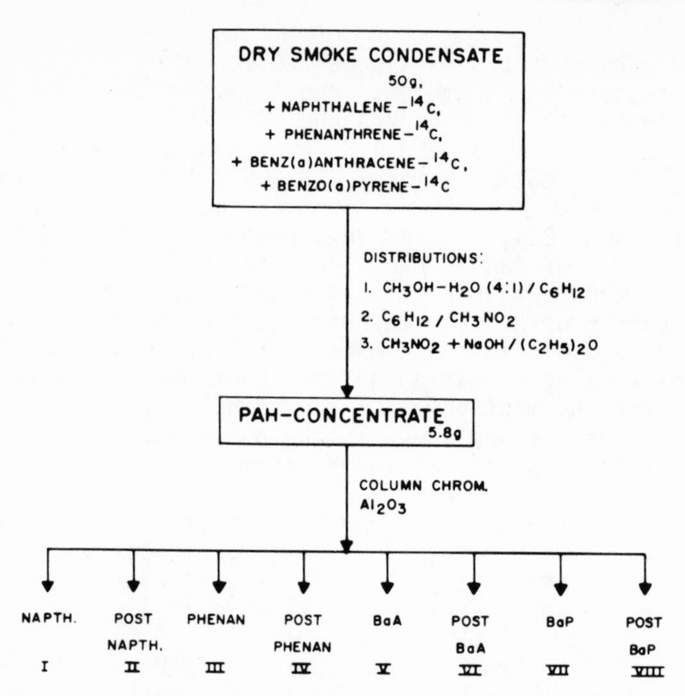

Figure 4. Enrichment method for polynuclear aromatic hydrocarbons from combustion products.

In summary, chemical analysis of the mainstream smoke of marijuana and tobacco has shown that, with the exception of the cannabinoids and tobacco alkaloids, both inhalants are qualitatively similar. Quantitatively, tobacco smoke contains about 3 times the concentration of isoprene and 50% more of the weakly tumor-promoting volatile phenols. Due to its poorer combustibility, marijuana smoke contains about 50% more of the carcinogenic hydrocarbons.

Bioassays

In Figure 6 we have summarized our findings on the mouse skin bioassay for the complete tumorigenicity of marijuana smoke condensate and for tobacco smoke condensate, the positive control. The chronic toxicity of tobacco "tar" was somewhat higher as indicated by the lower survival rate of the mice after about 6 months of treatment. After 74

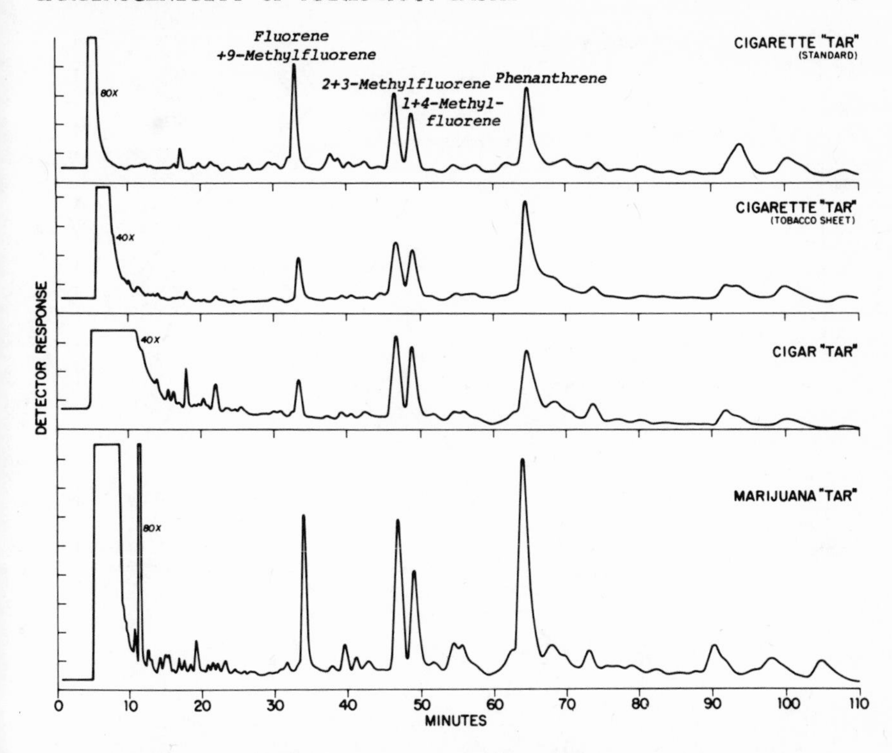

Figure 5. Gas chromatograms of phenanthrene fractions of condensates from various smoke products (each chromatogram was a concentrate from 48 mg condensate).

weeks of "tar" applications, 34% of the "marijuana mice" had survived as compared with only 20% of the "tobacco mice". The first tumor appeared after 17 weeks of "tar" application with 99 mice at risk. At the end of the experiment (74 weeks "tar" application), 6 out of 100 mice in the marijuana groups had developed 7 skin tumors and 14 out of 100 mice of the tobacco group had developed 18 skin tumors (first tumor, 18 weeks; at risk, 97 mice). All of the skin tumors observed after marijuana "tar" application were benign tumors (papillomas) whereas 2 out of the 14 tumor-bearing mice of the tobacco group had malignant tumors (carcinomas), and the

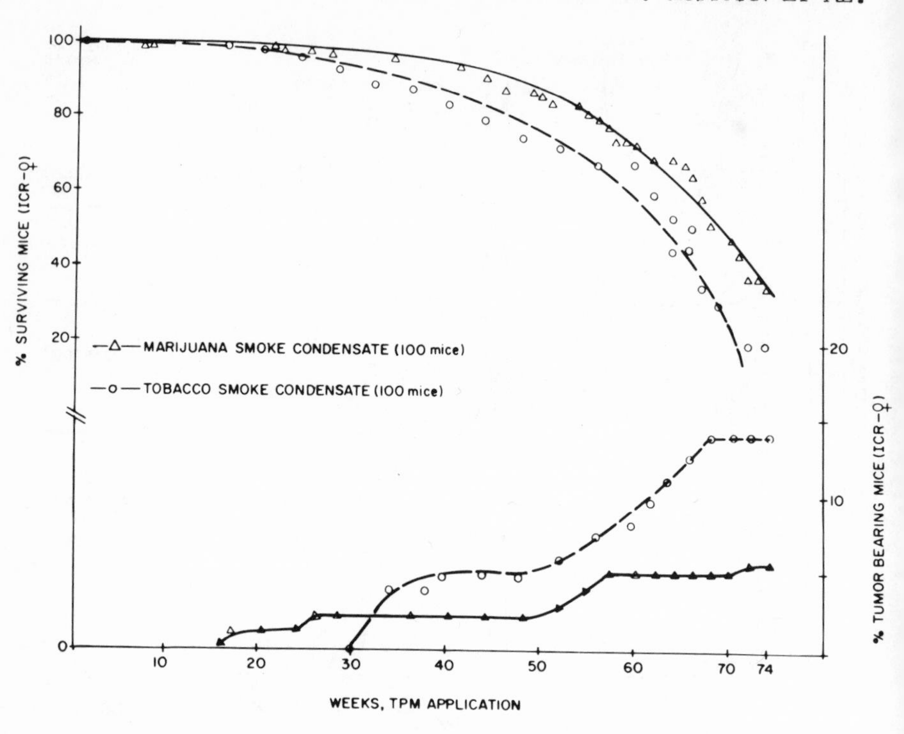

Figure 6. Survival rate and tumor response to cigarette smoke condensate: Complete carcinogenicity.

rest had papillomas. For this strain of Swiss mice we rarely observed skin tumors after treatment with the solvent acetone alone[11]. We can conclude, therefore, that in the mouse skin bioassay marijuana smoke condensate is a weak tumorigenic agent. The activity of marijuana "tar", however, is significantly lower than that of smoke condensate from tobacco cigarettes. In addition to the skin tumors, we observed, in the marijuana group, 25 mice with 34 lung adenomas and 1 mouse with lymphoma compared to 22 mice with 30 lung adenomas and 2 mice with lymphomas for the tobacco group. In the negative control we observed 31 mice with 34 lung adenomas.

The mice which were initiated with 75 μg DMBA were treated for 56 weeks with 50% suspensions of marijuana "tar" or tobacco "tar", respectively. At the end of this time 26 of the 60 mice tested with marijuana "tar" developed 48 skin papillomas and 3 carcinomas, 3 mice had fibrosarcomas, and 8 mice developed 14 lung adenomas (Fig. 7). In the tobacco group we observed that 34 out of 60 mice had 72 skin papillomas, 4 mice had 6 carcinomas and 4 mice had developed 6 lung adenomas. In the negative control group, with 75 μg DMBA as initiator and acetone as "promoter", 5 out of 60 mice developed 8 papillomas, and 1 mouse out of the 5 tumor-bearing mice had 2 carcinomas. In the control

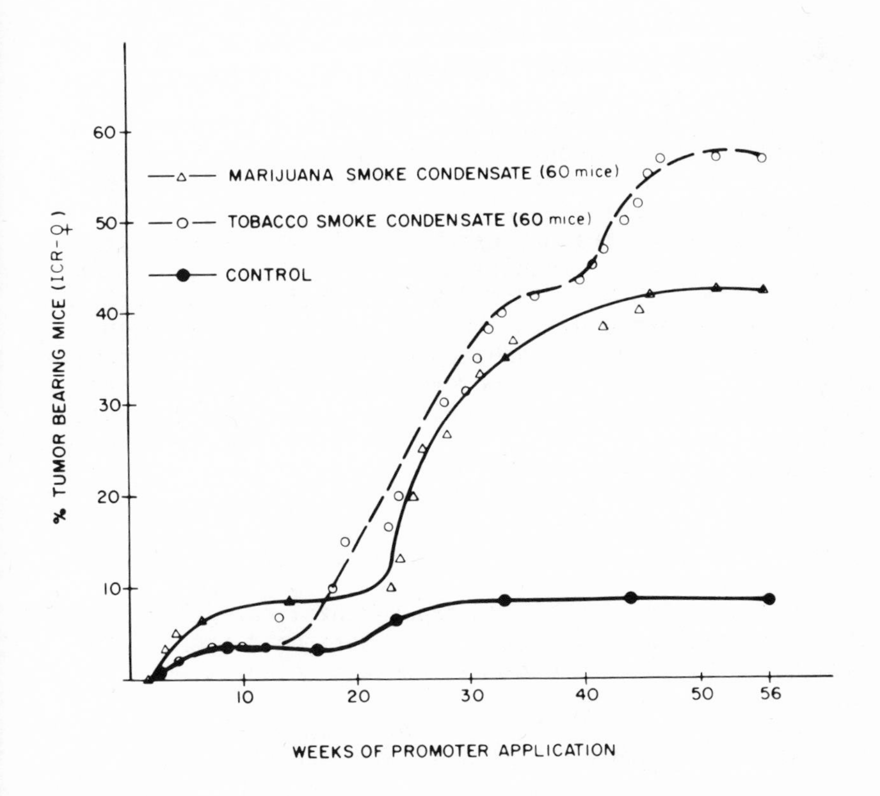

Figure 7. Tumor promoting activity of smoke condensates. Initiator: 75 μg DMBA

group we also found 7 mice with 9 adenomas and 1 mouse with a lymphoma. From these results we conclude that marijuana "tar" has a significant tumor-promoting activity, although lesser in degree than cigarette "tar" ($p<0.05$). Presently, we are engaged in identifying the tumor-promoting agents in marijuana smoke. Based on short-term tests it appears that the majority of the tumor-promoter resides in the weakly acidic fraction within a subfraction which contains non-volatile phenols and long-chain fatty acids.

We conclude from the chemical analytical data and the bioassays that marijuana smoke contains some volatile *N*-nitrosamines as well as some polycyclic hydrocarbon carcinogens. The latter carcinogens may serve in the experimental setting, as used in the present study, as tumor initiators. On mouse skin, marijuana "tar" is a weak tumorigenic agent and has tumor-promoting activity. Both activities, however, are significantly lower than the respective biologic activity of tobacco "tar".

Although we consider these findings of more than academic interest, we should not attempt to relate these bioassays to the human setting without some qualifications. Firstly, no epidemiological data relating marijuana smoke as a possible respiratory carcinogen to man are at hand; secondly, the daily exposure to marijuana smoke, even in extreme cases, is below the dose level of most tobacco cigarette smokers and, furthermore, marijuana has been smoked for fewer years than tobacco cigarettes. It remains as a reasonable question, however, to determine whether, particularly because of its relatively high PAH content, marijuana smoke is capable of potentiating the carcinogenic effects of cigarette smoke.

SUMMARY

The mainstream smoke of experimental marijuana cigarettes was analyzed for toxic and tumorigenic agents and compared with smoke of standard tobacco cigarettes.

Both aerosols are qualitatively comparable, except for differences in cannabinoids and tobacco alkaloids.

Quantitatively we found higher concentrations in the tobacco smoke for carbon monoxide, isoprene, volatile phenols, and pyridines and in the marijuana smoke higher concentrations for some polynuclear aromatic hydrocarbons. The transfer rate into the mainstream smoke of the major cannabinoid, Δ^1-tetrahydrocannabinol was about 20%, resulting in a significant enrichment of this toxic agent in the particulate matter of this inhalant.

The bioassays on mouse skin for tumorigenicity and tumor-promoting activity. Both activities, however, are significantly below those exchibited by tobacco smoke condensate.

The chemical analytical findings and the results from the bioassays for tumorigenicity are discussed. These data cannot be extrapolated to the human setting without qualification.

Acknowledgements

We express our appreciation to Dr. A.R. Patel from the Meloy Laboratories, Inc., Springfield, Va., for producing the marijuana cigarettes and for the large-scale preparation of marijuana "tar" for the mouse skin bioassays. We thank Dr. T. Okamoto of the American Health Foundation for the preparation and reading of histological slides. This study was supported by the National Cancer Institute Contract NO1-CP-33305.

REFERENCES

1. Pillard, R.C. 1970. Marijuana in New Engl. J. Med. 283: 294-303.
2. Singer, A.J. 1971. Marijuana: Chemistry, Pharmacology, and Pattern of Social Use. N.Y. Acad. Sci. 191: 1-269.
3. Nahas, G.H. 1973. "Marijuana, Deceptive Weed." p. 325. Raven Press, New York.
4. Mechoulam, R. 1973. "Marijuana: Chemistry, Pharmacology, Metabolism, and Clinical Effects". p. 409. Academic Press, New York.
5. Waller, C.W., J.J. Denny & M.A. Waltz. 1971. "Annotated Bibliography of Marijuana." University of Mississippi, University, Miss., 1964-1970, p. 301.
6. ______, J.J. Denny & M.A. Waltz. 1972 "Annotated Bibliography of Marijuana. University of Mississippi,

University, Miss., 1971 Suppl. p. 155.
7. Magus, R.D. & L.S. Harris. 1971. Carcinogenic potential of marijuana smoke condensate. Federation Proc. 30, Abstr.: 279.
8. Cottrell, J.C., S.S. Sohn, & W.H. Vogel. 1973. Toxic effects of marijuana tar on mouse skin. Arch. Environ Health, 26: 277-278.
9. Leuchtenberger, C. & R. Leuchtenberger. 1971. Enhancement of abnormal proliferation in lung explants after marijuana cigarette smoke. Experientia, 27: 737-738.
10. _____, R. Leuchtenberger, & A. Schneider. 1973. Effects of marijuana and tobacco smoke on lung physiology. Nature, 241: 137-139.
11. Wynder, E.L. & D. Hoffmann. 1967. "Tobacco and Tobacco Smoke. Studies in Experimental Carcinogenesis." p. 732. Academic Press, New York.
12. Lipp, G. 1969. Vergleich der Rauchausbeuten von drei verschiedenen Rauchmachinen I. Beitr. Tabakforsch. 5: 39-42.
13. Von Bethmann, M., G. Lipp, & H. Van Nooy. 1961. Feuchtigkeitsbestimmung in Tabak. Beitr. Tabakforsch. 1: 19-29.
14. Rathkamp, G., T.C. Tso, & D. Hoffmann. 1973. Chemical studies on tobacco smoke. XX. Smoke analysis of cigarettes made from Bright tobaccos differing in variety and stalk position. Beitr. Tabakforsch. 7: 179-189.
15. Hoffmann, D. & E.L. Wynder. 1967. The reduction of the tumorigenicity of cigarette smoke condensate by addition of sodium nitrate to tobacco. Cancer Res. 27: 172-174.
16. Brunnemann, K. D. & D. Hoffmann. 1974. Chemical studies on tobacco smoke. XXIV. Quantitative method for carbon monoxide and carbon dioxide in cigarette and cigar smoke. J. Chrom. Sci. 12: 70-75.
17. _____ & D. Hoffmann, unpublished.
18. Artho, A. & P. Koch. 1969. Uber die Bestimmung von Cyanwasserstoff im Cigarettenrauch, Beitr. Tabakforsc 5: 58-63.
19. Hoffmann, D., G. Rathkamp, & Y.Y. Liu. 1974. Chemical studies on tobacco smoke. XXVI. On the isolation and identification of volatile and nonvolatile *N*-nitrosamines and hydrazines in cigarette smoke. WHO Monograph (in print).
20. Brunnemann, K.D. & D. Hoffmann. 1974. Chemical studies

on tobacco smoke. XXV. The pH of tobacco smoke. Food Cosmet. Toxicol. 12: 115-124.
21. Pillsbury, H., C.C. Bright, K.J. O'Connor, & F.W.J. Irish. 1969. Tar and nicotine in cigarette smoke. J. Assoc. Offic. Anal. Chemists. 52: 458-462.
22. Hoffmann, D., G. Rathkamp, K.D. Brunnemann, & E.L. Wynder. 1973. Chemical studies on tobacco smoke. XXII. On the profile analysis of tobacco smoke. Sci. Total Environ. 2: 157-171.
23. Wynder, E.L. & D. Hoffmann. 1968. Selected laboratory method in tobacco carcinogenesis. Methods Cancer Res. 4: 3-52.
24. Druckrey, H., R. Preussmann, S. Ivankovic, & D. Schmähl. 1967. Organotrope carcinogene Wirkungen bei 65 verschiedenen *N*-Nitrosoverbindungen an BD-Ratten. Z. Krebsforsch. 69: 103-201.
25. Magee, P.N. & J.M. Barnes. 1967. Carcinogenic nitroso compounds. Advan. Cancer Res. 10: 163-246.
26. Federal Trade Commission: Tar-Nicotine Report, March 1974, Tobacco Reporter, 101 (4): 26 (1974).
27. Hoffmann, D. & E.L. Wynder. 1968. Chemical analysis and carcinogenic bioassays of organic particulate pollutants. In "Air Pollution" (A.C. Stern, Ed.), Vol. 2. pp. 187-247. Academic Press, New York.

Chapter Four

CHEMISTRY AND FOLK MEDICINE

JAMES A. DUKE

Plant Taxonomy Laboratory
Plant Genetics & Germplasm Institute
U.S. Department of Agriculture
Beltsville, Maryland

INTRODUCTION

The Plant Taxonomy Laboratory of the Plant Genetics and Germplasm Institute in Beltsville, Md., has a matrix devised for screening exotic diversification crops for underdeveloped areas. Repeated in tables here, the matrix contains abbreviations for scientific names of 1,000 crop species.

In this chapter I shall attempt to show how the matrix can be used to suggest correlations or lack of correlations between crop chemistry, mostly nutritional, and folk medicine, mostly Amerindian. In order to do this I will discuss some of the nutritional analyses of crop species, Amerindian ethnomedicinal uses of these same species, and modern medicinal recommendations for nutritional "therapy." Since others in this volume write of teratogens and antitumor agents, my examples are biased in that direction.

My personal physician (who chooses to remain anonymous) says that many of the people who visit him should not be there, that many have self-healing ailments that would mend without medication, given a little time, and that many are psychological cases who might "heal" as well with a placebo as with a pill. About 10% of the folk medicine plants of Mexico are said to be effective for what the natives use them. My interest in folk medicine began with my ethnobotanical work with the Choco (an endangered "species" of Indians), Cuna Indians, Blacks, and Castillians of Panama and Colombia. I had an ugly tropical ulcer on my ankle, blamed on dumbcane, *Dieffenbachia seguine*, which North American doctors, calling my ailment "cellulitis," and Special Forces doctors, calling it "jungle rot", stated would lead to my losing my foot unless I returned to the United States for treatment.

Then came "Swamp Fox II," a military mobility study in tropical Panama, with which I worked as a consultant on the botany of the Bayano area of Panama. Jungle Blacks said I might heal, if I treated the ulcer periodically with powdered sulfur after washing it with peroxide. Off I went to the jungle, laden with peroxide and sulfur, limping across the Rio Mamoni. After a total of three years in Panama, I still have my foot, thanks to luck, jungle medicine, or both.

After that experience, I more readily listened to native medicine men. I used *Piper darienensis* to treat, effectively, a toothache; I poulticed castor leaves on to sores, and wild fig juice on to cuts, with no obvious harm, with no proof of help either. I recorded the native uses of the flora of Panama and Colombia.[1] The Choco, Cuna and Blacks are rapidly being acculturated; their folk uses of plants may wane before they have all been documented.

More recently I made two trips to Bolivia and became interested in "Callahuaya" Indians. Today in Bolivia, a more prevalent meaning seems to be witch doctor or medicine man, be he Quechua, Aymara, or Castillian.

Drug plants, many if not all originally identified from folk medicinal practices, are important even in the United States. Of a sample of 2.5 million prescriptions written from 1956 through 1960, over 47% contained drugs of natural origin.[2] Between 1959 and 1960, use of drugs of natural origin increased 7.7%. Drugs containing plant products represented over 34% of all drugs of natural origin. Krochmal cited 125 medicinal plants of Appalachia for which there was a demand.[2] Among those nearly 10% are in our crop matrix: *Acorus calamus*, *Chenopodium ambrosioides*, *Fragaria virginiana*, *Gaultheria procumbens*, *Juglans cinerea*, *J. nigra*, *Juniperus communis*, *Lobelia inflata*, *Marrubium vulgare*, *Mentha piperita*, *spicata*, *Nasturtium officinale*, *Nepeta cateria*, *Panax*, *Phytolacca americana*, *Salvia officinalis*, and *Tanacetum vulgare*.

NUTRITIONAL CHEMISTRY

In the crop matrix, 1,000 potential crop species were bibliographically scored from selected Food Composition Tables.[3,4,5,6] Complete indices to scientific names, authorities and ecological tolerances are being published elsewhere[7], but Tables 2-16 (which are located in the Appendix at the end of this chapter) show examples of the matrix. For example, Table 2 presents the complete matrix of common names, while, in Table 3, I have underscored abbreviations of those species reported to have a low calorie content (<25/100g edible portion). The other three categories for calories are fair (25-100/100 g), moderate (100-300/100 g), and high (>300/100 g) (Table 4). The four data groups for protein are arbitrarily set at low (<1 g/100 g uncooked) (Table 5), fair (1-3 g/100 g uncooked), moderate (3-10 g/100 g uncooked), and high (>10 g/100 g uncooked) (Table 6). If interested in low protein crops, the reader will note from Table 5 that C-17 (ACO) has been underscored. By referring to Table 2, he finds that C-17 is the pineapple, *Ananas comosus* (L.) Merr.

Similar categories have been established for fat, carbohydrates, fiber, ash, calcium, phosphorus, magnesium, iron, sodium, potassium, vitamin A (Tables 7,8), thiamin, riboflavin, niacin, and vitamin C (Tables 9, 10). This enables a considerable reduction of much of the analytical data given in the four food-analysis books, and enables one rapidly to search for items high or low in a given category.

Someone forced to assume a diet high or low in any of these nutritional elements could consult the appropriate table and find which of the 1,000 species are reported to have the desired dietary attribute. Further research might still be in order. Some plants, like *Phaseolus vulgaris* (PVU at position c-14), have broad ranges in their chemical analyses, because the ripe seeds (e.g., pinto beans) differ widely from the green pods (e.g., string beans). Fruits may vary by a factor of three in mineral composition, depending on the substrate. Still, if one wished a diet low in calories and calcium, yet high in vitamins A and C, he could consult the appropriate tables and find plants reported to share all these attributes.

In addition to these chemical tables, information on ecological range (life zone), salt pH tolerance, and tolerances to a wide variety of environmental conditions such as salt, drought, waterlogging, frost, heat, poor soil, shade, insect attack, and disease have also been prepared in matrix form for the 1000 species.

FOLK MEDICINE

Many crop species are used as folk medicine by various ethnic groups. In our matrix, we have underscored those of the 1,000 species in the Blackfoot (Table 11), Callahuaya (Table 12), Choco (Table 13), Cuna (Table 14), Kumaon, Samoan, Santal, and Tongan pharmacopoeias, based on the sources cited in the references.[1,9,10,11,12,13,14,15,16,17]

We have also underscored species abbreviations if they were reported to treat abscesses, arthritic complaints, asthma, bronchitis, cancer, cholera, colds, dysentery, fever, hemorrhage, neuralgia, sores, soroche, stomach, stones, tumors (Table 15), or worms (Table 16).

Choco and Cuna Indians respect ipecac (CIP at position J-14)[1,11], and modern medicine respects ipecac. All ethnomedicinal lore should be studied before ethnic groups or the plants they use become genetically swamped, endangered, or extinct. Cuna and Choco Indians use stinging nettles (Urticaceae) to treat arthritic complaints, and rural people in the United States sometimes recommend bee stings for similar complaints. These might seem strange differences, though formic acid is active in both. There are dozens of species of stinging nettles, but only one, *Urtica dioica* (UDI at position m-5), is considered in the crop matrix. Although it is sometimes consumed as a potherb, no analytical data are available to me. It has been reported as a folk cure for arthritic complaints and for colds, and it is listed among the cancer actives in the cancer-screening program at Beltsville. It appears in the Callahuaya pharmacopoeia as an antitumor agent, but this stinging nettle is not reported in the Tropical Moist Forest Life Zone occupied by the Choco and Cuna Indians. Its place is taken, ecologically and pharmacologically, by a humid tropical *Urera* species.

About 15% of the species in our crop matrix are primarily drug plants. There is no fine line between drug plants, folk medicine plants, medicinal resins, and plants that have other uses equally as important as their drug uses, such as the masticatory coca, from which cocaine is derived, the fiber hemp, from which tetrahydrocannabinol is derived, and the poppyseed poppy, from which opium can be obtained. Many of the essential oils, fumitories, resins, spices, and latices are also used as drugs.

Probably all plants have been used medicinally, at one time or another. To tribes like the Cuna, all plants and animals have medicinal uses. One may also find pages from the Bible, bits of broken glass, tin cans, etc., in their medicine kit. Although I have reported plants for which they gave me specific uses, many curanderos said they used all plants.

Species diversity increases toward the tropics. Only 5 Blackfoot Indian plants occurred in our matrix of 1,000 crop species, in contrast to more than 60 Choco Indian medicinal plants. More than 100 Cuna plants for which I obtained specific uses are listed. The Callahuaya have closer to 200 (some of which occur in Bolivia) or 20% of the crops of the world in their pharmacopoeia (Table 12). Street venders in La Paz sell everything from llama foetuses to lima beans for medicines.

The Callahuaya are a subtribe of the Aymara who live in the Munecas and Caupolican districts of La Paz. They speak fluent Aymara but consider themselves ethnically distinct. Their primary tongue, since the days of the Inca, is Quechua. They are the famous traveling herb doctors of the Andes, most of the males traveling far and wide with herbal remedies and amulets.

Cardenas discusses the Callahuaya.[9] The first therapeutic plant mentioned in his medicinal plant chapter is coca, *Erythroxylum coca*, which is found in tombs of the Inca, Indians skilled with trepanation. Coincidentally ECO at position Q-17 is underscored in Tables 12, 15 and 16, meaning that coca is one of the few species in the Callahuaya pharmacopoeia which is reported to be an anti-tumor agent, and a vermifuge.

Cardenas says that another fountain of knowledge, in addition to that passed down from the extinct societies like the Inca, derived from Indians of Tahuantinsuyu, whose pharmacopoeia has been expanded and advanced to the present herb doctors of the vicinity of Charazani known as Callahuayas. Until recently these inhabitants of Kanlaya and Chajaya came to La Paz with their knapsacks on their backs, containing their precious drugs. However, Cardenas' appraisal of the Farmacopea Callahuaya is not high.[14] On the other hand, many crops, used with good phytochemical reason, occur in the recent Farmacopea Callahuaya.[14]

Callahuaya medicinal plants in our crop matrix include basil, castor, chamomile, coca, ephedra, fenugreek, garlic, hemp, ipecac, mandrake, poppy, quinine, wormseed, etc. Without speculating as to whether the Callahuaya gave some of these to the world pharmacopoeia, or vice versa, it is worth noting that most of these are now grown in Bolivia for medicinal purposes. Coca and quinine, at least, have been used since Inca times. Modern medicine still uses cocaine and quinine. Although there is a trend away from cocaine, natural quinine is indicated for certain types of malaria.

DISCUSSION

Calorie-rich plant foods in our crop matrix include such things as almonds, macadams, pecans, and walnuts, all >600 calories/100 g. Hemp seeds have 421 calories, and betel nuts have nearly 400 calories. Among low-calorie foods are bottle gourds, ceylon spinach, chayote, chinese cabbage, chrysanthemum, colza, cucumber, lettuce, snake gourd, spinach, towel gourd, turnip, udo, water cress, and waxgourd, etc. Among the low-calorie plants, red pepper, onion, and lemon, are also entered in our matrix for folk cancer cures. Horsetail probably does not occur in lowland

Panama, but the other three are components of sauces in the otherwise bland diets of the Cuna and Choco Indians. Three of the high-calorie foods enter the matrix as folk "remedies" for cancer: turmeric, fenugreek, and quinoa; none known to the Cuna and Choco, two however in the Callahuaya pharmacopoeia of Poblete.[14] Low-calorie intake is indicated for arteriosclerosis (if overweight), diabetes mellitis, gout, obesity, and for cardiovascular, gall bladder, and renal diseases; high-calorie intake for malnutrition, fever, and early stages of cystic fibrosis.

Commenting on nutritional management of the anemic geriatric patient, Jukes and Borsook state that providing needed vitamins and iron from food sources alone would require a diet too high in calories, too expensive, and too monotonous.[18] Using the matrix, however, I found that only watermelons and squashes showed through as high iron, low calorie, but this was a "false drop." Their seeds are high iron, high calorie, but the fruit pulp is low in calories and iron.

Protein-rich crops (>25 g protein/100 g, e.g., broadbeans, cucurbit seeds, goabean, hemp seed, horsegram, and, of course, soybean) are more and more in demand today. Ripe seeds of legumes are rich in protein; tender green beans in several genera, even the goabean, contain less protein than the leaves. Tops of cassava, radish, sweet potato, taro, and turnip, contain proportionately more protein than the roots. Non-leguminous leafy vegetables, and watery tropical roots and fruits are low in protein. Cucurbit leaves are often 4-10 times richer in protein than the fruits. Seeds of cucurbits are protein-rich, fruits are protein-poor. Hemp seed has 27% protein, poppyseed 24%, betel nut 6%, and betel leaf 3%. High-protein diets might be indicated for alcoholism, burn patients, kwashiorkor, malnourishment, and marasmus. Low-protein intake might be indicated for hepatic failures due to cirrhosis, hepatitis, and terminal alcoholism. Four syndromes result, in part, from geriatric protein deficiency; hunger edema, pellagra, nutritional liver disease, and nutritional macrocytic anemia.[19] There are five cancer "remedies" among our protein-poor foods, imbu, lemon, peach, pepper, tuna, and three among the protein-rich, fenugreek, opium poppy, and quinoa.

Those wishing to avoid oil in their diet might enjoy

Jerusalem artichoke and lotus seed which have only 1/10 g oil or fat/100 g. The betel nut is 10% oil, the betel leaf less than 1%, hemp seed closer to 25%. Some seeds containing more than 50% oil are the almond, brazilnut, candlenut, filbert, macadam, poppy, sesame, walnut, and watermelon. High oil might be indicated for constipation and malnourishment; low oil intake for arteriosclerosis, early stages of cystic fibrosis, hypercholesterolemia, obesity, pancreatic diseases, and sprue. Low oil is indicated to rest the gall bladder.[8] Corn, peanut, safflower, and sunflower oils, like whale oil, lower cholesterol levels, while cocoa and palm oils may raise it.

Among low-oil plants in our matrix, burdock, fenugreek, horsetail, lemon, onion, peach, peppers, sodom-apple, tuna, and turmeric are reported as folk cancer "cures," while only the opium poppy finds itself among the high-oil plants in the same matrix.

In this early stage of my compilation there are only the following folk "remedies" for cancer:

A-11 - Sweetflag	J-1 - Jackbean	a-5 - Tuna
B-19 - Onion	J-11 - Periwinkle	b-2 - Poppy
C-1 - Aloe	J-21 - Quinoa	e-14 - Peach
D-12 - Burdock	K-22 - Lemon	f-21 - Rosemary
D-21 - Absinth	M-2 - Ti Palm	h-7 - Jojoba
H-19 - Calendula	M-7 - Turmeric	h-15 - Sodomapple
I-7 - Pepper, green	Q-5 - Horsetail	j-14 - Java Plum
I-10 - Pepper, hot	Y-19 - Balsam apple	I-8 - Fenugreek
I-11 - Locoto	Z-20 - Tobacco	

The following are revealed as folk antitumor agents:

B-19 *Allium cepa* L.	J-20 *Chenopodium ambrosioides* L
C-23 *Annona cherimola* Mill.	J-21 *Chenopodium quinoa* Willd.
D-4 *Anthemis nobilis* L.	L-11 *Coffee arabica* L.
D-8 *Apium graveolens* L.	M-3 *Coriandrum sativum* L.
I-4 *Canna edulis* Ker	N-10 *Cyclanthera pedata* (L.) Schrad.

N-11 *Cydonia oblonga* Mill.
O-18 *Dioscorea bulbifera* L.
Q-17 *Erythroxylum coca* Lam.
R-12 *Ficus carica* L.
T-14 *Hordeum vulgare* L.
T-18 *Hyoscyamus niger* L.
T-22 *Hyssopus officinalis* L.
U-8 *Ipomoea batatas* (L.) Lam.
W-6 *Linum usitatissimum* L.
X-9 *Mandragora officinarum* L.
X-12 *Manihot esculenta* Crantz
X-17 *Marrubium vulgare* L.
Z-7 *Musa* x *paradisiaca* L.
Z-20 *Nicotiana tabacum* L.
Z-23 *Ocimum basilicum* L.
a-5 *Opuntia ficus-indica* (L.) Mill.
a-7 *Origanum vulgare* L.
a-13 *Oxalis tuberosa* Mol.
b-15 *Passiflora quadrangularis* L.
c-4 *Petroselinum crispum* (Mill.) A.W. Hill
c-14 *Phaseolus vulgaris* L.
c-16 *Phoenix dactylifera* L.
d-11 *Pisum sativum* L.
d-25 *Portulaca oleracea* L.
e-5 *Prunus armeniaca* L.
e-8 *Prunus cerasus* L.
f-19 *Ricinus communis* L.
f-21 *Rosmarinus officinalis* L.
g-20 *Secale cereale* L.
h-22 *Solanum nigrum* L.
1-8 *Trigonella foenum-graecum* L.
m-2 *Ullucus tuberosus* Caldas
m-5 *Urtica dioica* L.
m-22 *Vicia faba* L.

The only antihemorrhagic, antitumor vermifuge I know that is reported to be good for arthritis, asthma, and soroche is the coca plant, which has recently been added to the list of cancer actives. The coca plant has a wider ecological range than might be expected, occurring in Tropical Moist Forest, Tropical Wet Forest, Subtropical Moist Forest, and Warm Temperate Moist Forest Life Zones, the latter close to the ecological habitat of the Callahuaya, the former the home of the Cuna and Choco of Panama.

Of the folk antitumor agents in our crop matrix, cassava, parsley, purslane, and sweet potato are high in vitamin C, while banana, barley, flax, green bean, horsebean, and rye are low. Banana, barley, cassava, cherimoya, flax, green bean, horsebean, oca, and onion are low in vitamin A while apricot, basil, cassava (leaves), celery, coffee, coriander, green bean, onion, pennyroyal, purslane, sweet potato, and wonderberry are high.

All the Asian matrix cereals are high in carbohydrates (>60 g/100 g)[6]. Many roots and legume seeds fall in the moderate category (30-60 g). Many starchy fruits, roots, green legume fruits and cucurbit seeds fall in the fair category (10-30 g) while cucurbit fruits and leafy vegetables are frequent in the low carbohydrate category (<10 g). In legumes, low carbohydrates often correlate with high protein. Asian analyses of *Canavalia* are higher than readings from other continents. The dried betel nut is 70% carbohydrate, hearts of the betel palm 10%, but the betel leaf is only 1% carbohydrate. Hemp seeds are 28% carbohydrate. High carbohydrate intake might be indicated for early stages of cystic fibrosis, marasmus caused by poor appetite, or underweight. Low carbohydrate intake might be indicated for coeliac disease, diabetes, dyspepsia, hypoglycemia, insulinonoma, and obesity.

Because of low fiber content, American foods and the carcinogens associated with them remain in the intestinal tract longer than if we had a high-fiber diet. Many health fadists ingest sunflower seed to add fiber to their diet. However, there is proportionately more fiber in hemp, perilla, safflower, and sesame seed. Like sunflower seeds, cucurbit seeds, high in ash, are not so high in fiber. High fiber might be indicated for several types of diarrhea, as well as prophylactically in diverticulosis and cancer of the lower bowel. Low fiber might be indicated for ulcerative colitis.

Andean Indians may or may not benefit from calcium ingested with their coca. In spite of several inquiries, I have been unable to find a modern food analysis of coca. Instead I referred to Del Granado for analyses of Bolivian coca.[10] I mention this because my program was generated to help provide intelligent alternatives to countries which wish to phase out such narcotic-producing plants as coca and poppy.

There are two main coca-producing regions in Bolivia, Los Yungas and the Chapare. Coca from the drier, cooler, higher Yungas region is preferred by the Indians in La Paz to that from the humid tropical Chapare region; they claim that the Yungas leaves are sweeter. To me, sweet implies less cocaine (or more sugar). If the Indian chews coca for something other than cocaine, and if neither the beverage industry nor the medical industry has an irreplaceable need for cocaine, is there any legitimate reason for continued cultivation of this plant?

Although Bolivians recently assured me that Yungas coca was higher in cocaine than Chapare coca, reportedly based on three replications, this differs from the data of Del Granado[10]:

	Coca from Chapare	Coca from Totora	Coca from Yungas
cocaine	2.40%	1.00%	0.25%
starch	24.50	23.90	36.19
dextrin	5.00	1.20	1.16
sugar	1.40	17.50	11.46
pectic acid	0.80	2.20	1.82
fiber	8.80	7.50	7.80
wood	35.00	29.70	28.57
chlorophyll & oil & wax	3.70	1.00	0.25
ash	5.00	5.00	6.00
water	9.00	0.00	0.00
unknown	4.40	0.00	0.00
	100.00	100.00	100.00

Since there are discrepancies between the old and new reports, since I know of no modern food analysis for coca, and since unwashed coca leaves are often treated with insecticide, new analytical work is in order. Coca has been reported to be rich in vitamins, but I cannot document this as well as I would like. FDA analyses indicated that thiamin ranges from about 0.5-0.75 mg/100 g; riboflavin 0.9-1.2 mg/100 g, and vitamin C 7-23 mg/100 g. The monotonous Indian diet of potatoes, quinoa, barley, oca, etc., might benefit somewhat from coca and ash ingestion.

Vitamin A is often low (< IU/100 g) in colorless cereals, roots, pulses, and unpigmented seeds, fair in some tropical fruits, moderate in green vegetables, and temperate fruits, high (>1000 IU) in red and/or leafy aerial vegetables. Beets, like radish, though red, are devoid of beta-carotene, while beet greens contain ca 3000 IU. Calabash gourd has only 15 IU while the leaves have closer to 15,000, paralleled by gourds and chayotes. The betel piper leaf has ca 20,000 IU while hemp seed contain only 8. Low vitamin A intake might be indicated in hypervitaminosis A and Xanthosis cutis (carotenemia), high vitamin A for avitaminosis A, cystic fibrosis, and night blindness. Geriatric patients may exhibit Bitot eye spots, hyperkeratosis, night blindness, and xerophthalmia as a result of prolonged vitamin A deficiency. In one survey, 1.5% of Latin American children showed ocular lesions due to hypovitaminosis A.[20]

To test folk medicines, I independently prepared Table 1 to see what dietary elements were high or low in the folk medicines of the aforementioned ethnic groups. After going through the Food Composition Tables, I found that nearly 60 of the crops had been reported as low-calorie foods, 150 as high-calorie foods, establishing a ratio of expectation of about 1 low calorie to every 2½ high-calorie foods, based on our arbitrary classification. The farthest deviation from the expected ratio for calories was in the folk treatments for hemorrhages where 5 low-calorie items (artichoke, horsetail, lemon, onion, purslane) were used for hemorrhage. All except purslane, *Portulaca oleracea*, are in the Callahuaya pharmacopoeia. I know of no obvious reason why low- or high-calorie intake should affect hemorrhage on the spot, although high-calorie would be indicated to compensate for blood loss.

There was a departure from the norm on high-calorie intake among the folk medicinal remedies for dysentery: baobob, betel, breadfruit, burdock, cacao, chestnut, cinnamon, coconut, corn, cumin, flax, garbanzo, grass pea, Indian almond, Job's tear, lentil, pea, quinoa, rice, sesame, sunflower, and wheat. Ten of these, including rice and corn, are in the Callahuaya pharmacopoeia.

From the protein data of Table 1, we see that high protein departs from the norm in folk cures for dysentery: baobob, breadfruit, cacao, corn, cumin, flax, garbanzo, grass pea, Indian almond, Job's tear, lentil, opium, pea, quinoa, sesame, sunflower, and wheat. High-protein is logically indicated in dysentery.

Low-protein foods, among them annatto, apricot, banana, date, lemon, mango, persimmon, raspberry, and tuna were among the antihemorrhagics, seven of the nine in the Callahuaya pharmacopoeia. Here I think we can see a correlation

Table 1. Relative Incidence of Characters and Coincidence of Paired Characters (Parenthetical numbers indicate the number of the 1000 matrix species reported in a given category).

	TREAT ABSCESS (58)	TREAT ARTHRITIS (102)	TREAT ASTHMA (57)	TREAT BRONCHITIS (50)	TREAT CANCER (24)	TREAT CHOLERA (32)	TREAT COLD (101)	TREAT DYSENTERY (84)	TREAT FEVER (117)	TREAT HEMORRHAGE (33)	TREAT NEURALGIA (39)	TREAT SORES (106)	TREAT SOROCHE (9)	TREAT STOMACH (101)	TREAT STONES (47)	TREAT WORMS (85)
LOW CALORIE (58)	4	13	6	8	3	2	14	3	10	5	6	12	1	10	10	7
HIGH CALORIE (143)	13	16	8	8	3	8	14	22	21	4	5	20	2	16	9	16
LOW PROTEIN (122)	12	19	7	9	5	3	22	16	22	9	6	23	0	20	16	14
HIGH PROTEIN (114)	10	13	5	5	2	3	11	17	18	3	6	16	2	15	8	12
LOW OIL (280)	28	40	17	20	12	14	46	38	50	16	11	53	2	41	30	37
HIGH OIL (49)	4	6	3	5	1	0	8	7	11	2	1	10	0	6	2	10
LOW CARBOHYDRATE(164)	17	30	17	17	9	8	32	21	33	9	11	33	2	34	28	32
HIGH CARBOHYDRATE(64)	9	9	4	3	2	6	7	14	9	2	3	8	0	9	6	3
LOW FIBER (48)	5	8	4	6	2	1	10	6	11	2	4	6	1	8	6	12
HIGH FIBER (93)	13	14	6	4	4	3	14	18	16	3	6	17	2	17	5	12
LOW ASH (99)	8	13	2	6	6	1	18	13	16	4	3	12	1	16	12	13
HIGH ASH (152)	20	24	13	12	4	8	19	29	26	7	8	29	1	24	12	22
LOW CALCIUM (185)	18	25	9	9	5	8	28	23	32	7	9	29	2	25	14	26
HIGH CALCIUM (45)	5	5	3	1	0	5	5	8	5	0	4	5	0	7	6	7
LOW PHOSPHORUS (239)	25	36	11	14	8	7	41	27	45	12	10	43	2	35	28	32
HIGH PHOSPHORUS (91)	10	10	2	3	1	2	9	14	14	2	3	13	2	15	6	13
LOW MAGNESIUM (15)	2	1	0	1	0	1	2	0	2	0	0	1	0	2	1	0
HIGH MAGNESIUM (25)	5	4	1	2	1	2	4	2	2	0	3	5	1	5	2	3
LOW IRON (188)	22	28	8	11	7	6	30	28	36	10	9	29	2	29	21	23
HIGH IRON (71)	9	10	5	3	1	5	7	14	12	2	4	17	2	14	11	16
LOW SODIUM (190)	2	29	9	14	8	11	34	31	34	10	10	36	3	35	23	30
HIGH SODIUM (8)	0	2	1	0	0	0	1	2	2	0	1	1	0	1	2	1
LOW POTASSIUM (18)	0	0	0	0	0	0	3	0	2	0	0	3	0	1	2	1
HIGH POTASSIUM (8)	2	0	0	1	0	0	2	1	0	0	0	1	0	2	2	0
LOW VITAMIN A (117)	19	17	6	10	6	6	22	22	29	4	6	23	3	23	7	14
HIGH VITAMIN A (115)	12	21	11	13	4	8	21	15	22	8	10	20	2	24	20	27
LOW THIAMIN (48)	4	4	1	3	4	2	5	4	4	3	1	5	0	4	4	3
HIGH THIAMIN (81)	24	31	12	13	5	9	26	29	32	8	12	35	4	32	18	31
LOW RIBOFLAVIN (59)	6	11	3	6	5	2	13	8	11	3	2	9	1	7	6	7
HIGH RIBOFLAVIN (95)	15	17	9	13	4	10	20	22	21	5	8	23	1	24	19	21
LOW NIACIN (45)	3	8	2	4	3	1	11	9	12	4	1	6	1	6	7	6
HIGH NIACIN (52)	8	8	3	4	1	2	10	9	10	4	3	9	0	11	6	5
LOW VITAMIN C (74)	12	14	5	4	1	7	11	14	15	2	4	14	2	13	9	10
HIGH VITAMIN C (36)	6	6	5	4	2	4	7	6	4	3	2	9	0	6	4	7

between astringency and antihemorrhagic rather than a dietary correlation. Low-protein might be indicated for internal hemorrhage associated with hepatic ailments.

Low-oil was disproportionately high among folk remedies for stones. Since gall stones consist mostly of cholesterol, this might be a good correlation. High oil was disproportionately high among folk vermifuges. Except for laxative oils, there is no immediately obvious reason for this correlation. High carbohydrate, which might well be indicated for cholera, was disproportionately high in cholera remedies. I see no medical reason, however, why low carbohydrate might be recommended for worms.

High fiber could be indicated to stop the diarrhea associated with cholera, but I see no reason except coincidence for low fiber to be characteristic of bronchitis folk remedies!

Low ash, at least low calcium and low phosphorus, might be indicated in metastatic cancer, where blood levels of calcium are high; and high ash might help replenish electrolytes in cholera patients. On the other hand, I see no reason for low iron for cancer, or low or high phosphorus for soroche. High iron for soroche makes sense. Low magnesium might or might not be indicated for fever, depending on the cause of the fever, while high magnesium, like high zinc oxide, is good for healing sores. High-sodium for asthma makes no sense. High-potassium intake might be indicated for stomach patients whose electrolytic balance was upset. Low-potassium and/or low-sodium might be recommended for ulcers which heal slowly due to edema. Low vitamin A should be contraindicated for abscesses, but high vitamin A and stones seem unrelated.

I know of no reason why low thiamin might be indicated for cancer or high thiamin for asthma, although high thiamin would be indicated for neuralgia. Further I know no reason why low riboflavin might help cancer, but high riboflavin, like most of the B complex, might be indicated for cholera. I know not why low niacin should be good for cancer, but there is good reason for high niacin diets among neuralgics. Low vitamin C, if anything should be contraindicated for fever, while high vitamin C might be indicated for the cachexia associated with some types of cancer.

Scoring a plus for a logical positive correlation, a zero for a correlation with no obvious beneficial or deleterious effects, and a minus for a correlation which is contraindicated, we find that the folk pharmacopoeia, based on nutritional chemistry, is right more often than wrong and/or neutral.

For the second test of the folk medicines, I sought the help of the Medicinal Plant Resources Laboratory under the leadership of Dr. R. E. Perdue. Of their extensive and rather proprietary list of actives in their cancer screening program, 103 occur in our crop matrix. Dr. Perdue noted an apparent correlation between vermifugal activity and anticancer activity in African plants.

Assuming that all our data were complete and that coincidences of uncorrelated characters were random, one would expect five of our antitumor agents to show activity in the cancer screening program. Superposing the two matrices, we see coincidence at the expected 5 points, suggesting no correlation between our folk antitumor agents and tested cancer actives. Our data corroborate Perdue's vermifuge correlation. On basis of random distribution, we expect nine of our vermifuges to show up among the cancer actives. There are 15. An even higher ratio shows up for our "cancer remedies" and the cancer active file. Based on chance, there should be only three, but, in fact, there are seven, suggesting either that the Callahuaya Indians might have been right with some of their "cancer cures" or that cancer screening has been more intensive among reported folk cures. Of the seven, jute, jojoba, and periwinkle are exotic to the Callahuaya, while the other four folk "cancer cures" that have proven active in the cancer screening program, (calendula, lemon, peach, and senna) are important in the Callahuaya pharmacopoeia.

CONCLUSIONS

The crop matrix shows some potential new correlations between nutritional chemistry and folk medicine. Dietary manipulation of diseases, even in modern medicine, is often so slow that primitive man would not be able to see the cause and effect relationship.

A manual information retrieval can show correlations between some folk remedies and nutritional chemistry, but a computerized system will be necessary to test adequately the significance, if any, of these minor correlations.

Folk medicines must have evolved through trial and

error. Those that were tried and showed immediate benefits survived as the fittest; those that had no effect or negative effects were eliminated. Those with delayed results were more difficult to recognize. The primitive applying an oakgall to a cut is doing the same as modern man applying a styptic after a shaving cut. Results are immediately apparent. When an Indian chews coca leaves, he almost immediately knows that it numbs the mouth, helps a toothache, and curbs the appetite. It takes longer to learn that the bitter principle in quinine can cure malaria. But it would take long and detailed medical research to show that the calcium consumed with coca in Colombia, or with betel in Burma, might prevent or alleviate osteoporosis.

Folk responses to alkaloids like quinine and cocaine may be much more immediate than responses to nutritional chemistry. It might take months for man to realize that he was responding to good diet. But if primitive man could learn the correlation between sexual intercourse and childbirth nine months later, perhaps he could notice physical responses to dietary changes, nine months earlier.

While obviously laden with taxonomic errors, the Callahuaya pharmacopoeia contains about 2000 crops, many of them only recently showing new medicinal promise.

ACKNOWLEDGMENTS

Dr. J.T. Best reviewed the text and contributed medical opinions introduced in the text. Dr. E.E. Terrell, Dr. C.R. Reed, and Mr. S. Hurst helped with nomenclature of the crops.

REFERENCES

1. Duke, J.A. 1972. "Isthmian Ethnobotanical Dictionary." Harrod & Co., Fulton, Md.
2. Krochmal, A. 1968. Medicinal Plants and Appalachia. Econ. Bot. 22: 332.
3. Leung, W.W. & M. Flores. 1961. "Tabla de composición de alimentos para uso en America Latina." INCAP. Guatemala City, Guatemala.
4. Watt, B.K. & A.L. Merrill. 1963. "Composition of Foods." Agric. Handbook No. 8. U.S. Government Printing Office, Washington, D.C.
5. Leung, W.W., F. Busson, & C. Jardin. 1968. "Food composition table for use in Africa." FAO, Rome.
6. _____, R.R. Butrum, & F.H. Chang. 1972. "Food composition table for use in East Asia. FAO and U.S. Dept. HEW, Rome.
7. Duke, J.A. & E.E. Terrell. 1974. Crop diversification matrix: Introduction. Taxon. 23: 73.
8. Bender, A.E. 1973. "Nutrition and Dietetic Foods " (2nd edition of Dietetic Foods). Chemical Publishing Co., New York.
9. Cardenas, M. 1969. "Manual of Plantas Económicas de Bolivia." Imprenta Icthus, Cochabamba, Bolivia.
10. Del Granado, J.T. 1931. "Plantas Boliviana." Arno Hermanos, La Paz.
11. Dominguez, X.A., P. Rojas, V. Collins, and Ma. Del Refugio Morales. 1960. A phytochemical study of eight Mexican plants. Econ. Bot. 44: 157.
12. Jain, S.K. & C.R. Tarafder. 1970. Medicinal Plant-lore of the Santals. Econ. Bot. 24: 241.
13. Johnston, A. 1970. Blackfoot Indian utilization of the flora of the Northwestern Great Plains. Econ. Bot. 24: 301.
14. Poblete, E.O. 1969. "Plantas medicinales de Bolivia. Farmacopea Callawaya." Editorial Los Amigos del Libro, Cochabamba, La Paz.
15. Shah, N.C. & M.C. Joshi. 1971. An ethnobotanical study of the Kumaon Region of India. Econ. Bot. 25: 414.
16. Uhe, George. 1974. Medicinal plants of Samoa. A preliminary survey of the use of plants for medicinal purposes in the Samoan Islands. Econ. Bot. 28: 1.
17. Wiener, M.A. 1971. Ethnomedicine in Tonga. Econ. Bot. 25: 423.

18. Jukes, T.H., and H. Borsook. 1974. Nutritional management of the anemic geriatric patient. Geriatrics 29: 147.
19. Dreizen, S. 1974. Clinical manifestations of malnutrition. Geriatrics 29: 97.
20. Rueda-Williamson, R. 1972. Vitamin A malnutrition in Latin America and in Caribbean. Proc. Western Hemisp. Nutrition Congress III. 1971: 56.

APPENDIX OF CROP MATRIX TABLES

Table 2. Common Names of

	1	2	3	4	5	6	7	8	9	10	11	12	13
A	OKRA	MUSK OKRA	CHINA JUTE	HUISACHE	BLACK WATTLE	BABUL	GOLDEN WATTLE	GUM ARABIC	SHITTIM WOOD	SUGAR MAPLE	ACONITE	SWEET FLAG	NILE GRASS
B	CRESTED WHEATGR	TALL WHEATGR	INTERMED WHEATGR	QUACK-GRASS	STREAMBK WHEATGR	SIBERIAN WHEATGR	WESTERN WHEATGR	BLUEBUNCH WHEATGR	SLENDER WHEATGR	VELVET BENTGRASS	REDTOP	CREEPING BENTGRASS	COLONIAL BENTGRASS
C	BARBADOS ALOE	CAPE ALOE	SOCOTRINE ALOE	CREEPING FOXTAIL	MEADOW FOXTAIL	LESSER GALANGAL	ALYCE CLOVER	INCA WHEAT	SPANISH GREENS	PRINCESS-FEATHER	LIVID AMARANTH	CHINESE AMARANTH	SPANISH CARROT
D	SOURSOP	BULLOCK'S HEART	SUGAR APPLE	ROMAN CHAMOMILE	CHERVIL	KIDNEY VETCH	BIGNAY	CELERY	CELERIAC	PEANUT	SPIKENARD	BURDOCK	BETEL NUT
E	BREAD-FRUIT	CHEMPEDAK	JACKFRUIT	ARUNDIN-ARIA	GIANT REED	WOOLLY PAPAW	PAPAW	SPRENGER ASPARAGUS	ASPARAGUS	SWEET WOODRUFF	TRAGA-CANTH	SALTBUSH	BUTTER LEAVES
F	KAPUNDUNG	BACCAUREA RAMIFLORA	PEACH PALM	UDJUNG ATUP	DESERT DATE	BAMBUSA	CAMWOOD	UPLAND CRESS	BUCHU	MALABAR SPINACH	CAMEL'S FOOT	SLOUGH-GRASS	WAXGOURD
G	YELLOW BLUESTEM	COMAGUE-YANA	SILVER BLUESTEM	SIDEOATS GRAMA	BLACK GRAMA	BLUE GRAMA	SIGNAL GRASS	YAGUE	BRACHIAR-IA EMINII	PARAGRASS	BROWNTOP MILLET	BIRD RAPE	PAKCHOI
H	QUAKING GRASS	CALIF BROMEGR	SMOOTH BROMEGR	MOUNTAIN BROMEGR	RESCUE GRASS	RAMON	COW TREE	CUDDAPAH ALMOND	BUFFALO GRASS	BALSAMO	SHEA BUTTER	DIVI-DIVI	BRAZIL WOOD
I	JACKBEAN	SWORD-BEAN	OBLIQUE JACKBEAN	EDIBLE CANNA	MARI-JUANA	CAPER	MANGO PEPPER	PIMENT-CHIEN	CAPSICUM CHINENSE	TABASCO	GOATCHILI	PAPAYA	MOUNTAIN PAPAYA
J	SENNA COFFEE	ALEXANDR SENNA	JAPANESE CHESTNUT	AMERICAN CHESTNUT	CHINESE CHESTNUT	CHINQUA-PIN	EUROPEAN CHESTNUT	CASTILLOA RUBBER	CAUCHO RUBBER	KHAT	PERI-WINKLE	KAPOK	BUFFEL-GRASS
K	PYRETHRUM	PERSIAN INSECT FL	GARLAND CHRYSANT	COSTMARY	CROWN DAISY	CAINITO	AFRICAN STARAPPLE	GARBANZO	ENDIVE	CHICORY	QUININE	PADANG CASSIA	CAMPHOR
L	ORANGE	TACHIBANA	BOLOBOLO	CLAPPER POLYANDRA	CLAUSENA DENTATA	WAMPI	BUTTERFLY PEA	BLUE PEA	SEA GRAPE	COCONUT	COFFEE	BENGAL COFFEE	ROBUSTA COFFEE
M	TUSSA JUTE	TI PALM	CORIANDER	CROWN-VETCH	COROZO	EUROPEAN FILBERT	CHINESE HAZELNUT	TURKISH HAZELNUT	BEAKED HAZELNUT	HIMALAYAN HAZELNUT	SIBERIAN HAZELNUT	GIANT FILBERT	TIBETAN HAZELNUT
N	FIGLEAF GOURD	BOSTON MARROW	MIXTA SQUASH	CROOKNECK PUMPKIN	VEGETABLE MARROW	CUMIN	TURMERIC	ZEDOARY	GUAR	ARCHUCHA	QUINCE	SERE GRASS	MALABAR GRASS
O	ORCHARD GRASS	CROWFOOT GRASS	CARROT	DENDRO-CALAMUS	TUBA PUTCH	TUBA ROOT	BEGGAR-LICE	DIAZ BLUESTEM	ANGLETON BLUESTEM	DICHONDRA	DITTANY	YELLOW FOXGLOVE	DIGITALIS
P	BLACK SAPOTE	KAKI	PERSIMMON	CUMARU	ENG TONKA BEAN	HORSE GRAM	KETEM-BILLA	ARG PHEAS-ANT TREE	WINTER'S BARK	DURIAN	JELUTONG	SQUIRTING CUCUMBER	BARNYARD-GRASS
Q	BLUE WILD RYE	RUSSIAN WILD RYE	PAKISTANI EPHEDRA	CHINESE EPHEDRA	FIELD HORSETAIL	BOER LOVEGRASS	WEEPING LOVEGRASS	LEHMANN LOVEGRASS	TEF	SAND LOVEGRASS	CENTIPEDE GRASS	AUSTRAL DEST LIME	LOQUAT
R	EUROPEAN BEECH	FEIJOA	IND WOOD APPLE	TALL FESCUE	IDAHO FESCUE	HARD FESCUE	SHEEP FESCUE	MEADOW FESCUE	RED FESCUE	CHEWING'S FESCUE	HUANG T'ENG	COMMON FIG	RUBBER FIG
S	BANITI	KARIIS	IMBE	MANGO-STEEN	BIRA TAI	GAMBOGE TREE	WINTER-GREEN	HUCKLE-BERRY	DYER'S GENISTA	YELLOW GENTIAN	AVENS	MPEMPO	SOYBEAN
T	SUNCHOKE	JERUSALEM ARTICHOKE	KAMRAJ	LIMPO GRASS	RUBBER	PARA-RUBBER	KENAF	ROSELLE	CURLY MESQUITE	GALLETA GRASS	TOBOSA GRASS	BULBOUS BARLEY	TWO-ROWED BARLEY
U	NATAL-INDIGO	HAIRY INDIGO	SPICATA INDIGO	COMMON INDIGO	ICE CREAM BEAN	INULA	WATER SPINACH	SWEET POTATO	GHIABATO	JASMINE	CHILE COCO	HEARTNUT	BUTTERNUT
V	DRAGON'S HEAD	GUINEA GUM VINE	KIRK'S GUM VINE	W AFRICAN GUM VINE	GREATER GALANGA	LANGSAT	ROUGH PEA	GRASS PEA	SWEET BAY	BROAD LF LAVENDER	COMMON LAVENDER	FRENCH LAVENDER	HENNA
W	LOVAGE	OITICICA OIL	WHITE MEADWFOAM	BAKER'S MEADWFOAM	DOUGLAS'S MEADWFOAM	FLAX	SPANISH THYME	LITCHI NUT	INDIAN TOBACCO	ANNUAL RYEGRASS	PERENNIAL RYEGRASS	MACAQUIN-HO TIMBO	URUCU TIMBO
X	MACADAMIA NUT	MAHUA	SWEET MARJORAM	POT MARJORAM	BARBADOS CHERRY	SIBERIAN CRABAPPLE	APPLE	MAMEY APPLE	MANDRAKE	MANGO	MANICOBA	CASSAVA	CEARA RUBBER
Y	SNAIL MEDIC	CAJUPUT TREE	BURMESE LACQUER	SPANISH LIME	WHITE MELILOT	INDIAN MELILOT	YELLOW MELILOT	SWEET-CLOVER	MOLASSES GRASS	SWEET BALM	CORN MINT	SCOTCH SPEARMINT	PEPPER-MINT
Z	BLACK MULBERRY	RED MULBERRY	VELVET-BEAN	CURRY-LEAF TREE	DWARF BANANA	ENSETE	BANANA	ABACA	BRAZILIAN GRAPE TR	NUTMEG	TOLU BALSAM	PERU BALSAM	MYRRH
a	WATER DROPWORT	OLIVE	SAINFOIN	LILY TURF	INDIAN FIG	COHUNE PALM	OREGANO	SERRA-DELLA	AFRICAN RICE	RICE	INDIAN RICEGRASS	SMILO-GRASS	OCA
b	SCARLET POPPY	OPIUM POPPY	AFRICAN LOCUST BN	CANDLE-TREE	CUAJILOTE	GUAYULE	DALLIS-GRASS	BAHIA-GRASS	KODO MILLET	VASEY-GRASS	PASSION FRUIT	WATER-LEMON	SWT GRAN-ADILLA
c	PEREJIL	PERILLA	AVOCADO	PARSLEY	WATER CANARYGR	REED CANARYGR	TRUE CANARYGR	SUNOL-GRASS	HARDING GRASS	TEPARY BEAN	SIRATRO	SCARLET RUNNER BN	LIMA BEAN
d	BAY RUM TREE	ANISE	SILVER PINE	PINYON	BETEL PEPPER	IND LONG PEPPER	KAVA	BLACK PEPPER	JAVANESE LG PEPPER	PISTACHIO	PEA	KAFFIR POTATO	BIG BLUEGR
e	SPEKBOOM	EGG-FRUIT TREE	MAMMEE SAPOTE	POUTERIA VIRIDIS	APRICOT	SWEET CHERRY	MYROBALAN PLUM	SOUR CHERRY	COMMON PLUM	DAMSON PLUM	ALMOND	CHERRY LAUREL	BEACH PLUM
f	KUDZU	TROPICAL KUDZU	POME-GRANATE	PEAR	SAND PEAR	CORK-OAK	RADISH	RAUWOLFIA	SWEET RESEDA	CANTON RHUBARB	CHINESE RHUBARB	RHUBARB	CHINESE LACQUER
g	PURPLE RASPBERRY	GARDEN RUE	SUGARCANE	CHIA	SAGE	CLARY	EASTERN ELDERBRY	WESTERN ELDERBRY	BURNET	SNAKE PLANT	CHINESE TALLOW TR	SUMMER SAVORY	WINTER SAVORY
h	DANICHA	SESBANIA	ITALIAN MILLET	GOLDEN TIMOTHY	CASSABANA	ACEITUNA	JOJOBA	WHITE MUSTARD	SARSA-PARILLA	CUP-EGG-PLANT	GUNYANG	RAM-BEGUN	GILO
i	SUDAN POTATO	INDIAN-GRASS	ALMUM	SORGHUM	JOHNSON-GRASS	SUDAN-GRASS	SPANISH BROOM	GIRIGIRI	PARA CRESS	SPINACH	AMBARELLA	HOG PLUM	RED MOMBIN
j	KOMBE	ARROW POISON	STRYCH-NINE	BARBA-TIMAO	NIGERIAN LUCERNE	BRAZILIAN LUCERNE	TOWNSVIL LUCERNE	STYRAX	SIAM BENZOIN	OURICURU PALM	COMFREY	SWEET BERRY	CLOVE
k	VOGEL TEPHROSIA	INDIAN ALMOND	N ZEALAND SPINACH	ASPARAGUS PEA	KATEMFE	BACAO	CACAO	CUPUACU	WILD THYME	COMMON THYME	AMMI	SALSIFY	HORNED CHESTNUT
l	WHITE CLOVER	PERSIAN CLOVER	SUB CLOVER	WHITETIP CLOVER	ARROWLEAF CLOVER	SEASIDE CLOVER	HOP CLOVER	FENUGREEK	LIMEBERRY	TRITICALE	WHEAT	PERSIAN WHEAT	CLUB WHEAT
m	COLTS-FOOT	ULLUCU	GAMBIER	ARAMINA	STINGING NETTLE	LOWBUSH BLUEBERRY	RABBITEYE BLUEBERRY	HIGHBUSH BLUEBERRY	CRANBERRY	LIGON-BERRY	CORN SALAD	SPANISH TAMARIND	VANILLA
n	HAIRY VETCH	MOTH BEAN	ADZUKI BEAN	BLACK GRAM	MUNG BEAN	RICE BEAN	COW PEA	CATJAN	YARD-LONG BEAN	VIGNA VEXILLATA	VIROLA FAT	FOX GRAPE	MUSCADINE GRAPE

ALTERNATIVE

Crops in 1000 Crop Matrix

	14	15	16	17	18	19	20	21	22	23	24	25
A	KIWI	BAOBAB	MALABAR NUT	JOINTED GOAT GR	BAEL	MELEGUETA PEPPER	BABOON SPICE	HENEQUEN	LECHE-GUILLA	SISAL	FAIRWAY WHEATGR	THICKSPK WHEATGR
B	TUNG-OIL TREE	CANDLENUT	MU-OIL TREE	CAMEL-THORN	ELEPHANT GARLIC	ONION	RAKKYO	WELSH ONION	GARLIC	CHIVES	CHINESE CHIVES	GIANT TARO
C	EUROPEAN BEACHGR	AMERICAN BEACHGR	CASHEW	PINEAPPLE	GAMBA GRASS	BIG BLUESTEM	SAND BLUESTEM	DILL	ANGELICA	CHERIMOYA	ILAMA	CIMARRONA
D	SUGAR PALM	BACURY	HORSE-RADISH	EUROPEAN ARNICA	PERUVIAN CARROT	TALL OATGRASS	SOUTHERN WOOD	ABSINTH	RUSSIAN WORMSEED	TARRAGON	MARITIME WORMWOOD	MUGWORT
E	BELLA-DONNA	ABYSSIN-ICA OAT	RED OAT	WILD OAT	COMMON OAT	STERILE OAT	BRISTLE OAT	BILIMBI	CARAMBOLA	CARPET-GRASS	TROPICAL CARPETGR	RAMBAI
F	BRAZIL NUT	BEET	ANNATTO	AKEE	RAMIE	RED SILK COTTON	BORAGE	PALMYRA PALM	FRANKIN-CENSE	CAUCASIAN BLUESTEM	SWEET PITTED GR	AUSTRAL BLUESTEM
G	BROWN MUSTARD	RAPE	RUTABAGA	BLACK MUSTARD	COLLARDS	BROCCOLI	CABBAGE	BRUSSELS SPROUTS	KOHLRABI	ASPARAGUS BROCCOLI	CHINESE CABBAGE	TURNIP
H	SAPPAN WOOD	TARA	PIGEON PEA	RATTAN	CORN ROOT	CALENDULA	FRISOL-ILLA	COWSLIP	TEA	RAMPION	YLANG-YLANG	JAVA ALMOND
I	KARANDA	EGYPTIAN CARISSA	NATAL PALM	SAFFLOWER	CARAWAY	PECAN	HICKORY NUTS	SUARINUT	FISHTAIL PALM	MEXICAN APPLE	RINGWORM BUSH	AVARAM
J	IPECAC	CAROB	BUNGU	WAX PALM	TURNIP CHERVIL	LAMB'S QUARTER	WORMSEED GOOSEFOOT	QUINOA	RHODES-GRASS	FEATHER FINGER GR	AFRICAN OAK	FUSTIC MULBERRY
K	SAIGON CINNAMON	INDIAN BARK	CINNAMON	COLOCYNTH	WATER-MELON	LIME	SOUR ORANGE	PUMMELO	LEMON	CITRON	GRAPE-FRUIT	TANGERINE
L	CONGO COFFEE	LIBERIAN COFFEE	JOB'S TEARS	ABATA KOLA	GBANJA KOLA	OWE KOLA	COLCHICUM	INDIAN BORAGE	RATALA	DASHEEN	COPAIBA	WHITE JUTE
M	CRAMBE	SEA KALE	SAFFRON	SLENDERLF CROTALARI	SUNN-HEMP	LANCELEAF CROTALARI	CROTALARIA PALLIDA	SHOWY CROTALARI	PURGING CROTON	GHERKIN	CANTA-LOUPE	CUCUMBER
N	ROSHA GRASS	NARDUS GRASS	WINTER'S GRASS	CARDOON	GLOBE ARTICHOKE	BERMUDA-GRASS	STAR GRASS	AFRICAN BERMUDAGR	CHUFA	NUTGRASS	TREE TOMATO	HUON PINE
O	PANGOLA-GRASS	HUNGRY RICE	BLACK FONIO	WINGED YAM	AIRPOTATO	COMPOSITA YAM	CONVOLVU-LACEA YAM	FLORI-BUNDA YAM	CUCUL-MECA	CHINESE YAM	EBOE YAM	SERENDIP-ITY BERRY
P	BILLON DOLLAR GR	ANTELOPE GRASS	PERENN VELDTGR	LINGARO	OIL PALM	WATERNUT	CARDAMON	FINGER MILLET	GOOSE-GRASS	CANADA WILD RYE	BASIN WILD RYE	GIANT WILD RYE
Q	CARIB-GRASS	ROQUETTE	FITWEED	COCA	EUCA-LYPTUS	KADJATOA	SURINAM CHERRY	CANDE-LILLA	LONGAN	BUCKWHEAT	TATARIAN BUCKWHEAT	AMERICAN BEECH
R	VOGEL FIC	TIKUG	GOVERN PLUM	FENNEL	KUMQUAT	GARDEN STRAWBRY	CHILEAN STRAWBRY	EUROPEAN STRAWBRY	VIRGINIA STRAWBRY	FUNTUMIA AFRICANA	FUNTUMIA ELASTICA	MAURITIUS HEMP
S	COMMON LICORICE	AMERICAN LICORICE	WILD COTTON	TREE COTTON	SEA ISLND COTTON	LEVANT COTTON	UPLAND COTTON	PHALSA	NIGER SEED	PANQUE	SPANISH SAINFOIN	SUNFLOWER
T	BARLEY	TSI	HOPS	LUKRABAO	HENBANE	HIRTA GRASS	JARAGUA GRASS	BENEFING	HYSSOP	CANDY-TUFT	YERBA MATE	STAR-ANISE
U	CALIF BLK WALNUT	AMER BLK WALNUT	ENGLISH WALNUT	SEA-RUSH	JUNIPER	KENTJOER	GEOCARPA	SUMMER CYPRESS	LABLAB	LETTUCE	BITTER LETTUCE	CALABASH
V	SAPUCAJA NUT	MONKEY POD	LENTIL	GARDEN CRESS	PEPPER GRASS	AUSTRAL SEATREE	SERICEA	KOREAN CLOVER	JAPANESE CLOVER	FENDLER-POD	GORDON-POD	LEAD TREE
W	BARBASCO	BIRDSFOOT TREFOIL	BIG TREFOIL	NARROW LF TREFOIL	ANGLED LUFFA	SMOOTH LUFFA	WHITE LUPINE	BLUE LUPINE	YELLOW LUPINE	GOW KEE	TOMATO	CURRANT TOMATO
X	BALATA	CHICLE	ARROW-ROOT	HORE-HOUND	GERMAN CHAMOMILE	BURITI PALM	SPOTTED BURCLOVER	SICKLE MEDIC	BLACK MEDIC	BUTTON CLOVER	CALIF BURCLOVER	ALFALFA
Y	PENNY-ROYAL	APPLEMINT	SPEARMINT	CHAMPAC	HUEMEGA	BALSAM APPLE	BALSAM PEAR	CERIMAN	WINTER PURSLANE	INDIAN MULBERRY	BENOIL TREE	WHITE MULBERRY
Z	MYRTLE	WATER CRESS	SACRED LOTUS	CATNIP	RAMBUTAN	PULASAN	TOBACCO	BLACK CUMIN	COCHINEAL	SWEET BASIL	HOARY BASIL	HOLY BASIL
a	EROSUS YAMBEAN	TUBEROUS YAMBEAN	GUTTA PERCHA	GINSENG	BLUE PANICGR	KLEIN-GRASS	MAIDEN-CANE	GUINEA-GRASS	PROSO MILLET	VINE MESQUITE	TORPEDO-GRASS	SWITCH-GRASS
b	BANANA FRUIT	BARBAR-DINE	PARSNIP	GUARANA	YOCO	HARMALA SHRUB	SOGA	PEARL MILLET	KIKUYU-GRASS	ELEPHANT GRASS	OWALA OIL	AUSTRAL SHEEPBUSH
c	GREEN BEAN	TIMOTHY	DATE PALM	N ZEALAND FLAX	REED GRASS	INDIAN GOOSEBRY	EMBLIC	PHYLLO-STACHYS	TOMATILLO	PERUVIAN CHERRY	POKEWEED	ALLSPICE
d	ANNUAL BLUEGR	TEXAS BLUEGR	BULBOUS BLUEGR	MUTTON BLUEGR	KENTUCKY BLUEGR	SANDBERG BLUEGR	ROUGHISH BLUEGR	PATCHOULI	TUBEROSE	POLYSCIAS RUMPHIANA	TRIFOL ORANGE	PURSLANE
e	PEACH	JAPANESE PLUM	SIERRA	CATTLEY GUAVA	WILD GUAVA	LEMON GUAVA	GUINEA GUAVA	MOUNTAIN GUAVA	GOA BEAN	RED BARWOOD	SANDERS-WOOD	BARWOOD
f	HAIRY GOOSEBRY	BLACK CURRANT	GARDEN CURRANT	COMMON RED RIBES	EUROPEAN GOOSEBRY	CASTOR BEAN	CHRIST-MAS POPS	ROSEMARY	BLACK-BERRIES	DEWBER-RIES	RED RASPBERRY	BLACK RASPBERRY
g	QUEBRACHO	RED QUE-BRACHO	PEPPER-TREE	LITTLE BLUESTEM	MALAY LAC-TREE	MARULA NUT	RYE	CHAYOTE	MARKING NUT	WING-SESAME	SESAME	WILD SESAME
h	COCONA	SODOM APPLE	TIBBATU	SOLANUM KHASIANUM	KANGAROO APPLE	NATIVE EGGPLANT	EGGPLANT	MELON PEAR	WONDER-BERRY	LULO	PLATE-BUSH	POTATO
i	IMBU	ALKALI SACATON	SAND DROPSEED	SACATON	CHINESE ARTICHOKE	ST AUGUST GRASS	KAA HE'E	N & T GRASS	ESPARTO	GREEN NEEDLE GR	CREAM FRUIT	UMTSUTI
j	JAVA PLUM	ROSE APPLE	MALAY APPLE	JAVA APPLE	TAMARIND	TANSY	CHAULMO-OGRA	KOK-SAGHYZ	DANDELION	OYSTER NUT TREE	ZANZIBAR OIL VINE	WHITE TEPHROSIA
k	JESUIT NUT	MUZINDA	CLUB GOURD	SNAKE GOURD	BERSEEM CLOVER	KURA CLOVER	STRAWBRY CLOVER	ROSE CLOVER	ALSIKE CLOVER	CRIMSON CLOVER	BALL CLOVER	RED CLOVER
l	EMMER	DURUM	EINKORN	POLISH WHEAT	SPELT	TIMOPHE-ERI	ORIENTAL WHEAT	POULARD WHEAT	BURWEED	CHIKTI FIBRE	MPUNGA	ANU
m	MA-HA-WA-SOO	KINKAOIL IRONWEED	VETIVER	NARROWLF VETCH	MONANTHA VETCH	PURPLE VETCH	WOOLLYPOD VETCH	BITTER VETCH	HORSE BEAN	BARD VETCH	HUNGARIAN VETCH	COMMON VETCH
n	WINE GRAPE	BAMBARRA GROUNDNUT	YAUTIA	OZOTE	CORN	TEOSINTE	GINGER	WILD RICE	CHINESE JUJUBE	INDIAN JUJUBE	JAPANESE LAWNGRASS	MAN[illegible] GRASS

CROP MATRIX

Table 3. Low Calorie Crops (<25/100 g)

	1	2	3	4	5	6	7	8	9	10	11	12	13	14	15	16	17	18	19	20	21	22	23	24	25	
.A.	AES	AMO	ATH	AFA	AME	ANI	APY	ASE	ASE	ASA	ANA	ACA	AMA	ACH	ADI	AVA	ACY	AMA	AME	ASC	AFO	ALE	ASI	ACR	ADA	.A.
.B.	ADE	AEL	AIN	ARE	ARI	ASI	ASM	ASP	ATR	ACA	AGI	AST	ATE	AFO	AMO	AMO	APS	AAM	ACE	ACH	AFI	ASA	ASC	ATU	AMA	.B.
.C.	ABA	AFE	APE	AAR	APR	AOF	AVA	ACA	ACR	AHY	ALI	ATR	AVI	AAR	ABR	AOC	ACO	AGA	AGE	AHA	AGR	AAR	ACH	ADI	AMO	.C.
.D.	AMU	ARE	ASQ	AMO	ACE	AVU	ABU	AGR	AGR	AHY	ACO	ALA	ACA	API	AES	ARU	AMO	AXA	AEL	AAB	AAB	ACI	ADR	AMA	AVU	.D.
.E.	AAL	ACH	AIN	ASP	ADO	AGR	ATR	ADE	AOF	AOD	AGU	ACA	AHO	ABE	AAB	ABY	AFA	ASA	AST	AST	ABI	ACA	AAF	ACO	BMO	.E.
.F.	BRA	BRA	BGA	BFR	BAE	BSP	BNI	BVU	BBE	BAL	BES	BSY	BHI	BEX	BVU	BOR	BSA	BNI	BCE	BOF	BFL	BCA	BCA	BIN	BIN	.F.
.G.	BIS	BPE	BSA	BCU	BER	BGR	BBR	BDE	BEM	BMU	BRA	BCA	BCH	BJU	BNA	BNA	BNI	BOL	BOL	BOL	BOL	BOL	BOL	BPE	BRA	.G.
.H.	BHU	BCA	BIN	BMA	BUN	BAL	BUT	BLA	BDA	BSA	BPA	CCO	CES	CSA	CSP	CCA	CRO	CAL	COF	CMU	CPA	CSI	CRA	COD	CIN	.H.
.I.	CEN	CGL	CPL	CED	CSA	CSP	CAN	CBA	CCH	CFR	CPU	CPA	CPU	CCA	CED	CGR	CTI	CCA	CIL	CSP	CAM	CUR	CED	CAL	CAU	.I.
.J.	CCC	CSE	CCR	CDE	CMO	CPU	CSA	CEL	CUL	CED	CRO	CPE	CCI	CIP	CSI	CSE	CAL	CBU	CAL	CAM	CQU	CGA	CVI	CEX	CTI	.J.
.K.	CCI	CCO	CCO	CMA	CSP	CCA	CDE	CAR	CEN	CIN	COF	CBU	CCA	CCA	CTA	CVE	CCO	CLA	CAU	CAU	CGR	CLI	CME	CPA	CRE	.K.
.L.	CSI	CTA	CFI	CPO	CDE	CLA	CLA	CTE	CUV	CNU	CAR	CBE	CCA	CCO	CLI	CLA	CAC	CNI	CVE	CAU	CAM	CPA	CES	COF	CCA	.L.
.M.	COL	CTE	CSA	CVA	COL	CAV	CCH	CCO	CCO	CFE	CHE	CMA	CTI	CAB	CMA	CSA	CBR	CJU	CLA	CPA	CSP	CTI	CAN	CME	CSA	.M.
.N.	CFI	CMA	CMI	CMO	CPE	CCY	CLO	CZE	CTE	CPE	COB	CCI	CFL	CMA	CNA	CWI	CCA	CSC	CDA	CPL	CTR	CES	CRO	CBE	DFR	.N.
.O.	DGL	DAE	DCA	DSP	DEL	DMA	DIN	DAN	DAR	DRE	DAL	DGR	DPU	DDE	DEX	DIB	DAL	DBU	DCO	DCO	DFL	DMA	DOP	DRO	DCU	.O.
.P.	DEB	DKA	DVI	DOD	DOP	DUN	DHE	DMA	DWI	DZI	DCO	EEL	ECR	ECR	EPY	ECA	EPH	EGU	EDU	ECA	ECO	EIN	ECA	ECI	ECO	.P.
.Q.	EGL	EJU	EGE	ESI	EAR	ECH	ECU	ELE	ETE	ETR	EOP	EGL	EJA	EPO	EVE	EFO	ECO	ESP	EUT	EUN	EAN	ELO	FES	FTA	FGR	.Q.
.R.	FSY	FSE	FLI	FAR	FID	FLO	FOV	FPR	FRU	FRU	FTI	FCA	FEL	FVO	FGL	FIN	FVU	FSP	FAN	FCH	FVE	FVI	FAF	FEL	FFO	.R.
.S.	GDU	GLA	GLI	GMA	GMU	GTI	GPR	GBA	GTI	GLU	GUR	GJA	GMA	GGL	GLE	GAN	GAR	GBA	GHE	GHI	GAS	GAB	GTI	HCO	HAN	.S.
.T.	HAN	HTU	HZE	HAL	HBE	HBR	HCA	HSA	HBE	HJA	HMU	HBU	HDI	HVU	HCO	HLU	HAN	HNI	HHI	HRU	HSP	HOF	IAM	IPA	IVE	.T.
.U.	IAR	IHI	ISP	ITI	IED	IHE	IAQ	IBA	IER	JOF	JCH	JAI	JCI	JHI	JNI	JRE	JMA	JCO	KGA	KGE	KSC	LPU	LSA	LVI	LSI	.U.
.V.	LIB	LHE	LKI	LOW	LGA	LDO	LHI	LSA	LNO	LLA	LSP	LST	LIN	LMI	LOL	LCU	LSA	LVI	LLA	LCU	LST	LST	LFE	LGO	LLE	.V.
.W.	LOF	LRI	LAL	LBA	LDO	LUS	LMI	LCH	LIN	LMU	LPE	LNI	LUR	LUT	LCO	LPE	LTE	LAC	LCY	LAL	LAN	LLU	LCH	LES	LPI	.W.
.X.	MSP	MLO	MHO	MON	MPU	MBA	MSY	MAM	MOF	MIN	MDI	MES	MGL	MBI	MZA	MAR	MVU	MCH	MVI	MAR	MFA	MLU	MOR	MPO	MSA	.X.
.Y.	MSC	MCA	MUS	MBI	MAL	MIN	MOF	MSU	MMI	MOF	MAR	MGE	MPI	MPU	MRO	MSP	MCH	MMI	MBA	MCH	MDE	MPE	MCI	MOL	MAL	.Y.
.Z.	MNI	MRU	MDE	MKO	MAC	MEN	MPA	MTE	MCA	MFR	MBA	MPE	MOD	MCO	NOF	NNU	NCA	NLA	NMU	NTA	NSA	NCO	OBA	OKI	OSA	.Z.
.a.	OJA	OEU	OVI	OJA	OFI	OCO	OVU	OSA	OGL	OSA	OHY	OMI	OTU	PER	PTU	PGU	PSP	PAN	PCO	PHE	PMA	PMI	POB	PRE	PVI	.a.
.b.	PBR	PSO	PFI	PCE	PED	PAR	PDI	PNO	PSC	PUR	PED	PLA	PLI	PMO	PQU	PSA	PCU	PYO	PHA	PPT	PAM	PCL	PPU	PMA	PIN	.b.
.c.	PPE	PFR	PAM	PCR	PAQ	PAR	PCA	PCO	PST	PAC	PAT	PCO	PLU	PVU	PPR	PDA	PTE	PAU	PAC	PEM	PSP	PIX	PPE	PAM	PDI	.c.
.d.	PRA	PAN	PED	PQU	PBE	PLO	PME	PNI	PRE	PVE	PSA	PES	PAM	PAN	PAR	PBU	PFE	PPR	PSE	PTR	PCA	PTU	PRU	PTR	POL	.d.
.e.	PAF	PCA	PSA	PVI	PAR	PAV	PCE	PCE	PDO	PDO	PDU	PLA	PMA	PPE	PSA	PSU	PCA	PFR	PGU	PGU	PMO	PTE	PER	PSA	PSO	.e.
.f.	PLO	PPH	PGR	PCO	PPY	QSU	RSA	RSE	ROD	ROF	RPA	RRH	RVE	RHI	RNI	RRU	RSA	RUV	RCO	RCO	ROF	RSP	RSP	RID	ROC	.f.
.g.	ROC	RGR	SOF	SCH	SOF	SSC	SCA	SCA	SMI	STR	SSE	SHO	SMO	SBA	SLO	SMO	SSC	SOL	SCA	SCE	SED	SAN	SAL	SIN	SRA	.g.
.h.	SBI	SEX	SIT	SSP	SOD	SGL	SCH	SAL	SAR	SAE	SAV	SFE	SGI	SHY	SIN	SIN	SKH	SLA	SMA	SME	SMU	SNI	SQU	STO	STU	.h.
.i.	SRO	SAV	SAL	SBI	SHA	SSU	SJU	SST	SAC	SOL	SCY	SMO	SPU	STU	SAI	SCR	SWR	SSI	SSE	SRE	SCO	STE	SVI	SGR	SHI	.i.
.j.	SKO	SSA	SNU	SAD	SER	SGU	SHU	SBE	STO	SCO	SPE	SDU	SAR	SCU	SJA	SMA	SSA	TIN	TVU	TKU	TKO	TOF	TOC	TPE	TCA	.j.
.k.	TVO	TCA	TTE	TPU	TDA	TBI	TCA	TGR	TSE	TVU	TAM	TPO	TBI	TNA	TAF	TAN	TCU	TAL	TAM	TFR	THI	THY	TIN	TNI	TPR	.k.
.l.	TRE	TRE	TSU	TVA	TVE	TWI	TSP	TFO	TTR	XTR	TAE	TCA	TCO	TDI	TDU	TMO	TPO	TSP	TTI	TTU	TTU	TCO	TRH	TTO	TTU	.l.
.m.	TFA	UTU	UGA	ULO	UDI	VAN	VAS	VCO	VMA	VVI	VLO	VMA	VPL	VCA	VAN	VZI	VAN	VAR	VBE	VDA	VER	VFA	VMO	VPA	VSA	.m.
.n.	VVI	VAC	VAN	VMU	VRA	VUM	VUN	VUN	VUN	VVE	VSE	VLA	VRO	VVI	VSU	XSA	YEL	ZMA	ZMA	ZOF	ZAQ	ZJU	ZMA	ZJA	ZMA	.n.

ALTERNATIVE CROP MATRIX

Table 4. High Calorie Crops (>300/100 g)

	1	2	3	4	5	6	7	8	9	10	11	12	13	14	15	16	17	18	19	20	21	22	23	24	25	
.A.	AES	AMO	ATH	AFA	AME	ANI	APY	ASE	ASE	ASA	ANA	ACA	AMA	ACH	ADI	AVA	ACY	AMA	AME	ASC	AFO	ALE	ASI	ACR	ADA	.A.
.B.	ADE	AEL	AIN	ARE	ARI	ASI	ASM	ASP	ATR	ACA	AGI	AST	ATE	AFO	AMO	AMO	APS	AAM	ACE	ACH	AFI	ASA	ASC	ATU	AMA	.B.
.C.	ABA	AFE	APE	AAR	APR	AOF	AVA	ACA	ACR	AHY	ALI	ATR	AVI	AAR	ABR	AOC	ACO	AGA	AGE	AHA	AGR	AAR	ACH	ADI	AMO	.C.
.D.	AMU	ARE	ASQ	AMO	ACE	AVU	ABU	AGR	AGR	AHY	ACO	ALA	ACA	API	AES	ARU	AMO	AXA	AEL	AAB	AAB	ACI	ADR	AMA	AVU	.D.
.E	AAL	ACH	AIN	ASP	ADO	AGR	ATR	ADE	AOF	AOD	AGU	ACA	AHO	ABE	AAB	ABY	AFA	ASA	AST	AST	ABI	ACA	AAF	ACO	BMO	E.
.F.	BRA	BRA	BGA	BFR	BAE	BSP	BNI	BVU	BBE	BAL	BES	BSY	BHI	BEX	BVU	BOR	BSA	BNI	BCE	BOF	BFL	BCA	BCA	BIN	BIN	.F.
.G.	BIS	BPE	BSA	BCU	BER	BGR	BBR	BDE	BEM	BMU	BRA	BCA	BCH	BJU	BNA	BNA	BNI	BOL	BOL	BOL	BOL	BOL	BOL	BPE	BRA	.G.
.H.	BHU	BCA	BIN	BMA	BUN	BAL	BUT	BLA	BDA	BSA	BPA	CCO	CES	CSA	CSP	CCA	CRO	CAL	COF	CMU	CPA	CSI	CRA	COD	CIN	.H.
.I.	CEN	CGL	CPL	CED	CSA	CSP	CAN	CBA	CCH	CFR	CPU	CPA	CPU	CCA	CED	CGR	CTI	CCA	CIL	CSP	CAM	CUR	CED	CAL	CAU	.I.
.J	CCC	CSE	CCR	CDE	CMO	CPU	CSA	CEL	CUL	CED	CRO	CPE	CCI	CIP	CSI	CSE	CAL	CBU	CAL	CAM	CQU	CGA	CVI	CEX	CTI	J.
.K.	CCI	CCO	CCO	CMA	CSP	CCA	CDE	CAR	CEN	CIN	COF	CBU	CCA	CCA	CTA	CVE	CCO	CLA	CAU	CAU	CGR	CLI	CME	CPA	CRE	.K.
.L.	CSI	CTA	CFI	CPO	CDE	CLA	CLA	CTE	CUV	CNU	CAR	CBE	CCA	CCO	CLI	CLA	CAC	CNI	CVE	CAU	CAM	CPA	CES	COF	CCA	.L.
.M.	COL	CTE	CSA	CVA	COL	CAV	CCH	CCO	CCO	CFE	CHE	CMA	CTI	CAB	CMA	CSA	CBR	CJU	CLA	CPA	CSP	CTI	CAN	CME	CSA	.M.
.N.	CFI	CMA	CMI	CMO	CPE	CCY	CLO	CZE	CTE	CPE	COB	CCI	CFL	CMA	CNA	CWI	CCA	CSC	CDA	CPL	CTR	CES	CRO	CBE	DFR	.N.
.O	DGL	DAE	DCA	DSP	DEL	DMA	DIN	DAN	DAR	DRE	DAL	DGR	DPU	DDE	DEX	DIB	DAL	DBU	DCO	DCO	DFL	DMA	DOP	DRO	DCU	O.
.P.	DEB	DKA	DVI	DOD	DOP	DUN	DHE	DMA	DWI	DZI	DCO	EEL	ECR	ECR	EPY	ECA	EPH	EGU	EDU	ECA	ECO	EIN	ECA	ECI	ECO	.P.
.Q.	EGL	EJU	EGE	ESI	EAR	ECH	ECU	ELE	ETE	ETR	EOP	EGL	EJA	EPO	EVE	EFO	ECO	ESP	EUT	EUN	EAN	ELO	FES	FTA	FGR	.Q.
.R.	FSY	FSE	FLI	FAR	FID	FLO	FOV	FPR	FRU	FRU	FTI	FCA	FEL	FVO	FGL	FIN	FVU	FSP	FAN	FCH	FVE	FVI	FAF	FEL	FFO	.R.
.S.	GDU	GLA	GLI	GMA	GMU	GTI	GPR	GBA	GTI	GLU	GUR	GJA	GMA	GGL	GLE	GAN	GAR	GBA	GHE	GHI	GAS	GAB	GTI	HCO	HAN	.S.
.T	HAN	HTU	HZE	HAL	HBE	HBR	HCA	HSA	HBE	HJA	HMU	HBU	HDI	HVU	HCO	HLU	HAN	HNI	HHI	HRU	HSP	HOF	IAM	IPA	IVE	T.
.U.	IAR	IHI	ISP	ITI	IED	IHE	IAQ	IBA	IER	JOF	JCH	JAI	JCI	JHI	JNI	JRE	JMA	JCO	KGA	KGE	KSC	LPU	LSA	LVI	LSI	.U.
.V.	LIB	LHE	LKI	LOW	LGA	LDO	LHI	LSA	LNO	LLA	LSP	LST	LIN	LMI	LOL	LCU	LSA	LVI	LLA	LCU	LST	LST	LFE	LGO	LLE	.V.
.W.	LOF	LRI	LAL	LBA	LDO	LUS	LMI	LCH	LIN	LMU	LPE	LNI	LUR	LUT	LCO	LPE	LTE	LAC	LCY	LAL	LAN	LLU	LCH	LES	LPI	.W.
.X.	MSP	MLO	MHO	MON	MPU	MBA	MSY	MAM	MOF	MIN	MDI	MES	MGL	MBI	MZA	MAR	MVU	MCH	MVI	MAR	MFA	MLU	MOR	MPO	MSA	.X.
.Y	HSC	MCA	MUS	MBI	MAL	MIN	MOF	MSU	MMI	MOF	MAR	MGE	MPI	MPU	MRO	MSP	MCH	MMI	MBA	MCH	MDE	MPE	MCI	MOL	MAL	Y.
.Z.	MNI	MRU	MDE	MKO	MAC	MEN	MPA	MTE	MCA	MFR	MBA	MPE	MOD	MCO	NOF	NNU	NCA	NLA	NMU	NTA	NSA	NCO	OBA	OKI	OSA	.Z.
.a.	OJA	OEU	OVI	OJA	OFI	OCO	OVU	OSA	OGL	OSA	OHY	OMI	OTU	PER	PTU	PGU	PSP	PAN	PCO	PHE	PMA	PMI	POB	PRE	PVI	.a.
.b.	PBR	PSO	PFI	PCE	PED	PAR	PDI	PNO	PSC	PUR	PED	PLA	PLI	PMO	PQU	PSA	PCU	PYO	PHA	PPT	PAM	PCL	PPU	PMA	PIN	.b.
.c.	PPE	PFR	PAM	PCR	PAQ	PAR	PCA	PCO	PST	PAC	PAT	PCO	PLU	PVU	PPR	PDA	PTE	PAU	PAC	PEM	PSP	PIX	PPE	PAM	PDI	.c.
.d	PRA	PAN	PED	PQU	PBE	PLO	PME	PNI	PRE	PVE	PSA	PES	PAM	PAN	PAR	PBU	PFE	PPR	PSE	PTR	PCA	PTU	PRU	PTR	POL	d.
.e.	PAF	PCA	PSA	PVI	PAR	PAV	PCE	PCE	PDO	PDO	PDU	PLA	PMA	PPE	PSA	PSU	PCA	PFR	PGU	PGU	PMO	PTE	PER	PSA	PSO	.e.
.f.	PLO	PPH	PGR	PCO	PPY	QSU	RSA	RSE	ROD	ROF	RPA	RRH	RVE	RHI	RNI	RRU	RSA	RUV	RCO	RCO	ROF	RSP	RSP	RID	ROC	.f.
.g.	ROC	RGR	SOF	SCH	SOF	SSC	SCA	SCA	SMI	STR	SSE	SHO	SMO	SBA	SLO	SMO	SSC	SOL	SCA	SCE	SED	SAN	SAL	SIN	SRA	.g.
.h.	SBI	SEX	SIT	SSP	SOD	SGL	SCH	SAL	SAR	SAE	SAV	SFE	SGI	SHY	SIN	SIN	SKH	SLA	SMA	SME	SMU	SNI	SQU	STO	STU	.h.
.i	SRO	SAV	SAL	SBI	SHA	SSU	SJU	SST	SAC	SOL	SCY	SMO	SPU	STU	SAI	SCR	SWR	SSI	SSE	SRE	SCO	STE	SVI	SGR	SHI	i.
.j.	SKO	SSA	SNU	SAD	SER	SGU	SHU	SBE	STO	SCO	SPE	SDU	SAR	SCU	SJA	SMA	SSA	TIN	TVU	TKU	TKO	TOF	TOC	TPE	TCA	.j.
.k.	TVO	TCA	TTE	TPU	TDA	TBI	TCA	TGR	TSE	TVU	TAM	TPO	TBI	TNA	TAF	TAN	TCU	TAL	TAM	TFR	THI	THY	TIN	TNI	TPR	.k.
.l.	TRE	TRE	TSU	TVA	TVE	TWI	TSP	TFO	TTR	XTR	TAE	TCA	TCO	TDI	TDU	TMO	TPO	TSP	TTI	TTU	TTU	TCO	TRH	TTO	TTU	.l.
.m.	TFA	UTU	UGA	ULO	UDI	VAN	VAS	VCO	VMA	VVI	VLO	VMA	VPL	VCA	VAN	VZI	VAN	VAR	VBE	VDA	VER	VFA	VMO	VPA	VSA	.m.
.n	VVI	VAC	VAN	VMU	VRA	VUM	VUN	VUN	VUN	VVE	VSE	VLA	VRO	VVI	VSU	XSA	YEL	ZMA	ZMA	ZOF	ZAQ	ZJU	ZMA	ZJA	ZMA	n.

ALTERNATIVE CROP MATRIX

Table 5. Low Protein Crops (<1 g/100 g)

	1	2	3	4	5	6	7	8	9	10	11	12	13	14	15	16	17	18	19	20	21	22	23	24	25	
.A.	AES	AMO	ATH	AFA	AME	ANI	APY	ASE	ASE	ASA	ANA	ACA	AMA	ACH	ADI	AVA	ACY	AMA	AME	ASC	AFO	ALE	ASI	ACR	ADA	.A.
.B.	ADE	AEL	AIN	ARE	ARI	ASI	ASM	ASP	ATR	ACA	AGI	AST	ATE	AFO	AMO	AMO	APS	AAM	ACE	ACH	AFI	ASA	ASC	ATU	AMA	.B.
.C.	ABA	AFE	APE	AAR	APR	AOF	AVA	ACA	ACR	AHY	ALI	ATR	AVI	AAR	ABR	AOC	ACO	AGA	AGE	AHA	AGR	AAR	ACH	ADI	AMO	.C.
.D.	AMU	ARE	ASQ	AMO	ACE	AVU	ABU	AGR	AGR	AHY	ACO	ALA	ACA	API	AES	ARU	AMO	AXA	AEL	AAB	AAB	ACI	ADR	AMA	AVU	.D.
.E	AAL	ACH	AIN	ASP	ADO	AGR	ATR	ADE	AOF	AOD	AGU	ACA	AHO	ABE	AAB	ABY	AFA	ASA	AST	AST	ABI	ACA	AAF	ACO	BMO	E.
.F.	BRA	BRA	BGA	BFR	BAE	BSP	BNI	BVU	BBE	BAL	BES	BSY	BHI	BEX	BVU	BOR	BSA	BNI	BCE	BOF	BFL	BCA	BCA	BIN	BIN	.F.
.G.	BIS	BPE	BSA	BCU	BER	BGR	BBR	BDE	BEM	BMU	BRA	BCA	BCH	BJU	BNA	BNA	BNI	BOL	BOL	BOL	BOL	BOL	BOL	BPE	BRA	.G.
.H.	BHU	BCA	BIN	BMA	BUN	BAL	BUT	BLA	BDA	BSA	BPA	CCO	CES	CSA	CSP	CCA	CRO	CAL	COF	CMU	CPA	CSI	CRA	COD	CIN	.H.
.I.	CEN	CGL	CPL	CED	CSA	CSP	CAN	CBA	CCH	CFR	CPU	CPA	CPU	CCA	CED	CGR	CTI	CCA	CIL	CSP	CAM	CUR	CED	CAL	CAU	.I.
.J	COC	CSE	CCR	CDE	CMO	CPU	CSA	CEL	CUL	CED	CRO	CPE	CCI	CIP	CSI	CSE	CAL	CBU	CAL	CAM	CQU	CGA	CVI	CEX	CTI	J.
.K.	CCI	CCO	CCO	CMA	CSP	CCA	CDE	CAR	CEN	CIN	COF	CBU	CCA	CCA	CTA	CVE	CCO	CLA	CAU	CAU	CGR	CLI	CME	CPA	CRE	.K.
.L.	CSI	CTA	CFI	CPO	CDE	CLA	CLA	CTE	CUV	CNU	CAR	CBE	CCA	CCO	CLI	CLA	CAC	CNI	CVE	CAU	CAM	CPA	CES	COF	CCA	.L.
.M.	COL	CTE	CSA	CVA	COL	CAV	CCH	CCO	CCO	CFE	CHE	CMA	CTI	CAB	CMA	CSA	CBR	CJU	CLA	CPA	CSP	CTI	CAN	CME	CSA	.M.
.N.	CFI	CMA	CMI	CMO	CPE	CCY	CLO	CZE	CTE	CPE	COB	CCI	CFL	CMA	CNA	CWI	CCA	CSC	CDA	CPL	CTR	CES	CRO	CBE	DFR	.N.
.O	DGL	DAE	DCA	DSP	DEL	DMA	DIN	DAN	DAR	DRE	DAL	DGR	DPU	DDE	DEX	DIB	DAL	DBU	DCO	DCO	DFL	DMA	DOP	DRO	DCU	O.
.P.	DEB	DKA	DVI	DOD	DOP	DUN	DHE	DMA	DWI	DZI	DCO	EEL	ECR	ECR	EPY	ECA	EPH	EGU	EDU	ECA	ECO	EIN	ECA	ECI	ECO	.P.
.Q.	EGL	EJU	EGE	ESI	EAR	ECH	ECU	ELE	ETE	ETR	EOP	EGL	EJA	EPO	EVE	EFO	ECO	ESP	EUT	EUN	EAN	ELO	FES	FTA	FGR	.Q.
.R.	FSY	FSE	FLI	FAR	FID	FLO	FOV	FPR	FRU	FRU	FTI	FCA	FEL	FVO	FGL	FIN	FVU	FSP	FAN	FCH	FVE	FVI	FAF	FEL	FFO	.R.
.S.	GDU	GLA	GLI	GMA	GMU	GTI	GPR	GBA	GTI	GLU	GUR	GJA	GMA	GGL	GLE	GAN	GAR	GBA	GHE	GHI	GAS	GAB	GTI	HCO	HAN	.S.
.T	HAN	HTU	HZE	HAL	HBE	HBR	HCA	HSA	HBE	HJA	HMU	HBU	HDI	HVU	HCO	HLU	HAN	HNI	HHI	HRU	HSP	HOF	IAM	IPA	IVE	T.
.U.	IAR	IHI	ISP	ITI	IED	IHE	IAQ	IBA	IER	JOF	JCH	JAI	JCI	JHI	JNI	JRE	JMA	JCO	KGA	KGE	KSC	LPU	LSA	LVI	LSI	.U.
.V.	LIB	LHE	LKI	LOW	LGA	LDO	LHI	LSA	LNO	LLA	LSP	LST	LIN	LMI	LOL	LCU	LSA	LVI	LLA	LCU	LST	LST	LFE	LGO	LLE	.V.
.W.	LOF	LRI	LAL	LBA	LDO	LUS	LMI	LCH	LIN	LMU	LPE	LNI	LUR	LUT	LCO	LPE	LTE	LAC	LCY	LAL	LAN	LLU	LCH	LES	LPI	.W.
.X.	MSP	MLO	MHO	MON	MPU	MBA	MSY	MAM	MOF	MIN	MDI	MES	MGL	MBI	MZA	MAR	MVU	MCH	MVI	MAR	MFA	MLU	MOR	MPO	MSA	.X.
.Y	MSC	MCA	MUS	MBI	MAL	MIN	MOF	MSU	MMI	MOF	MAR	MGE	MPI	MPU	MRO	MSP	MCH	MMI	MBA	MCH	MDE	MPE	MCI	MOL	MAL	Y.
.Z.	MNI	MRU	MDE	MKO	MAC	MEN	MPA	MTE	MCA	MFR	MBA	MPE	MOD	MCO	NOF	NNU	NCA	NLA	NMU	NTA	NSA	NCO	OBA	OKI	OSA	.Z.
.a.	OJA	OEU	OVI	OJA	OFI	OCO	OVU	OSA	OGL	OSA	OHY	OMI	OTU	PER	PTU	PGU	PSP	PAN	PCO	PHE	PMA	PMI	POB	PRE	PVI	.a.
.b.	PBR	PSO	PFI	PCE	PED	PAR	PDI	PNO	PSC	PUR	PED	PLA	PLI	PMO	PQU	PSA	PCU	PYO	PHA	PPT	PAM	PCL	PPU	PMA	PIN	.b.
.c.	PPE	PFR	PAM	PCR	PAQ	PAR	PCA	PCO	PST	PAC	PAT	PCO	PLU	PVU	PPR	PDA	PTE	PAU	PAC	PEM	PSP	PIX	PPE	PAM	PDI	.c.
.d	PRA	PAN	PED	PQU	PBE	PLO	PME	PNI	PRE	PVE	PSA	PES	PAM	PAN	PAR	PBU	PFE	PPR	PSE	PTR	PCA	PTU	PRU	PTR	POL	d.
.e.	PAF	PCA	PSA	PVI	PAR	PAV	PCE	PCE	PDO	PDO	PDU	PLA	PMA	PPE	PSA	PSU	PCA	PFR	PGU	PGU	PMO	PTE	PER	PSA	PSO	.e.
.f.	PLO	PPH	PGR	PCO	PPY	QSU	RSA	RSE	ROD	ROF	RPA	RRH	RVE	RHI	RNI	RRU	RSA	RUV	RCO	RCO	ROF	RSP	RSP	RID	ROC	.f.
.g.	ROC	RGR	SOF	SCH	SOF	SSC	SCA	SCA	SMI	STR	SSE	SHO	SMO	SBA	SLO	SMO	SSC	SOL	SCA	SCE	SED	SAN	SAL	SIN	SRA	.g.
.h.	SBI	SEX	SIT	SSP	SOD	SGL	SCH	SAL	SAR	SAE	SAV	SFE	SGI	SHY	SIN	SIN	SKH	SLA	SMA	SME	SMU	SNI	SQU	STO	STU	.h.
.i	SRO	SAV	SAL	SBI	SHA	SSU	SJU	SST	SAC	SOL	SCY	SMO	SPU	STU	SAI	SCR	SWR	SSI	SSE	SRE	SCO	STE	SVI	SGR	SHI	i.
.j.	SKO	SSA	SNU	SAD	SER	SGU	SHU	SBE	STO	SCO	SPE	SDU	SAR	SCU	SJA	SMA	SSA	TIN	TVU	TKU	TKO	TOF	TOC	TPE	TCA	.j.
.k.	TVO	TCA	TTE	TPU	TDA	TBI	TCA	TGR	TSE	TVU	TAM	TPO	TBI	TNA	TAF	TAN	TCU	TAL	TAM	TFR	THI	THY	TIN	TNI	TPR	.k.
.l.	TRE	TRE	TSU	TVA	TVE	TWI	TSP	TFO	TTR	XTR	TAE	TCA	TCO	TDI	TDU	TMO	TPO	TSP	TTI	TTU	TTU	TCO	TRH	TTO	TTU	.l.
.m.	TFA	UTU	UGA	ULO	UDI	VAN	VAS	VCO	VMA	VVI	VLO	VMA	VPL	VCA	VAN	VZI	VAN	VAR	VBE	VDA	VER	VFA	VMO	VPA	VSA	.m.
.n	VVI	VAC	VAN	VMU	VRA	VUM	VUN	VUN	VUN	VVE	VSE	VLA	VRO	VVI	VSU	XSA	YEL	ZMA	ZMA	ZOF	ZAQ	ZJU	ZMA	ZJA	ZMA	n.

ALTERNATIVE CROP MATRIX

Table 6. High Protein Crops (>10 g/100 g)

	1	2	3	4	5	6	7	8	9	10	11	12	13	14	15	16	17	18	19	20	21	22	23	24	25	
A	AES	AMO	ATH	AFA	AME	ANI	APY	ASE	ASE	ASA	ANA	ACA	AMA	ACH	ADI	AVA	ACY	AMA	AME	ASC	AFO	ALE	ASI	ACR	ADA	A
B	ADE	AEL	AIN	ARE	ARI	ASI	ASM	ASP	ATR	ACA	AGI	AST	ATE	AFO	AMO	AMO	APS	AAM	ACE	ACH	AFI	ASA	ASC	ATU	AMA	B
C	ABA	AFE	APE	AAR	APR	AOF	AVA	ACA	ACR	AHY	ALI	ATR	AVI	AAR	ABR	AOC	ACO	AGA	AGE	AHA	AGR	AAR	ACH	ADI	AMO	C
D	AMU	ARE	ASQ	AMO	ACE	AVU	ABU	AGR	AGR	AHY	ACO	ALA	ACA	API	AES	ARU	AMO	AXA	AEL	AAB	AAB	ACI	ADR	AMA	AVU	D
E	AAL	ACH	AIN	ASP	ADO	AGR	ATR	ADE	AOF	AOD	AGU	ACA	AHO	ABE	AAB	ABY	AFA	ASA	AST	AST	ABI	ACA	AAF	ACO	BMO	E
F	BRA	BRA	BGA	BFR	BAE	BSP	BNI	BVU	BBE	BAL	BES	BSY	BHI	BEX	BVU	BOR	BSA	BNI	BCE	BOF	BFL	BCA	BCA	BIN	BIN	F
G	BIS	BPE	BSA	BCU	BER	BGR	BBR	BDE	BEM	BMU	BRA	BCA	BCH	BJU	BNA	BNA	BNI	BOL	BOL	BOL	BOL	BOL	BOL	BPE	BRA	G
H	BHU	BCA	BIN	BMA	BUN	BAL	BUT	BLA	BDA	BSA	BPA	CCO	CES	CSA	CSP	CCA	CRO	CAL	COF	CMU	CPA	CSI	CRA	COD	CIN	H
I	CEN	CGL	CPL	CED	CSA	CSP	CAN	CBA	CCH	CFR	CPU	CPA	CPU	CCA	CED	CGR	CTI	CCA	CIL	CSP	CAM	CUR	CED	CAL	CAU	I
J	COC	CSE	CCR	CDE	CMO	CPU	CSA	CEL	CUL	CED	CRO	CPE	CCI	CIP	CSI	CSE	CAL	CBU	CAL	CAM	CQU	CGA	CVI	CEX	CTI	J
K	CCI	CCO	CCO	CMA	CSP	CCA	CDE	CAR	CEN	CIN	COF	CBU	CCA	CCA	CTA	CVE	CCO	CLA	CAU	CAU	CGR	CLI	CME	CPA	CRE	K
L	CSI	CTA	CFI	CPO	CDE	CLA	CLA	CTE	CUV	CNU	CAR	CBE	CCA	CCO	CLI	CLA	CAC	CNI	CVE	CAU	CAM	CPA	CES	COF	CCA	L
M	COL	CTE	CSA	CVA	COL	CAV	CCH	CCO	CCO	CFE	CHE	CMA	CTI	CAB	CMA	CSA	CBR	CJU	CLA	CPA	CSP	CTI	CAN	CME	CSA	M
N	CFI	CMA	CMI	CMO	CPE	CCY	CLO	CZE	CTE	CPE	COB	CCI	CFL	CMA	CNA	CWI	CCA	CSC	CDA	CPL	CTR	CES	CRO	CBE	DFR	N
O	DGL	DAE	DCA	DSP	DEL	DMA	DIN	DAN	DAR	DRE	DAL	DGR	DPU	DDE	DEX	DIB	DAL	DBU	DCO	DCO	DFL	DMA	DOP	DRO	DCU	O
P	DEB	DKA	DVI	DOD	DOP	DUN	DHE	DMA	DWI	DZI	DCO	EEL	ECR	ECR	EPY	ECA	EPH	EGU	EDU	ECA	ECO	EIN	ECA	ECI	ECO	P
Q	EGL	EJU	EGE	ESI	EAR	ECH	ECU	ELE	ETE	ETR	EOP	EGL	EJA	EPO	EVE	EFO	ECO	ESP	EUT	EUN	EAN	ELO	FES	FTA	FGR	Q
R	FSY	FSE	FLI	FAR	FID	FLO	FOV	FPR	FRU	FRU	FTI	FCA	FEL	FVO	FGL	FIN	FVU	FSP	FAN	FCH	FVE	FVI	FAF	FEL	FFO	R
S	GDU	GLA	GLI	GMA	GMU	GTI	GPR	GBA	GTI	GLU	GUR	GJA	GMA	GGL	GLE	GAN	GAR	GBA	GHE	GHI	GAS	GAB	GTI	HCO	HAN	S
T	HAN	HTU	HZE	HAL	HBE	HBR	HCA	HSA	HBE	HJA	HMU	HBU	HDI	HVU	HCO	HLU	HAN	HNI	HHI	HRU	HSP	HOF	IAM	IPA	IVE	T
U	IAR	IHI	ISP	ITI	IED	IHE	IAQ	IBA	IER	JOF	JCH	JAI	JCI	JHI	JNI	JRE	JMA	JCO	KGA	KGE	KSC	LPU	LSA	LVI	LSI	U
V	LIB	LHE	LKI	LOW	LGA	LDO	LHI	LSA	LNO	LLA	LSP	LST	LIN	LMI	LOL	LCU	LSA	LVI	LLA	LCU	LST	LST	LFE	LGO	LLE	V
W	LOF	LRI	LAL	LBA	LDO	LUS	LMI	LCH	LIN	LMU	LPE	LNI	LUR	LUT	LCO	LPE	LTE	LAC	LCY	LAL	LAN	LLU	LCH	LES	LPI	W
X	MSP	MLO	MHO	MON	MPU	MBA	MSY	MAM	MOF	MIN	MDI	MES	MGL	MBI	MZA	MAR	MVU	MCH	MVI	MAR	MFA	MLU	MOR	MPO	MSA	X
Y	MSC	MCA	MUS	MBI	MAL	MIN	MOF	MSU	MMI	MOF	MAR	MGE	MPI	MPU	MRO	MSP	MCH	MMI	MBA	MCH	MDE	MPE	MCI	MOL	MAL	Y
Z	MNI	MRU	MDE	MKO	MAC	MEN	MPA	MTE	MCA	MFR	MBA	MPE	MOD	MCO	NOF	NNU	NCA	NLA	NMU	NTA	NSA	NCO	OBA	OKI	OSA	Z
a	OJA	OEU	OVI	OJA	OFI	OCO	OVU	OSA	OGL	OSA	OHY	OMI	OTU	PER	PTU	PGU	PSP	PAN	PCO	PHE	PMA	PMI	POB	PRE	PVI	a
b	PBR	PSO	PFI	PCE	PED	PAR	PDI	PNO	PSC	PUR	PED	PLA	PLI	PMO	PQU	PSA	PCU	PYO	PHA	PPT	PAM	PCL	PPU	PMA	PIN	b
c	PPE	PFR	PAM	PCR	PAQ	PAR	PCA	PCO	PST	PAC	PAT	PCO	PLU	PVU	PPR	PDA	PTE	PAU	PAC	PEM	PSP	PIX	PPE	PAM	PDI	c
d	PRA	PAN	PED	PQU	PBE	PLO	PME	PNI	PRE	PVE	PSA	PES	PAM	PAN	PAR	PBU	PFE	PPR	PSE	PTR	PCA	PTU	PRU	PTR	POL	d
e	PAF	PCA	PSA	PVI	PAR	PAV	PCE	PCE	PDO	PDO	PDU	PLA	PMA	PPE	PSA	PSU	PCA	PFR	PGU	PGU	PMO	PTE	PER	PSA	PSO	e
f	PLO	PPH	PGR	PCO	PPY	QSU	RSA	RSE	ROD	ROF	RPA	RRH	RVE	RHI	RNI	RRU	RSA	RUV	RCO	RCO	ROF	RSP	RSP	RID	ROC	f
g	ROC	RGR	SOF	SCH	SOF	SSC	SCA	SCA	SMI	STR	SSE	SHO	SMO	SBA	SLO	SMO	SSC	SOL	SCA	SCE	SED	SAN	SAL	SIN	SRA	g
h	SBI	SEX	SIT	SSP	SOD	SGL	SCH	SAL	SAR	SAE	SAV	SFE	SGI	SHY	SIN	SIN	SKH	SLA	SMA	SME	SMU	SNI	SQU	STO	STU	h
i	SRO	SAV	SAL	SBI	SHA	SSU	SJU	SST	SAC	SOL	SCY	SMO	SPU	STU	SAI	SCR	SWR	SSI	SSE	SRE	SCO	STE	SVI	SGR	SHI	i
j	SKO	SSA	SNU	SAD	SER	SGU	SHU	SBE	STO	SCO	SPE	SDU	SAR	SCU	SJA	SMA	SSA	TIN	TVU	TKU	TKO	TOF	TOC	TPE	TCA	j
k	TVO	TCA	TTE	TPU	TDA	TBI	TCA	TGR	TSE	TVU	TAM	TPO	TBI	TNA	TAF	TAN	TCU	TAL	TAM	TFR	THI	THY	TIN	TNI	TPR	k
l	TRE	TRE	TSU	IVA	TVE	TWI	TSP	TFO	TTR	XTR	TAE	TCA	TCO	TDI	TDU	TMO	TPO	TSP	TTI	TTU	TTU	TCO	TRH	TTO	TTU	l
m	TFA	UTU	UGA	ULO	UDI	VAN	VAS	VCO	VMA	VVI	VLO	VMA	VPL	VCA	VAN	VZI	VAN	VAR	VBE	VDA	VER	VFA	VMO	VPA	VSA	m
n	VVI	VAC	VAN	VMU	VRA	VUM	VUN	VUN	VUN	VVE	VSE	VLA	VRO	VVI	VSU	XSA	YEL	ZMA	ZMA	ZOF	ZAQ	ZJU	ZMA	ZJA	ZMA	n

ALTERNATIVE CROP MATRIX

Table 7. Low Vitamin A Crops (<2 IU/100 g)

	1	2	3	4	5	6	7	8	9	10	11	12	13	14	15	16	17	18	19	20	21	22	23	24	25	
.A.	AES	AMO	ATH	AFA	AME	ANI	APY	ASE	ASE	ASA	ANA	ACA	AMA	ACH	ADI	AVA	ACY	AMA	AME	ASC	AFO	ALE	ASI	ACR	ADA	.A.
.B.	ADE	AEL	AIN	ARE	ARI	ASI	ASM	ASP	ATR	ACA	AGI	AST	ATE	AFO	AMO	AMO	APS	AAM	ACE	ACH	AFI	ASA	ASC	ATU	AMA	.B.
.C.	ABA	AFE	APE	AAR	APR	AOF	AVA	ACA	ACR	AHY	ALI	ATR	AVI	AAR	ABR	AOC	ACO	AGA	AGE	AHA	AGR	AAR	ACH	ADI	AMO	.C.
.D.	AMU	ARE	ASQ	AMO	ACE	AVU	ABU	AGR	AGR	AHY	ACO	ALA	ACA	API	AES	ARU	AMO	AXA	AEL	AAB	AAB	ACI	ADR	AMA	AVU	.D.
.E.	AAL	ACH	AIN	ASP	ADO	AGR	ATR	ADE	AOF	AOD	AGU	ACA	AHO	ABE	AAB	ABY	AFA	ASA	AST	AST	ABI	ACA	AAF	ACO	BMO	.E.
.F.	BRA	BRA	BGA	BFR	BAE	BSP	BNI	BVU	BBE	BAL	BES	BSY	BHI	BEX	BVU	BOR	BSA	BNI	BCE	BOF	BFL	BCA	BCA	BIN	BIN	.F.
.G.	BIS	BPE	BSA	BCU	BER	BGR	BBR	BDE	BEM	BMU	BRA	BCA	BCH	BJU	BNA	BNA	BNI	BOL	BOL	BOL	BOL	BOL	BOL	BPE	BRA	.G.
.H.	BHU	BCA	BIN	BMA	BUN	BAL	BUT	BLA	BDA	BSA	BPA	CCO	CES	CSA	CSP	CCA	CRO	CAL	COF	CMU	CPA	CSI	CRA	COD	CIN	.H.
.I.	CEN	CGL	CPL	CED	CSA	CSP	CAN	CBA	CCH	CFR	CPU	CPA	CPU	CCA	CED	CGR	CTI	CCA	CIL	CSP	CAM	CUR	CED	CAL	CAU	.I.
.J.	CCC	CSE	CCR	CDE	CMO	CPU	CSA	CEL	CUL	CED	CRO	CPE	CCI	CIP	CSI	CSE	CAL	CBU	CAL	CAM	CQU	CGA	CVI	CEX	CTI	.J.
.K.	CCI	CCO	CCO	CMA	CSP	CCA	CDE	CAR	CEN	CIN	COF	CBU	CCA	CCA	CTA	CVE	CCO	CLA	CAU	CAU	CGR	CLI	CME	CPA	CRE	.K.
.L.	CSI	CTA	CFI	CPO	CDE	CLA	CLA	CTE	CUV	CNU	CAR	CBE	CCA	CCO	CLI	CLA	CAC	CNI	CVE	CAU	CAM	CPA	CES	COF	CCA	.L.
.M.	COL	CTE	CSA	CVA	COL	CAV	CCH	CCO	CCO	CFE	CHE	CMA	CTI	CAB	CMA	CSA	CBR	CJU	CLA	CPA	CSP	CTI	CAN	CME	CSA	.M.
.N.	CFI	CMA	CMI	CMO	CPE	CCY	CLO	CZE	CTE	CPE	COB	CCI	CFL	CMA	CNA	CWI	CCA	CSC	CDA	CPL	CTR	CES	CRO	CBE	DFR	.N.
.O.	DGL	DAE	DCA	DSP	DEL	DMA	DIN	DAN	DAR	DRE	DAL	DGR	DPU	DDE	DEX	DIB	DAL	DBU	DCO	DCO	DFL	DMA	DOP	DRO	DCU	.O.
.P.	DEB	DKA	DVI	DOD	DOP	DUN	DHE	DMA	DWI	DZI	DCO	EEL	ECR	ECR	EPY	ECA	EPH	EGU	EDU	ECA	ECO	EIN	ECA	ECI	ECO	.P.
.Q.	EGL	EJU	EGE	ESI	EAR	ECH	ECU	ELE	ETE	ETR	EOP	EGL	EJA	EPO	EVE	EFO	ECO	ESP	EUT	EUN	EAN	ELO	FES	FTA	FGR	.Q.
.R.	FSY	FSE	FLI	FAR	FID	FLO	FOV	FPR	FRU	FRU	FTI	FCA	FEL	FVO	FGL	FIN	FVU	FSP	FAN	FCH	FVE	FVI	FAF	FEL	FFO	.R.
.S.	GDU	GLA	GLI	GMA	GMU	GTI	GPR	GBA	GTI	GLU	GUR	GJA	GMA	GGL	GLE	GAN	GAR	GBA	GHE	GHI	GAS	GAB	GTI	HCO	HAN	.S.
.T.	HAN	HTU	HZE	HAL	HBE	HBR	HCA	HSA	HBE	HJA	HMU	HBU	HDI	HVU	HCO	HLU	HAN	HNI	HHI	HRU	HSP	HOF	IAM	IPA	IVE	.T.
.U.	IAR	IHI	ISP	ITI	IED	IHE	IAQ	IBA	IER	JOF	JCH	JAI	JCI	JHI	JNI	JRE	JMA	JCO	KGA	KGE	KSC	LPU	LSA	LVI	LSI	.U.
.V.	LIB	LHE	LKI	LOW	LGA	LDO	LHI	LSA	LNO	LLA	LSP	LST	LIN	LMI	LOL	LCU	LSA	LVI	LLA	LCU	LST	LST	LFE	LGO	LLE	.V.
.W.	LOF	LRI	LAL	LBA	LDO	LUS	LMI	LCH	LIN	LMU	LPE	LNI	LUR	LUT	LCO	LPE	LTE	LAC	LCY	LAL	LAN	LLU	LCH	LES	LPI	.W.
.X.	MSP	MLO	MHO	MON	MPU	MBA	MSY	MAM	MOF	MIN	MDI	MES	MGL	MBI	MZA	MAR	MVU	MCH	MVI	MAR	MFA	MLU	MOR	MPO	MSA	.X.
.Y.	MSC	MCA	MUS	MBI	MAL	MIN	MOF	MSU	MMI	MOF	MAR	MGE	MPI	MPU	MRO	MSP	MCH	MMI	MBA	MCH	MDE	MPE	MCI	MOL	MAL	.Y.
.Z.	MNI	MRU	MDE	MKO	MAC	MEN	MPA	MTE	MCA	MFR	MBA	MPE	MOD	MCO	NOF	NNU	NCA	NLA	NMU	NTA	NSA	NCO	OBA	OKI	OSA	.Z.
.a.	OJA	OEU	OVI	OJA	OFI	OCO	OVU	OSA	OGL	OSA	OHY	OMI	OTU	PER	PTU	PGU	PSP	PAN	PCO	PHE	PMA	PMI	POB	PRE	PVI	.a.
.b.	PBR	PSO	PFI	PCE	PED	PAR	PDI	PNO	PSC	PUR	PED	PLA	PLI	PMO	PQU	PSA	PCU	PYO	PHA	PPT	PAM	PCL	PPU	PMA	PIN	.b.
.c.	PPE	PFR	PAM	PCR	PAQ	PAR	PCA	PCO	PST	PAC	PAT	PCO	PLU	PVU	PPR	PDA	PTE	PAU	PAC	PEM	PSP	PIX	PPE	PAM	PDI	.c.
.d.	PRA	PAN	PED	PQU	PBE	PLO	PME	PNI	PRE	PVE	PSA	PES	PAM	PAN	PAR	PBU	PFE	PPR	PSE	PTR	PCA	PTU	PRU	PTR	POL	.d.
.e.	PAF	PCA	PSA	PVI	PAR	PAV	PCE	PCE	PDO	PDO	PDU	PLA	PMA	PPE	PSA	PSU	PCA	PFR	PGU	PGU	PMO	PTE	PER	PSA	PSO	.e.
.f.	PLO	PPH	PGR	PCO	PPY	QSU	RSA	RSE	ROD	ROF	RPA	RRH	RVE	RHI	RNI	RRU	RSA	RUV	RCO	RCO	ROF	RSP	RSP	RID	ROC	.f.
.g.	ROC	RGR	SOF	SCH	SOF	SSC	SCA	SCA	SMI	STR	SSE	SHO	SMO	SBA	SLO	SMO	SSC	SOL	SCA	SCE	SED	SAN	SAL	SIN	SFA	.g.
.h.	SBI	SEX	SIT	SSP	SOD	SGL	SCH	SAL	SAR	SAE	SAV	SFE	SGI	SHY	SIN	SIN	SKH	SLA	SMA	SME	SMU	SNI	SQU	STO	STU	.h.
.i.	SRO	SAV	SAL	SBI	SHA	SSU	SJU	SST	SAC	SOL	SCY	SMO	SPU	STU	SAI	SCR	SWR	SSI	SSE	SRE	SCO	STE	SVI	SGR	SHI	.i.
.j.	SKO	SSA	SNU	SAD	SER	SGU	SHU	SBE	STO	SCO	SPE	SDU	SAR	SCU	SJA	SMA	SSA	TIN	TVU	TKU	TKO	TOF	TOC	TPE	TCA	.j.
.k.	TVO	TCA	TTE	TPU	TDA	TBI	TCA	TGR	TSE	TVU	TAM	TPO	TBI	TNA	TAF	TAN	TCU	TAL	TAM	TFR	THI	THY	TIN	TNI	TPR	.k.
.l.	TRE	TRE	TSU	TVA	TVE	TWI	TSP	TFO	TTR	XTR	TAE	TCA	TCO	TDI	TDU	TMO	TPO	TSP	TTI	TTU	TTU	TCO	TRH	TTO	TTU	.l.
.m.	TFA	UTU	UGA	ULO	UDI	VAN	VAS	VCO	VMA	VVI	VLO	VMA	VPL	VCA	VAN	VZI	VAN	VAR	VBE	VDA	VER	VFA	VMO	VPA	VSA	.m.
.n.	VVI	VAC	VAN	VMU	VRA	VUM	VUN	VUN	VUN	VVE	VSE	VLA	VRO	VVI	VSU	XSA	YEL	ZMA	ZMA	ZOF	ZAQ	ZJU	ZMA	ZJA	ZMA	.n.

ALTERNATIVE CROP MATRIX

Table 8. High Vitamin A Crops (>1000 IU/100 g)

	1	2	3	4	5	6	7	8	9	10	11	12	13	14	15	16	17	18	19	20	21	22	23	24	25	
.A.	AES	AMO	ATH	AFA	AME	ANI	APY	ASE	ASE	ASA	ANA	ACA	AMA	ACH	ADI	AVA	ACY	AMA	AME	ASC	AFO	ALE	ASI	ACR	ADA	.A.
.B.	ADE	AEL	AIN	ARE	ARI	ASI	ASM	ASP	ATR	ACA	AGI	AST	ATE	AFO	AMO	AMO	APS	AAM	ACE	ACH	AFI	ASA	ASC	ATU	AMA	.B.
.C.	ABA	AFE	APE	AAR	APR	AOF	AVA	ACA	ACR	AHY	ALI	ATR	AVI	AAR	ABR	AOC	ACO	AGA	AGE	AHA	AGR	AAR	ACH	ADI	AMO	.C.
.D.	AMU	ARE	ASQ	AMO	ACE	AVU	ABU	AGR	AGR	AHY	ACO	ALA	ACA	API	AES	ARU	AMO	AXA	AEL	AAB	AAB	ACI	ADR	AMA	AVU	.D.
.E.	AAL	ACH	AIN	ASP	ADO	AGR	ATR	ADE	AOF	AOD	AGU	ACA	AHO	ABE	AAB	ABY	AFA	ASA	AST	AST	ABI	ACA	AAF	ACO	BMO	.E.
.F.	BRA	BRA	BGA	BFR	BAE	BSP	BNI	BVU	BBE	BAL	BES	BSY	BHI	BEX	BVU	BOR	BSA	BNI	BCE	BOF	BFL	BCA	BCA	BIN	BIN	.F.
.G.	BIS	BPE	BSA	BCU	BER	BGR	BBR	BDE	BEM	BMU	BRA	BCA	BCH	BJU	BNA	BNA	BNI	BOL	BOL	BOL	BOL	BOL	BOL	BPE	BRA	.G.
.H.	BHU	BCA	BIN	BMA	BUN	BAL	BUT	BLA	BDA	BSA	BPA	CCO	CES	CSA	CSP	CCA	CRO	CAL	COF	CMU	CPA	CSI	CRA	COD	CIN	.H.
.I.	CEN	CGL	CPL	CED	CSA	CSP	CAN	CBA	CCH	CFR	CPU	CPA	CPU	CCA	CED	CGR	CTI	CCA	CIL	CSP	CAM	CUR	CED	CAL	CAU	.I.
.J.	COC	CSE	CCR	CDE	CMO	CPU	CSA	CEL	CUL	CED	CRO	CPE	CCI	CIP	CSI	CSE	CAL	CBU	CAL	CAM	CQU	CGA	CVI	CEX	CTI	.J.
.K.	CCI	CCO	CCO	CMA	CSP	CCA	CDE	CAR	CEN	CIN	COF	CBU	CCA	CCA	CTA	CVE	CCO	CLA	CAU	CAU	CGR	CLI	CME	CPA	CRE	.K.
.L.	CSI	CTA	CFI	CPO	CDE	CLA	CLA	CTE	CUV	CNU	CAR	CBE	CCA	CCO	CLI	CLA	CAC	CNI	CVE	CAU	CAM	CPA	CES	COF	CCA	.L.
.M.	COL	CTE	CSA	CVA	COL	CAV	CCH	CCO	CCO	CFE	CHE	CMA	CTI	CAB	CMA	CSA	CBR	CJU	CLA	CPA	CSP	CTI	CAN	CME	CSA	.M.
.N.	CFI	CMA	CMI	CMO	CPE	CCY	CLO	CZE	CTE	CPE	COB	CCI	CFL	CMA	CNA	CWI	CCA	CSC	CDA	CPL	CTR	CES	CRO	CBE	DFR	.N.
.O.	DGL	DAE	DCA	DSP	DEL	DMA	DIN	DAN	DAR	DRE	DAL	DGR	DPU	DDE	DEX	DIB	DAL	DBU	DCO	DCO	DFL	DMA	DOP	DRO	DCU	.O.
.P.	DEB	DKA	DVI	DOD	DOP	DUN	DHE	DMA	DWI	DZI	DCO	EEL	ECR	ECR	EPY	ECA	EPH	EGU	EDU	ECA	ECO	EIN	ECA	ECI	ECO	.P.
.Q.	EGL	EJU	EGE	ESI	EAR	ECH	ECU	ELE	ETE	ETR	EOP	EGL	EJA	EPO	EVE	EFO	ECO	ESP	EUT	EUN	EAN	ELO	FES	FTA	FGR	.Q.
.R.	FSY	FSE	FLI	FAR	FID	FLO	FOV	FPR	FRU	FRU	FTI	FCA	FEL	FVO	FGL	FIN	FVU	FSP	FAN	FCH	FVE	FVI	FAF	FEL	FFO	.R.
.S.	GDU	GLA	GLI	GMA	GMU	GTI	GPR	GBA	GTI	GLU	GUR	GJA	GMA	GGL	GLE	GAN	GAR	GBA	GHE	GHI	GAS	GAB	GTI	HCO	HAN	.S.
.T.	HAN	HTU	HZE	HAL	HBE	HBR	HCA	HSA	HBE	HJA	HMU	HBU	HDI	HVU	HCO	HLU	HAN	HNI	HHI	HRU	HSP	HOF	IAM	IPA	IVE	.T.
.U.	IAR	IHI	ISP	ITI	IED	IHE	IAQ	IBA	IER	JOF	JCH	JAI	JCI	JHI	JNI	JRE	JMA	JCO	KGA	KGE	KSC	LPU	LSA	LVI	LSI	.U.
.V.	LIB	LHE	LKI	LOW	LGA	LDO	LHI	LSA	LNO	LLA	LSP	LST	LIN	LMI	LOL	LCU	LSA	LVI	LLA	LCU	LST	LST	LFE	LGO	LLF	.V.
.W.	LOF	LRI	LAL	LBA	LDO	LUS	LMI	LCH	LIN	LMU	LPE	LNI	LUR	LUT	LCO	LPE	LTE	LAC	LCY	LAL	LAN	LLU	LCH	LES	LPI	.W.
.X.	MSP	MLO	MHO	MON	MPU	MBA	MSY	MAM	MOF	MIN	MDI	MES	MGL	MBI	MZA	MAR	MVU	MCH	MVI	MAR	MFA	MLU	MOR	MPO	MSA	.X.
.Y.	MSC	MCA	MUS	MBI	MAL	MIN	MOF	MSU	MMI	MOF	MAR	MGE	MPI	MPU	MRO	MSP	MCH	MMI	MBA	MCH	MDE	MPE	MCI	MOL	MAL	.Y.
.Z.	MNI	MRU	MDE	MKO	MAC	MEN	MPA	MTE	MCA	MFR	MBA	MPE	MOD	MCO	NOF	NNU	NCA	NLA	NMU	NTA	NSA	NCO	OBA	OKI	OSA	.Z.
.a.	OJA	OEU	OVI	OJA	OFI	OCO	OVU	OSA	OGL	OSA	OHY	OMI	OTU	PER	PTU	PGU	PSP	PAN	PCO	PHE	PMA	PMI	POB	PRE	PVI	.a.
.b.	PBR	PSO	PFI	PCE	PED	PAR	PDI	PNO	PSC	PUR	PED	PLA	PLI	PMO	PQU	PSA	PCU	PYO	PHA	PPT	PAM	PCL	PPU	PMA	PIN	.b.
.c.	PPE	PFR	PAM	PCR	PAQ	PAR	PCA	PCO	PST	PAC	PAT	PCO	PLU	PVU	PPR	PDA	PTE	PAU	PAC	PEM	PSP	PIX	PPE	PAM	PDI	.c.
.d.	PRA	PAN	PED	PQU	PBE	PLO	PME	PNI	PRE	PVE	PSA	PES	PAM	PAN	PAR	PBU	PFE	PPR	PSE	PTR	PCA	PTU	PRU	PTR	POL	.d.
.e.	PAF	PCA	PSA	PVI	PAR	PAV	PCE	PCE	PDO	PDO	PDU	PLA	PMA	PPE	PSA	PSU	PCA	PFR	PGU	PGU	PMO	PTE	PER	PSA	PSO	.e.
.f.	PLO	PPH	PGR	PCO	PPY	QSU	RSA	RSE	ROD	ROF	RPA	RRH	RVE	RHI	RNI	RRU	RSA	RUV	RCO	RCO	ROF	RSP	RSP	RID	ROC	.f.
.g.	ROC	RGR	SOF	SCH	SOF	SSC	SCA	SCA	SMI	STR	SSE	SHO	SMO	SBA	SLO	SMO	SSC	SOL	SCA	SCE	SED	SAN	SAL	SIN	SRA	.g.
.h.	SBI	SEX	SIT	SSP	SOD	SGL	SCH	SAL	SAR	SAE	SAV	SFE	SGI	SHY	SIN	SIN	SKH	SLA	SMA	SME	SMU	SNI	SQU	STO	STU	.h.
.i.	SRO	SAV	SAL	SBI	SHA	SSU	SJU	SST	SAC	SOL	SCY	SMO	SPU	STU	SAI	SCR	SWR	SSI	SSE	SRE	SCO	STE	SVI	SGR	SHI	.i.
.j.	SKO	SSA	SNU	SAD	SER	SGU	SHU	SBE	STO	SCO	SPE	SDU	SAR	SCU	SJA	SMA	SSA	TIN	TVU	TKU	TKO	TOF	TCC	TPE	TCA	.j.
.k.	TVO	TCA	TTE	TPU	TDA	TBI	TCA	TGR	TSE	TVU	TAM	TPO	TBI	TNA	TAF	TAN	TCU	TAL	TAM	TFR	THI	THY	TIN	TNI	TPR	.k.
.l.	TRE	TRE	TSU	TVA	TVE	TWI	TSP	TFO	TTR	XTR	TAE	TCA	TCO	TDI	TDU	TMO	TPO	TSP	TTI	TTU	TTU	TCO	TRH	TTO	TTU	.l.
.m.	TFA	UTU	UGA	ULO	UDI	VAN	VAS	VCO	VMA	VVI	VLO	VMA	VPL	VCA	VAN	VZI	VAN	VAR	VBE	VDA	VER	VFA	VMO	VPA	VSA	.m.
.n.	VVI	VAC	VAN	VMU	VRA	VUM	VUN	VUN	VUN	VVE	VSE	VLA	VRO	VVI	VSU	XSA	YEL	ZMA	ZMA	ZOF	ZAQ	ZJU	ZMA	ZJA	ZMA	.n.

ALTERNATIVE CROP MATRIX

Table 9. Low Ascorbic Acid Crops (<2 mg/ 100 g)

	1	2	3	4	5	6	7	8	9	10	11	12	13	14	15	16	17	18	19	20	21	22	23	24	25	
.A.	AES	AMO	ATH	AFA	AME	ANI	APY	ASE	ASE	ASA	ANA	ACA	AMA	ACH	ADI	AVA	ACY	AMA	AME	ASC	AFO	ALE	ASI	ACR	ADA	.A.
.B.	ADE	AEL	AIN	ARE	ARI	ASI	ASM	ASP	ATR	ACA	AGI	AST	ATE	AFO	AHO	AMO	APS	AAM	ACE	ACH	AFI	ASA	ASC	ATU	AMA	.B.
.C.	ABA	AFE	APE	AAR	APR	AOF	AVA	ACA	ACR	AHY	ALI	ATR	AVI	AAR	ABR	AOC	ACO	AGA	AGE	AHA	AGR	AAR	ACH	ADI	AMO	.C.
.D.	AMU	ARE	ASQ	AMO	ACE	AVU	ABU	AGR	AGR	AHY	ACO	ALA	ACA	API	AES	ARU	AMO	AXA	AEL	AAB	AAB	ACI	ADR	AMA	AVU	.D.
.E.	AAL	ACH	AIN	ASP	ADO	AGR	ATR	ADE	AOF	AOD	AGU	ACA	AHO	ABE	AAB	ABY	AFA	ASA	AST	AST	ABI	ACA	AAF	ACO	BMO	.E.
.F.	BRA	BRA	BGA	BFR	BAE	BSP	BNI	BVU	BBE	BAL	BES	BSY	BHI	BEX	BVU	BOR	BSA	BNI	BCE	BOF	BFL	BCA	BCA	BIN	BIN	.F.
.G.	BIS	BPE	BSA	BCU	BER	BGR	BBR	BDE	BEM	BMU	BRA	BCA	BCH	BJU	BNA	BNA	BNI	BOL	BOL	BOL	BOL	BOL	BOL	BPE	BRA	.G.
.H.	BHU	BCA	BIN	BMA	BUN	BAL	BUT	BLA	BDA	BSA	BPA	CCO	CES	CSA	CSP	CCA	CRO	CAL	COF	CMU	CPA	CSI	CRA	COD	CIN	.H.
.I.	CEN	CGL	CPL	CED	CSA	CSP	CAN	CBA	CCH	CFR	CPU	CPA	CPU	CCA	CED	CGR	CTI	CCA	CIL	CSP	CAM	CUR	CED	CAL	CAU	.I.
.J.	COC	CSE	CCR	CDE	CMO	CPU	CSA	CEL	CUL	CED	CRO	CPE	CCI	CIP	CSI	CSE	CAL	CBU	CAL	CAM	CQU	CGA	CVI	CEX	CTI	.J.
.K.	CCI	CCO	CCO	CMA	CSP	CCA	CDE	CAR	CEN	CIN	COF	CBU	CCA	CCA	CTA	CVE	CCO	CLA	CAU	CAU	CGR	CLI	CME	CPA	CRE	.K.
.L.	CSI	CTA	CFI	CPO	CDE	CLA	CLA	CTE	CUV	CNU	CAR	CBE	CCA	CCO	CLI	CLA	CAC	CNI	CVE	CAU	CAM	CPA	CES	COF	CCA	.L.
.M.	COL	CTE	CSA	CVA	COL	CAV	CCH	CCO	CCO	CFE	CHE	CMA	CTI	CAB	CMA	CSA	CBR	CJU	CLA	CPA	CSP	CTI	CAN	CME	CSA	.M.
.N.	CFI	CMA	CMI	CMO	CPE	CCY	CLO	CZE	CTE	CPE	COB	CCI	CFL	CMA	CNA	CWI	CCA	CSC	CDA	CPL	CTR	CFS	CRO	CBE	DFR	.N.
.O.	DGL	DAE	DCA	DSP	DEL	DMA	DIN	DAN	DAR	DRE	DAL	DGR	DPU	DDE	DEX	DIB	DAL	DBU	DCO	DCO	DFL	DMA	DOP	DRO	DCU	.O.
.P.	DEB	DKA	DVI	DOD	DOP	DUN	DHE	DMA	DWI	DZI	DCO	EEL	ECR	ECR	EPY	ECA	EPH	EGU	EDU	ECA	ECO	EIN	ECA	ECI	ECO	.P.
.Q.	EGL	EJU	EGE	ESI	EAR	ECH	ECU	ELE	ETE	ETR	EOP	EGL	EJA	EPO	EVE	EFO	ECO	ESP	EUT	EUN	EAN	ELO	FES	FTA	FGR	.Q.
.R.	FSY	FSE	FLI	FAR	FID	FLO	FOV	FPR	FRU	FRU	FTI	FCA	FEL	FVO	FGL	FIN	FVU	FSP	FAN	FCH	FVE	FVI	FAF	FEL	FFO	.R.
.S.	GDU	GLA	GLI	GMA	GMU	GTI	GPR	GBA	GTI	GLU	GUR	GJA	GMA	GGL	GLE	GAN	GAR	GBA	GHE	GHI	GAS	GAB	GTI	HCO	HAN	.S.
.T.	HAN	HTU	HZE	HAL	HBE	HBR	HCA	HSA	HBE	HJA	HMU	HBU	HDI	HVU	HCO	HLU	HAN	HNI	HHI	HRU	HSP	HOF	IAM	IPA	IVE	.T.
.U.	IAR	IHI	ISP	ITI	IED	IHE	IAQ	IBA	IER	JOF	JCH	JAI	JCI	JHI	JNI	JRE	JMA	JCO	KGA	KGE	KSC	LPU	LSA	LVI	LSI	.U.
.V.	LIB	LHE	LKI	LOW	LGA	LDO	LHI	LSA	LNO	LLA	LSP	LST	LIN	LMI	LOL	LCU	LSA	LVI	LLA	LCU	LST	LST	LFE	LGO	LLF	.V.
.W.	LOF	LRI	LAL	LBA	LDO	LUS	LMI	LCH	LIN	LMU	LPE	LNI	LUR	LUT	LCO	LPE	LTE	LAC	LCY	LAL	LAN	LLU	LCH	LES	LPI	.W.
.X.	MSP	MLO	MHO	MON	MPU	MBA	MSY	MAM	MOF	MIN	MDI	MES	MGL	MBI	MZA	MAR	MVU	MCH	MVI	MAR	MFA	MLU	MOR	MPO	MSA	.X.
.Y.	MSC	MCA	MUS	MBI	MAL	MIN	MOF	MSU	MMI	MOF	MAR	MGE	MPI	MPU	MRO	MSP	MCH	MMI	MBA	MCH	MDE	MPE	MCI	MOL	MAL	.Y.
.Z.	MNI	MRU	MDE	MKO	MAC	MEN	MPA	MTE	MCA	MFR	MBA	MPE	MOD	MCO	NOF	NNU	NCA	NLA	NMU	NTA	NSA	NCO	OBA	OKI	OSA	.Z.
.a.	OJA	OEU	OVI	OJA	OFI	OCO	OVU	OSA	OGL	OSA	OHY	OMI	OTU	PER	PTU	PGU	PSP	PAN	PCO	PHE	PMA	PMI	POB	PRE	PVI	.a.
.b.	PBR	PSO	PFI	PCE	PED	PAR	PDI	PNO	PSC	PUR	PED	PLA	PLI	PMO	PQU	PSA	PCU	PYO	PHA	PPT	PAM	PCL	PPU	PMA	PIN	.b.
.c.	PPE	PFR	PAM	PCR	PAQ	PAR	PCA	PCO	PST	PAC	PAT	PCO	PLU	PVU	PPR	PDA	PTE	PAU	PAC	PEM	PSP	PIX	PPE	PAM	PDI	.c.
.d.	PRA	PAN	PED	PQU	PBE	PLO	PME	PNI	PRE	PVE	PSA	PES	PAM	PAN	PAR	PBU	PFE	PPR	PSE	PTR	PCA	PTU	PRU	PTR	POL	.d.
.e.	PAF	PCA	PSA	PVI	PAR	PAV	PCE	PCE	PDO	PDO	PDU	PLA	PMA	PPE	PSA	PSU	PCA	PFR	PGU	PGU	PMO	PTE	PER	PSA	PSO	.e.
.f.	PLO	PPH	PGR	PCO	PPY	QSU	RSA	RSE	ROD	ROF	RPA	RRH	RVE	RHI	RNI	RRU	RSA	RUV	RCO	RCO	ROF	RSP	RSP	RID	ROC	.f.
.g.	ROC	RGR	SOF	SCH	SOF	SSC	SCA	SCA	SMI	STR	SSE	SHO	SMO	SBA	SLO	SMO	SSC	SOL	SCA	SCE	SED	SAN	SAL	SIN	SRA	.g.
.h.	SBI	SEX	SIT	SSP	SOD	SGL	SCH	SAL	SAR	SAE	SAV	SFE	SGI	SHY	SIN	SIN	SKH	SLA	SMA	SME	SMU	SNI	SQU	STO	STU	.h.
.i.	SRO	SAV	SAL	SBI	SHA	SSU	SJU	SST	SAC	SOL	SCY	SMO	SPU	STU	SAI	SCR	SWR	SSI	SSE	SRE	SCO	STE	SVI	SGR	SHI	.i.
.j.	SKO	SSA	SNU	SAD	SER	SGU	SHU	SBE	STO	SCO	SPE	SDU	SAR	SCU	SJA	SMA	SSA	TIN	TVU	TKU	TKO	TOF	TOC	TPE	TCA	.j.
.k.	TVO	TCA	TTE	TPU	TDA	TBI	TCA	TGR	TSE	TVU	TAM	TPO	TBI	TNA	TAF	TAN	TCU	TAL	TAM	TFR	THI	THY	TIN	TNI	TPR	.k.
.l.	TRE	TRE	TSU	TVA	TVE	TWI	TSP	TFO	TTR	XTR	TAE	TCA	TCO	TDI	TDU	TMO	TPO	TSP	TTI	TTU	TTU	TCO	TRH	TTO	TTU	.l.
.m.	TFA	UTU	UGA	ULO	UDI	VAN	VAS	VCO	VMA	VVI	VLO	VMA	VPL	VCA	VAN	VZI	VAN	VAR	VBE	VDA	VER	VFA	VMO	VPA	VSA	.m.
.n.	VVI	VAC	VAN	VMU	VRA	VUM	VUN	VUN	VUN	VVE	VSE	VLA	VRO	VVI	VSU	XSA	YEL	ZMA	ZMA	ZOF	ZAO	ZJU	ZMA	ZJA	ZMA	.n.

ALTERNATIVE CROP MATRIX

Table 10. High Ascorbic Acid Crops (>100 mg/100 g)

	1	2	3	4	5	6	7	8	9	10	11	12	13	14	15	16	17	18	19	20	21	22	23	24	25	
.A.	AES	AMO	ATH	AFA	AME	ANI	APY	ASE	ASE	ASA	ANA	ACA	AMA	ACH	ADI	AVA	ACY	AMA	AME	ASC	AFO	ALE	ASI	ACR	ADA	.A.
.B.	ADE	AEL	AIN	ARE	ARI	ASI	ASM	ASP	ATR	ACA	AGI	AST	ATE	AFO	AMO	AMO	APS	AAM	ACE	ACH	AFI	ASA	ASC	ATU	AMA	.B.
.C.	ABA	AFE	APE	AAR	APR	AOF	AVA	ACA	ACR	AHY	ALI	ATR	AVI	AAR	ABR	AOC	ACO	AGA	AGE	AHA	AGR	AAR	ACH	ADI	AMO	.C.
.D.	AMU	ARE	ASQ	AMO	ACE	AVU	ABU	AGR	AGR	AHY	ACO	ALA	ACA	API	AES	ARU	AMO	AXA	AEL	AAB	AAB	ACI	ADR	AMA	AVU	.D.
.E.	AAL	ACH	AIN	ASP	ADO	AGR	ATR	ADE	AOF	AOD	AGU	ACA	AHO	ABE	AAB	ABY	AFA	ASA	AST	AST	ABI	ACA	AAF	ACO	BMO	.E.
.F.	BRA	BRA	BGA	BFR	BAE	BSP	BNI	BVU	BBE	BAL	BES	BSY	BHI	BEX	BVU	BOR	BSA	BNI	BCE	BOF	BFL	BCA	BCA	BIN	BIN	.F.
.G.	BIS	BPE	BSA	BCU	BER	BGR	BBR	BDE	BEM	BMU	BRA	BCA	BCH	BJU	BNA	BNA	BNI	BOL	BOL	BOL	BOL	BOL	BOL	BPE	BRA	.G.
.H.	BHU	BCA	BIN	BMA	BUN	BAL	BUT	BLA	BDA	BSA	BPA	CCO	CES	CSA	CSP	CCA	CRO	CAL	COF	CMU	CPA	CSI	CRA	COD	CIN	.H.
.I.	CEN	CGL	CPL	CED	CSA	CSP	CAN	CBA	CCH	CFR	CPU	CPA	CPU	CCA	CED	CGR	CTI	CCA	CIL	CSP	CAM	CUR	CED	CAL	CAU	.I.
.J.	COC	CSE	CCR	CDE	CMO	CPU	CSA	CEL	CUL	CED	CRO	CPE	CCI	CIP	CSI	CSE	CAL	CBU	CAL	CAM	CQU	CGA	CVI	CEX	CTI	J.
.K.	CCI	CCO	CCO	CMA	CSP	CCA	CDE	CAR	CEN	CIN	COF	CBU	CCA	CCA	CTA	CVE	CCO	CLA	CAU	CAU	CGR	CLI	CME	CPA	CRE	.K.
.L.	CSI	CTA	CFI	CPO	CDE	CLA	CLA	CTE	CUV	CNU	CAR	CBE	CCA	CCO	CLI	CLA	CAC	CNI	CVE	CAU	CAM	CPA	CES	COF	CCA	.L.
.M.	COL	CTE	CSA	CVA	COL	CAV	CCH	CCO	CCO	CFE	CHE	CMA	CTI	CAB	CMA	CSA	CBR	CJU	CLA	CPA	CSP	CTI	CAN	CME	CSA	.M.
.N.	CFI	CMA	CMI	CMO	CPE	CCY	CLO	CZE	CTE	CPE	COB	CCI	CFL	CMA	CNA	CWI	CCA	CSC	CDA	CPL	CTR	CES	CRO	CBE	DFR	.N.
.O.	DGL	DAE	DCA	DSP	DEL	DMA	DIN	DAN	DAR	DRE	DAL	DGR	DPU	DDE	DEX	DIB	DAL	DBU	DCO	DCO	DFL	DMA	DOP	DRO	DCU	O.
.P.	DEB	DKA	DVI	DOD	DOP	DUN	DHE	DMA	DWI	DZI	DCO	EEL	ECR	ECR	EPY	ECA	EPH	EGU	EDU	ECA	ECO	EIN	ECA	ECI	ECO	.P.
.Q.	EGL	EJU	EGE	ESI	EAR	ECH	ECU	ELE	ETE	ETR	EOP	EGL	EJA	EPO	EVE	EFO	ECO	ESP	EUT	EUN	EAN	ELO	FES	FTA	FGR	.Q.
.R.	FSY	FSE	FLI	FAR	FID	FLO	FOV	FPR	FRU	FRU	FTI	FCA	FEL	FVO	FGL	FIN	FVU	FSP	FAN	FCH	FVE	FVI	FAF	FEL	FFO	.R.
.S.	GDU	GLA	GLI	GMA	GMU	GTI	GPR	GBA	GTI	GLU	GUR	GJA	GMA	GGL	GLE	GAN	GAR	GBA	GHE	GHI	GAS	GAB	GTI	HCO	HAN	.S.
.T.	HAN	HTU	HZE	HAL	HBE	HBR	HCA	HSA	HBE	HJA	HMU	HBU	HDI	HVU	HCO	HLU	HAN	HNI	HHI	HRU	HSP	HOF	IAM	IPA	IVE	T.
.U.	IAR	IHI	ISP	ITI	IED	IHE	IAQ	IBA	IER	JOF	JCH	JAI	JCI	JHI	JNI	JRE	JMA	JCO	KGA	KGE	KSC	LPU	LSA	LVI	LSI	.U.
.V.	LIB	LHE	LKI	LOW	LGA	LDO	LHI	LSA	LNO	LLA	LSP	LST	LIN	LMI	LOL	LCU	LSA	LVI	LLA	LCU	LST	LST	LFE	LGO	LLF	.V.
.W.	LOF	LRI	LAL	LBA	LDO	LUS	LMI	LCH	LIN	LMU	LPE	LNI	LUR	LUT	LCO	LPE	LTE	LAC	LCY	LAL	LAN	LLU	LCH	LES	LPI	.W.
.X.	MSP	MLO	MHO	MON	MPU	MBA	MSY	MAM	MOF	MIN	MDI	MES	MGL	MBI	MZA	MAR	MVU	MCH	MVI	MAR	MFA	MLU	MOR	MPO	MSA	.X.
.Y.	MSC	MCA	MUS	MBI	MAL	MIN	MOF	MSU	MMI	MOF	MAR	MGE	MPI	MPU	MRO	MSP	MCH	MMI	MBA	MCH	MDE	MPE	MCI	MOL	MAL	Y.
.Z.	MNI	MRU	MDE	MKO	MAC	MEN	MPA	MTE	MCA	MFR	MBA	MPE	MOD	MCO	NOF	NNU	NCA	NLA	NMU	NTA	NSA	NCO	OBA	OKI	OSA	.Z.
.a.	OJA	OEU	OVI	OJA	OFI	OCO	OVU	OSA	OGL	OSA	OHY	OMI	OTU	PER	PTU	PGU	PSP	PAN	PCO	PHE	PMA	PMI	POB	PRE	PVI	.a.
.b.	PBR	PSO	PFI	PCE	PED	PAR	PDI	PNO	PSC	PUR	PED	PLA	PLI	PMO	PQU	PSA	PCU	PYO	PHA	PPT	PAM	PCL	PPU	PMA	PIN	.b.
.c.	PPE	PFR	PAM	PCR	PAQ	PAR	PCA	PCO	PST	PAC	PAT	PCO	PLU	PVU	PPR	PDA	PTE	PAU	PAC	PEM	PSP	PIX	PPE	PAM	PDI	.c.
.d.	PRA	PAN	PED	PQU	PBE	PLO	PME	PNI	PRE	PVE	PSA	PES	PAM	PAN	PAR	PBU	PFE	PPR	PSE	PTR	PCA	PTU	PRU	PTR	POL	d.
.e.	PAF	PCA	PSA	PVI	PAR	PAV	PCE	PCE	PDO	PDO	PDU	PLA	PMA	PPE	PSA	PSU	PCA	PFR	PGU	PGU	PMO	PTE	PER	PSA	PSO	.e.
.f.	PLO	PPH	PGR	PCO	PPY	QSU	RSA	RSE	ROD	ROF	RPA	RRH	RVE	RHI	RNI	RRU	RSA	RUV	RCO	RCO	ROF	RSP	RSP	RID	ROC	.f.
.g.	ROC	RGR	SOF	SCH	SOF	SSC	SCA	SCA	SMI	STR	SSE	SHO	SMO	SBA	SLO	SMO	SSC	SOL	SCA	SCE	SED	SAN	SAL	SIN	SRA	.g.
.h.	SBI	SEX	SIT	SSP	SOD	SGL	SCH	SAL	SAR	SAE	SAV	SFE	SGI	SHY	SIN	SIN	SKH	SLA	SMA	SME	SMU	SNI	SQU	STO	STU	.h.
.i.	SRO	SAV	SAL	SBI	SHA	SSU	SJU	SST	SAC	SOL	SCY	SMO	SPU	STU	SAI	SCR	SWR	SSI	SSE	SRE	SCO	STE	SVI	SGR	SHI	i.
.j.	SKO	SSA	SNU	SAD	SER	SGU	SHU	SBE	STO	SCO	SPE	SDU	SAR	SCU	SJA	SMA	SSA	TIN	TVU	TKU	TKO	TOF	TOC	TPE	TCA	.j.
.k.	TVO	TCA	TTE	TPU	TDA	TBI	TCA	TGR	TSE	TVU	TAM	TPO	TBI	TNA	TAF	TAN	TCU	TAL	TAM	TFR	THI	THY	TIN	TNI	TPR	.k.
.l.	TRE	TRE	TSU	TVA	TVE	TWI	TSP	TFO	TTR	XTR	TAE	TCA	TCO	TDI	TDU	TMO	TPO	TSP	TTI	TTU	TTU	TCO	TRH	TTO	TTU	.l.
.m.	TFA	UTU	UGA	ULO	UDI	VAN	VAS	VCO	VMA	VVI	VLO	VMA	VPL	VCA	VAN	VZI	VAN	VAR	VBE	VDA	VER	VFA	VMO	VPA	VSA	.m.
.n.	VVI	VAC	VAN	VMU	VRA	VUM	VUN	VUN	VUN	VVE	VSE	VLA	VRO	VVI	VSU	XSA	YEL	ZMA	ZMA	ZOF	ZAQ	ZJU	ZMA	ZJA	ZMA	n.

ALTERNATIVE CROP MATRIX

Table 11. Crops Occurring in the Blackfoot Pharmacopoeia

	1	2	3	4	5	6	7	8	9	10	11	12	13	14	15	16	17	18	19	20	21	22	23	24	25	
.A.	AES	AMO	ATH	AFA	AME	ANI	APY	ASE	ASE	ASA	ANA	ACA	AMA	ACH	ADI	AVA	ACY	AMA	AME	ASC	AFO	ALE	ASI	ACR	ADA	.A.
.B.	ADE	AEL	AIN	ARE	ARI	ASI	ASM	ASP	ATR	ACA	AGI	AST	ATE	AFO	AMO	AMO	APS	AAM	ACE	ACH	AFI	ASA	ASC	ATU	AMA	.B.
.C.	ABA	AFE	APE	AAR	APR	AOF	AVA	ACA	ACR	AHY	ALI	ATR	AVI	AAR	ABR	AOC	ACO	AGA	AGE	AHA	AGR	AAR	ACH	ADI	AMO	.C.
.D.	AMU	ARE	ASQ	AMO	ACE	AVU	ABU	AGR	AGR	AHY	ACO	ALA	ACA	API	AES	ARU	AMO	AXA	AEL	AAB	AAB	ACI	ADR	AMA	AVU	.D.
.E	AAL	ACH	AIN	ASP	ADO	AGR	ATR	ADE	AOF	AOD	AGU	ACA	AHO	ABE	AAB	ABY	AFA	ASA	AST	AST	ABI	ACA	AAF	ACO	BMO	E.
.F.	BRA	BRA	BGA	BFR	BAE	BSP	BNI	BVU	BBE	BAL	BES	BSY	BHI	BEX	BVU	BOR	BSA	BNI	BCE	BOF	BFL	BCA	BCA	BIN	BIN	.F.
.G.	BIS	BPE	BSA	BCU	BER	BGR	BBR	BDE	BEM	BMU	BRA	BCA	BCH	BJU	BNA	BNA	BNI	BOL	BOL	BOL	BOL	BOL	BOL	BPE	BRA	.G.
.H.	BHU	BCA	BIN	BMA	BUN	BAL	BUT	BLA	BDA	BSA	BPA	CCO	CES	CSA	CSP	CCA	CRO	CAL	COF	CMU	CPA	CSI	CRA	COD	CIN	.H.
.I.	CEN	CGL	CPL	CED	CSA	CSP	CAN	CBA	CCH	CFR	CPU	CPA	CPU	CCA	CED	CGR	CTI	CCA	CIL	CSP	CAM	CUR	CED	CAL	CAU	.I.
.J	CCC	CSE	CCR	CDE	CMO	CPU	CSA	CEL	CUL	CED	CRO	CPE	CCI	CIP	CSI	CSE	CAL	CBU	CAL	CAM	CQU	CGA	CVI	CEX	CTI	J.
.K.	CCI	CCO	CCO	CMA	CSP	CCA	CDE	CAR	CEN	CIN	COF	CBU	CCA	CCA	CTA	CVE	CCO	CLA	CAU	CAU	CGR	CLI	CME	CPA	CRE	.K.
.L.	CSI	CTA	CFI	CPO	CDE	CLA	CLA	CTE	CUV	CNU	CAR	CBE	CCA	CCO	CLI	CLA	CAC	CNI	CVE	CAU	CAM	CPA	CES	COF	CCA	.L.
.M.	COL	CTE	CSA	CVA	COL	CAV	CCH	CCO	CCO	CFE	CHE	CMA	CTI	CAB	CMA	CSA	CBR	CJU	CLA	CPA	CSP	CTI	CAN	CME	CSA	.M.
.N.	CFI	CMA	CMI	CMO	CPE	CCY	CLO	CZE	CTE	CPE	COB	CCI	CFL	CMA	CNA	CWI	CCA	CSC	CDA	CPL	CTR	CES	CRO	CBE	DFR	.N.
.O	DGL	DAE	DCA	DSP	DEL	DMA	DIN	DAN	DAR	DRE	DAL	DGR	DPU	DDE	DEX	DIB	DAL	DBU	DCO	DCO	DFL	DMA	DOP	DRO	DCU	O.
.P.	DEB	DKA	DVI	DOD	DOP	DUN	DHE	DMA	DWI	DZI	DCO	EEL	ECR	ECR	EPY	ECA	EPH	EGU	EDU	ECA	ECO	EIN	ECA	ECI	ECO	.P.
.Q.	EGL	EJU	EGE	ESI	EAR	ECH	ECU	ELE	ETE	ETR	EOP	EGL	EJA	EPO	EVE	EFO	ECO	ESP	EUT	EUN	EAN	ELO	FES	FTA	FGR	.Q.
.R.	FSY	FSE	FLI	FAR	FID	FLO	FOV	FPR	FRU	FRU	FTI	FCA	FEL	FVO	FGL	FIN	FVU	FSP	FAN	FCH	FVE	FVI	FAF	FEL	FFO	.R.
.S.	GDU	GLA	GLI	GMA	GMU	GTI	GPR	GBA	GTI	GLU	GUR	GJA	GMA	GGL	GLE	GAN	GAR	GBA	GHE	GHI	GAS	GAB	GTI	HCO	HAN	.S.
.T	HAN	HTU	HZE	HAL	HBE	HBR	HCA	HSA	HBE	HJA	HMU	HBU	HDI	HVU	HCO	HLU	HAN	HNI	HHI	HRU	HSP	HOF	IAM	IPA	IVE	T.
.U.	IAR	IHI	ISP	ITI	IED	IHE	IAQ	IBA	IER	JOF	JCH	JAI	JCI	JHI	JNI	JRE	JMA	JCO	KGA	KGE	KSC	LPU	LSA	LVI	LSI	.U.
.V.	LIB	LHE	LKI	LOW	LGA	LDO	LHI	LSA	LNO	LLA	LSP	LST	LIN	LMI	LOL	LCU	LSA	LVI	LLA	LCU	LST	LST	LFE	LGO	LLE	.V.
.W.	LOF	LRI	LAL	LBA	LDO	LUS	LMI	LCH	LIN	LMU	LPE	LNI	LUR	LUT	LCO	LPE	LTE	LAC	LCY	LAL	LAN	LLU	LCH	LES	LPI	.W.
.X.	MSP	MLO	MHO	MON	MPU	MBA	MSY	MAM	MOF	MIN	MDI	MES	MGL	MBI	MZA	MAR	MVU	MCH	MVI	MAR	MFA	MLU	MOR	MPO	MSA	.X.
.Y	MSC	MCA	MUS	MBI	MAL	MIN	MOF	MSU	MMI	MOF	MAR	MGE	MPI	MPU	MRO	MSP	MCH	MMI	MBA	MCH	MDE	MPE	MCI	MOL	MAL	Y.
.Z.	MNI	MRU	MDE	MKO	MAC	MEN	MPA	MTE	MCA	MFR	MBA	MPE	MOD	MCO	NOF	NNU	NCA	NLA	NMU	NTA	NSA	NCO	OBA	OKI	OSA	.Z.
.a.	OJA	OEU	OVI	OJA	OFI	OCO	OVU	OSA	OGL	OSA	OHY	OMI	OTU	PER	PTU	PGU	PSP	PAN	PCO	PHE	PMA	PMI	POB	PRE	PVI	.a.
.b.	PBR	PSO	PFI	PCE	PED	PAR	PDI	PNO	PSC	PUR	PED	PLA	PLI	PMO	PQU	PSA	PCU	PYO	PHA	PPT	PAM	PCL	PPU	PMA	PIN	.b.
.c.	PPE	PFR	PAM	PCR	PAQ	PAR	PCA	PCO	PST	PAC	PAT	PCO	PLU	PVU	PPR	PDA	PTE	PAU	PAC	PEM	PSP	PIX	PPE	PAM	PDI	.c.
.d	PRA	PAN	PED	PQU	PBE	PLO	PME	PNI	PRE	PVE	PSA	PES	PAM	PAN	PAR	PBU	PFE	PPR	PSE	PTR	PCA	PTU	PRU	PTR	POL	d.
.e.	PAF	PCA	PSA	PVI	PAR	PAV	PCE	PCE	PDO	PDO	PDU	PLA	PMA	PPE	PSA	PSU	PCA	PFR	PGU	PGU	PMO	PTE	PER	PSA	PSO	.e.
.f.	PLO	PPH	PGR	PCO	PPY	QSU	RSA	RSE	ROD	ROF	RPA	RRH	RVE	RHI	RNI	RRU	RSA	RUV	RCO	RCO	ROF	RSP	RSP	RID	ROC	.f.
.g.	ROC	RGR	SOF	SCH	SOF	SSC	SCA	SCA	SMI	STR	SSE	SHO	SMO	SBA	SLO	SMO	SSC	SOL	SCA	SCE	SED	SAN	SAL	SIN	SRA	.g.
.h.	SBI	SEX	SIT	SSP	SOD	SGL	SCH	SAL	SAR	SAE	SAV	SFE	SGI	SHY	SIN	SIN	SKH	SLA	SMA	SME	SMU	SNI	SQU	STO	STU	.h.
.i	SRO	SAV	SAL	SBI	SHA	SSU	SJU	SST	SAC	SOL	SCY	SMO	SPU	STU	SAI	SCR	SWR	SSI	SSE	SRE	SCO	STE	SVI	SGR	SHI	i.
.j.	SKO	SSA	SNU	SAD	SER	SGU	SHU	SBE	STO	SCO	SPE	SDU	SAR	SCU	SJA	SMA	SSA	TIN	TVU	TKU	TKO	TOF	TOC	TPE	TCA	.j.
.k.	TVO	TCA	TTE	TPU	TDA	TBI	TCA	TGR	TSE	TVU	TAM	TPO	TBI	TNA	TAF	TAN	TCU	TAL	TAM	TFR	THI	THY	TIN	TNI	TPR	.k.
.l.	TRE	TRE	TSU	TVA	TVE	TWI	TSP	TFO	TTR	XTR	TAE	TCA	TCO	TDI	TDU	TMO	TPO	TSP	TTI	TTU	TTU	TCO	TRH	TTO	TTU	.l.
.m.	TFA	UTU	UGA	ULO	UDI	VAN	VAS	VCO	VMA	VVI	VLO	VMA	VPL	VCA	VAN	VZI	VAN	VAR	VBE	VDA	VER	VFA	VMO	VPA	VSA	.m.
.n	VVI	VAC	VAN	VMU	VRA	VUM	VUN	VUN	VUN	VVE	VSE	VLA	VRO	VVI	VSU	XSA	YEL	ZMA	ZMA	ZOF	ZAQ	ZJU	ZMA	ZJA	ZMA	n.

ALTERNATIVE CROP MATRIX

Table 12. Crops Occurring in the Callahuaya Pharmacopoeia

	1	2	3	4	5	6	7	8	9	10	11	12	13	14	15	16	17	18	19	20	21	22	23	24	25	
A	AES	AMO	ATH	AFA	AME	ANI	APY	ASE	ASE	ASA	ANA	ACA	AMA	ACH	ADI	AVA	ACY	AMA	AME	ASC	AFO	ALE	ASI	ACR	ADA	A
B	ADE	AEL	AIN	ARE	ARI	ASI	ASM	ASP	ATR	ACA	AGI	AST	ATE	AFO	AMO	AMO	APS	AAM	ACE	ACH	AFI	ASA	ASC	ATU	AMA	B
C	ABA	AFE	APE	AAR	APR	AOF	AVA	ACA	ACR	AHY	ALI	ATR	AVI	AAR	ABR	AOC	ACO	AGA	AGE	AHA	AGR	AAR	ACH	ADI	AMO	C
D	AMU	ARE	ASQ	AMO	ACE	AVU	ABU	AGR	AGR	AHY	ACO	ALA	ACA	API	AES	ARU	AMO	AXA	AEL	AAB	AAB	ACI	ADR	AMA	AVU	D
E	AAL	ACH	AIN	ASP	ADO	AGR	ATR	ADE	AOF	AOD	AGU	ACA	AHO	ABE	AAB	ABY	AFA	ASA	AST	AST	ABI	ACA	AAF	ACO	BMO	E
F	BRA	BRA	BGA	BFR	BAE	BSP	BNI	BVU	BBE	BAL	BES	BSY	BHI	BEX	BVU	BOR	BSA	BNI	BCE	BOF	BFL	BCA	BCA	BIN	BIN	F
G	BIS	BPE	BSA	BCU	BER	BGR	BBR	BDE	BEM	BMU	BRA	BCA	BCH	BJU	BNA	BNA	BNI	BOL	BOL	BOL	BOL	BOL	BOL	BPE	BRA	G
H	BHU	BCA	BIN	BMA	BUN	BAL	BUT	BLA	BDA	BSA	BPA	CCO	CES	CSA	CSP	CCA	CRO	CAL	COF	CMU	CPA	CSI	CRA	COD	CIN	H
I	CEN	CGL	CPL	CED	CSA	CSP	CAN	CBA	CCH	CFR	CPU	CPA	CPU	CCA	CED	CGR	CTI	CCA	CIL	CSP	CAM	CUR	CED	CAL	CAU	I
J	CCC	CSE	CCR	CDE	CMO	CPU	CSA	CEL	CUL	CED	CRO	CPE	CCI	CIP	CSI	CSE	CAL	CBU	CAL	CAM	CQU	CGA	CVI	CEX	CTI	J
K	CCI	CCO	CCO	CMA	CSP	CCA	CDE	CAR	CEN	CIN	COF	CBU	CCA	CCA	CTA	CVE	CCO	CLA	CAU	CAU	CGR	CLI	CME	CPA	CRE	K
L	CSI	CTA	CFI	CPO	CDE	CLA	CLA	CTE	CUV	CNU	CAR	CBE	CCA	CCO	CLI	CLA	CAC	CNI	CVE	CAU	CAM	CPA	CES	COF	CCA	L
M	COL	CTE	CSA	CVA	COL	CAV	CCH	CCO	CCO	CFE	CHE	CMA	CTI	CAB	CMA	CSA	CBR	CJU	CLA	CPA	CSP	CTI	CAN	CME	CSA	M
N	CFI	CMA	CMI	CMO	CPE	CCY	CLO	CZE	CTE	CPE	COB	CCI	CFL	CMA	CNA	CWI	CCA	CSC	CDA	CPL	CTR	CES	CRO	CBE	DFR	N
O	DGL	DAE	DCA	DSP	DEL	DMA	DIN	DAN	DAR	DRE	DAL	DGR	DPU	DDE	DEX	DIB	DAL	DBU	DCO	DCO	DFL	DMA	DOP	DRO	DCU	O
P	DEB	DKA	DVI	DOD	DOP	DUN	DHE	DMA	DWI	DZI	DCO	EEL	ECR	ECR	EPY	ECA	EPH	EGU	EDU	ECA	ECO	EIN	ECA	ECI	ECO	P
Q	EGL	EJU	EGE	ESI	EAR	ECH	ECU	ELE	ETE	ETR	EOP	EGL	EJA	EPO	EVE	EFO	ECO	ESP	EUT	EUN	EAN	ELO	FES	FTA	FGR	Q
R	FSY	FSE	FLI	FAR	FID	FLO	FOV	FPR	FRU	FRU	FTI	FCA	FEL	FVO	FGL	FIN	FVU	FSP	FAN	FCH	FVE	FVI	FAF	FEL	FFO	R
S	GDU	GLA	GLI	GMA	GMU	GTI	GPR	GBA	GTI	GLU	GUR	GJA	GMA	GGL	GLE	GAN	GAR	GBA	GHE	GHI	GAS	GAB	GTI	HCO	HAN	S
T	HAN	HTU	HZE	HAL	HBE	HBR	HCA	HSA	HBE	HJA	HMU	HBU	HDI	HVU	HCO	HLU	HAN	HNI	HHI	HRU	HSP	HOF	IAM	IPA	IVE	T
U	IAR	IHI	ISP	ITI	IED	IHE	IAQ	IBA	IER	JOF	JCH	JAI	JCI	JHI	JNI	JRE	JMA	JCO	KGA	KGE	KSC	LPU	LSA	LVI	LSI	U
V	LIB	LHE	LKI	LOW	LGA	LDO	LHI	LSA	LNO	LLA	LSP	LST	LIN	LMI	LOL	LCU	LSA	LVI	LLA	LCU	LST	LST	LFE	LGO	LLE	V
W	LOF	LRI	LAL	LBA	LDO	LUS	LMI	LCH	LIN	LMU	LPE	LNI	LUR	LUT	LCO	LPE	LTE	LAC	LCY	LAL	LAN	LLU	LCH	LES	LPI	W
X	MSP	MLO	MHO	MON	MPU	MBA	MSY	MAM	MOF	MIN	MDI	MES	MGL	MBI	MZA	MAR	MVU	MCH	MVI	MAR	MFA	MLU	MOR	MPO	MSA	X
Y	MSC	MCA	MUS	MBI	MAL	MIN	MOF	MSU	MMI	MOF	MAR	MGE	MPI	MPU	MRO	MSP	MCH	MMI	MBA	MCH	MDE	MPE	MCI	MOL	MAL	Y
Z	MNI	MRU	MDE	MKO	MAC	MEN	MPA	MTE	MCA	MFR	MBA	MPE	MOD	MCO	NOF	NNU	NCA	NLA	NMU	NTA	NSA	NCO	OBA	OKI	OSA	Z
a	OJA	OEU	OVI	OJA	OFI	OCO	OVU	OSA	OGL	OSA	OHY	OMI	OTU	PER	PTU	PGU	PSP	PAN	PCO	PHE	PMA	PMI	POB	PRE	PVI	a
b	PBR	PSO	PFI	PCE	PED	PAR	PDI	PNO	PSC	PUR	PED	PLA	PLI	PMO	PQU	PSA	PCU	PYO	PHA	PPT	PAM	PCL	PPU	PMA	PIN	b
c	PPE	PFR	PAM	PCR	PAQ	PAR	PCA	PCO	PST	PAC	PAT	PCO	PLU	PVU	PPR	PDA	PTE	PAU	PAC	PEM	PSP	PIX	PPE	PAM	PDI	c
d	PRA	PAN	PED	PQU	PBE	PLO	PME	PNI	PRE	PVE	PSA	PES	PAM	PAN	PAR	PBU	PFE	PPR	PSE	PTR	PCA	PTU	PRU	PTR	POL	d
e	PAF	PCA	PSA	PVI	PAR	PAV	PCE	PCE	PDO	PDO	PDU	PLA	PMA	PPE	PSA	PSU	PCA	PFR	PGU	PGU	PMO	PTE	PER	PSA	PSO	e
f	PLO	PPH	PGR	PCO	PPY	QSU	RSA	RSE	ROD	ROF	RPA	RRH	RVE	RHI	RNI	RRU	RSA	RUV	RCQ	RCO	ROF	RSP	RSP	RID	ROC	f
g	ROC	RGR	SOF	SCH	SOF	SSC	SCA	SCA	SMI	STR	SSE	SHO	SMO	SBA	SLO	SMO	SSC	SOL	SCA	SCE	SED	SAN	SAL	SIN	SRA	g
h	SBI	SEX	SIT	SSP	SOD	SGL	SCH	SAL	SAR	SAE	SAV	SFE	SGI	SHY	SIN	SIN	SKH	SLA	SMA	SME	SMU	SNI	SQU	STO	STU	h
i	SRO	SAV	SAL	SBI	SHA	SSU	SJU	SST	SAC	SOL	SCY	SMO	SPU	STU	SAI	SCR	SWR	SSI	SSE	SRE	SCO	STE	SVI	SGR	SHI	i
j	SKO	SSA	SNU	SAD	SER	SGU	SHU	SBE	STO	SCO	SPE	SDU	SAR	SCU	SJA	SMA	SSA	TIN	TVU	TKU	TKO	TOF	TOC	TPE	TCA	j
k	TVO	TCA	TTE	TPU	TDA	TBI	TCA	TGR	TSE	TVU	TAM	TPO	TBI	TNA	TAF	TAN	TCU	TAL	TAM	TFR	THI	THY	TIN	TNI	TPR	k
l	TRE	TRE	TSU	TVA	TVE	TWI	TSP	TFO	TTR	XTR	TAE	TCA	TCO	TDI	TDU	TMO	TPO	TSP	TTI	TTU	TTU	TCO	TRH	TTO	TTU	l
m	TFA	UTU	UGA	ULO	UDI	VAN	VAS	VCO	VMA	VVI	VLO	VMA	VPL	VCA	VAN	VZI	VAN	VAR	VBE	VDA	VER	VFA	VMO	VPA	VSA	m
n	VVI	VAC	VAN	VMU	VRA	VUM	VUN	VUN	VUN	VVE	VSE	VLA	VRO	VVI	VSU	XSA	YEL	ZMA	ZMA	ZOF	ZAQ	ZJU	ZMA	ZJA	ZMA	n

ALTERNATIVE CROP MATRIX

Table 13. Crops Occurring in the Choco Pharmacopoeia

	1	2	3	4	5	6	7	8	9	10	11	12	13	14	15	16	17	18	19	20	21	22	23	24	25	
.A.	AES	AMO	ATH	AFA	AME	ANI	APY	ASE	ASE	ASA	ANA	ACA	AMA	ACH	ADI	AVA	ACY	AMA	AME	ASC	AFO	ALE	ASI	ACR	ADA	.A.
.B.	ADE	AEL	AIN	ARE	ARI	ASI	ASM	ASP	ATR	ACA	AGI	AST	ATE	AFO	AMO	AMO	APS	AAM	ACE	ACH	AFI	ASA	ASC	ATU	AMA	.B.
.C.	ABA	AFE	APE	AAR	APR	AOF	AVA	ACA	ACR	AHY	ALI	ATR	AVI	AAR	ABR	AOC	ACO	AGA	AGE	AHA	AGR	AAR	ACH	ADI	AMO	.C.
.D.	AMU	ARE	ASQ	AMO	ACE	AVU	ABU	AGR	AGR	AHY	ACO	ALA	ACA	API	AES	ARU	AMO	AXA	AEL	AAB	AAB	ACI	ADR	AMA	AVU	.D.
.E.	AAL	ACH	AIN	ASP	ADO	AGR	ATR	ADE	AOF	AOD	AGU	ACA	AHO	ABE	AAB	ABY	AFA	ASA	AST	AST	ABI	ACA	AAF	ACO	BMO	.E.
.F.	BRA	BRA	BGA	BFR	BAE	BSP	BNI	BVU	BBE	BAL	BES	BSY	BHI	BEX	BVU	BOR	BSA	BNI	BCE	BOF	BFL	BCA	BCA	BIN	BIN	.F.
.G.	BIS	BPE	BSA	BCU	BER	BGR	BBR	BDE	BEM	BMU	BRA	BCA	BCH	BJU	BNA	BNA	BNI	BOL	BOL	BOL	BOL	BOL	BOL	BPE	BRA	.G.
.H.	BHU	BCA	BIN	BMA	BUN	BAL	BUT	BLA	BDA	BSA	BPA	CCO	CES	CSA	CSP	CCA	CRO	CAL	COF	CMU	CPA	CSI	CRA	COD	CIN	.H.
.I.	CEN	CGL	CPL	CED	CSA	CSP	CAN	CBA	CCH	CFR	CPU	CPA	CPU	CCA	CED	CGR	CTI	CCA	CIL	CSP	CAM	CUR	CED	CAL	CAU	.I.
.J.	CCC	CSE	CCR	CDE	CMO	CPU	CSA	CEL	CUL	CED	CRO	CPE	CCI	CIP	CSI	CSE	CAL	CBU	CAL	CAM	CQU	CGA	CVI	CEX	CTI	.J.
.K.	CCI	CCO	CCO	CMA	CSP	CCA	CDE	CAR	CEN	CIN	COF	CBU	CCA	CCA	CTA	CVE	CCO	CLA	CAU	CAU	CGR	CLI	CME	CPA	CRE	.K.
.L.	CSI	CTA	CFI	CPO	CDE	CLA	CLA	CTE	CUV	CNU	CAR	CBE	CCA	CCO	CLI	CLA	CAC	CNI	CVE	CAU	CAM	CPA	CES	COF	CCA	.L.
.M.	COL	CTE	CSA	CVA	COL	CAV	CCH	CCO	CCO	CFE	CHE	CMA	CTI	CAB	CMA	CSA	CBR	CJU	CLA	CPA	CSP	CTI	CAN	CME	CSA	.M.
.N.	CFI	CMA	CMI	CMO	CPE	CCY	CLO	CZE	CTE	CPE	COB	CCI	CFL	CMA	CNA	CWI	CCA	CSC	CDA	CPL	CTR	CES	CRO	CBE	DFR	.N.
.O.	DGL	DAE	DCA	DSP	DEL	DMA	DIN	DAN	DAR	DRE	DAL	DGR	DPU	DDE	DEX	DIB	DAL	DBU	DCO	DCO	DFL	DMA	DOP	DRO	DCU	.O.
.P.	DEB	DKA	DVI	DOD	DOP	DUN	DHE	DMA	DWI	DZI	DCO	EEL	ECR	ECR	EPY	ECA	EPH	EGU	EDU	ECA	ECO	EIN	ECA	ECI	ECO	.P.
.Q.	EGL	EJU	EGE	ESI	EAR	ECH	ECU	ELE	ETE	ETR	EOP	EGL	EJA	EPO	EVE	EFO	ECO	ESP	EUT	EUN	EAN	ELO	FES	FTA	FGR	.Q.
.R.	FSY	FSE	FLI	FAR	FID	FLO	FOV	FPR	FRU	FRU	FTI	FCA	FEL	FVO	FGL	FIN	FVU	FSP	FAN	FCH	FVE	FVI	FAF	FEL	FFO	.R.
.S.	GDU	GLA	GLI	GMA	GMU	GTI	GPR	GBA	GTI	GLU	GUR	GJA	GMA	GGL	GLE	GAN	GAR	GBA	GHE	GHI	GAS	GAB	GTI	HCO	HAN	.S.
.T.	HAN	HTU	HZE	HAL	HBE	HBR	HCA	HSA	HBE	HJA	HMU	HBU	HDI	HVU	HCO	HLU	HAN	HNI	HHI	HRU	HSP	HOF	IAM	IPA	IVE	.T.
.U.	IAR	IHI	ISP	ITI	IED	IHE	IAQ	IBA	IER	JOF	JCH	JAI	JCI	JHI	JNI	JRE	JMA	JCO	KGA	KGE	KSC	LPU	LSA	LVI	LSI	.U.
.V.	LIB	LHE	LKI	LOW	LGA	LDO	LHI	LSA	LNO	LLA	LSP	LST	LIN	LMI	LOL	LCU	LSA	LVI	LLA	LCU	LST	LST	LFE	LGO	LLE	.V.
.W.	LOF	LRI	LAL	LBA	LDO	LUS	LMI	LCH	LIN	LMU	LPE	LNI	LUR	LUT	LCO	LPE	LTE	LAC	LCY	LAL	LAN	LLU	LCH	LES	LPI	.W.
.X.	MSP	MLO	MHO	MON	MPU	MBA	MSY	MAM	MOF	MIN	MDI	MES	MGL	MBI	MZA	MAR	MVU	MCH	MVI	MAR	MFA	MLU	MOR	MPO	MSA	.X.
.Y.	MSC	MCA	MUS	MBI	MAL	MIN	MOF	MSU	MMI	MOF	MAR	MGE	MPI	MPU	MRO	MSP	MCH	MMI	MBA	MCH	MDE	MPE	MCI	MOL	MAL	.Y.
.Z.	MNI	MRU	MDE	MKO	MAC	MEN	MPA	MTE	MCA	MFR	MBA	MPE	MOD	MCO	NOF	NNU	NCA	NLA	NMU	NTA	NSA	NCO	OBA	OKI	OSA	.Z.
.a.	OJA	OEU	OVI	OJA	OFI	OCO	OVU	OSA	OGL	OSA	OHY	OMI	OTU	PER	PTU	PGU	PSP	PAN	PCO	PHE	PMA	PMI	POB	PRE	PVI	.a.
.b.	PBR	PSO	PFI	PCE	PED	PAR	PDI	PNO	PSC	PUR	PED	PLA	PLI	PMO	PQU	PSA	PCU	PYO	PHA	PPT	PAM	PCL	PPU	PMA	PIN	.b.
.c.	PPE	PFR	PAM	PCR	PAQ	PAR	PCA	PCO	PST	PAC	PAT	PCO	PLU	PVU	PPR	PDA	PTE	PAU	PAC	PEM	PSP	PIX	PPE	PAM	PDI	.c.
.d.	PRA	PAN	PED	PQU	PBE	PLO	PME	PNI	PRE	PVE	PSA	PES	PAM	PAN	PAR	PBU	PFE	PPR	PSE	PTR	PCA	PTU	PRU	PTR	POL	.d.
.e.	PAF	PCA	PSA	PVI	PAR	PAV	PCE	PCE	PDO	PDO	PDU	PLA	PMA	PPE	PSA	PSU	PCA	PFR	PGU	PGU	PMO	PTE	PER	PSA	PSO	.e.
.f.	PLO	PPH	PGR	PCO	PPY	QSU	RSA	RSE	ROD	ROF	RPA	RRH	RVE	RHI	RNI	RRU	RSA	RUV	RCO	RCO	ROF	RSP	RSP	RID	ROC	.f.
.g.	ROC	RGR	SOF	SCH	SOF	SSC	SCA	SCA	SMI	STR	SSE	SHO	SMO	SBA	SLO	SMO	SSC	SOL	SCA	SCE	SED	SAN	SAL	SIN	SRA	.g.
.h.	SBI	SEX	SIT	SSP	SOD	SGL	SCH	SAL	SAR	SAE	SAV	SFE	SGI	SHY	SIN	SIN	SKH	SLA	SMA	SME	SMU	SNI	SQU	STO	STU	.h.
.i.	SRO	SAV	SAL	SBI	SHA	SSU	SJU	SST	SAC	SOL	SCY	SMO	SPU	STU	SAI	SCR	SWR	SSI	SSE	SRE	SCO	STE	SVI	SGR	SHI	.i.
.j.	SKO	SSA	SNU	SAD	SER	SGU	SHU	SBE	STO	SCO	SPE	SDU	SAR	SCU	SJA	SMA	SSA	TIN	TVU	TKU	TKO	TOF	TOC	TPE	TCA	.j.
.k.	TVO	TCA	TTE	TPU	TDA	TBI	TCA	TGR	TSE	TVU	TAM	TPO	TBI	TNA	TAF	TAN	TCU	TAL	TAM	TFR	THI	THY	TIN	TNI	TPR	.k.
.l.	TRE	TRE	TSU	TVA	TVE	TWI	TSP	TFO	TTR	XTR	TAE	TCA	TCO	TDI	TDU	TMO	TPO	TSP	TTI	TTU	TTU	TCO	TRH	TTO	TTU	.l.
.m.	TFA	UTU	UGA	ULO	UDI	VAN	VAS	VCO	VMA	VVI	VLO	VMA	VPL	VCA	VAN	VZI	VAN	VAR	VBE	VDA	VER	VFA	VMO	VPA	VSA	.m.
.n.	VVI	VAC	VAN	VMU	VRA	VUM	VUN	VUN	VUN	VVE	VSE	VLA	VRO	VVI	VSU	XSA	YEL	ZMA	ZMA	ZOF	ZAQ	ZJU	ZMA	ZJA	ZMA	.n.

ALTERNATIVE CROP MATRIX

Table 14. Crops Occurring in the Cuna Pharmacopoeia

	1	2	3	4	5	6	7	8	9	10	11	12	13	14	15	16	17	18	19	20	21	22	23	24	25	
.A.	AES	AMO	ATH	AFA	AME	ANI	APY	ASE	ASE	ASA	ANA	ACA	AMA	ACH	ADI	AVA	ACY	AMA	AME	ASC	AFO	ALE	ASI	ACR	ADA	.A.
.B.	ADE	AEL	AIN	ARE	ARI	ASI	ASM	ASP	ATR	ACA	AGI	AST	ATE	AFO	AMO	AMO	APS	AAM	ACE	ACH	AFI	ASA	ASC	ATU	AMA	.B.
.C.	ABA	AFE	APE	AAR	APR	AOF	AVA	ACA	ACR	AHY	ALI	ATR	AVI	AAR	ABR	AOC	ACO	AGA	AGE	AHA	AGR	AAR	ACH	ADI	AMO	.C.
.D.	AMU	ARE	ASQ	AMO	ACE	AVU	ABU	AGR	AGR	AHY	ACO	ALA	ACA	API	AES	ARU	AMO	AXA	AEL	AAB	AAB	ACI	ADR	AMA	AVU	.D.
.E.	AAL	ACH	AIN	ASP	ADO	AGR	ATR	ADE	AOF	AOD	AGU	ACA	AHO	ABE	AAB	ABY	AFA	ASA	AST	AST	ABI	ACA	AAF	ACO	BMO	.E.
.F.	BRA	BRA	BGA	BFR	BAE	BSP	BNI	BVU	BBE	BAL	BES	BSY	BHI	BEX	BVU	BOR	BSA	BNI	BCE	BOF	BFL	BCA	BCA	BIN	BIN	.F.
.G.	BIS	BPE	BSA	BCU	BER	BGR	BBR	BDE	BEM	BMU	BRA	BCA	BCH	BJU	BNA	BNA	BNI	BOL	BOL	BOL	BOL	BOL	BOL	BPE	BRA	.G.
.H.	BHU	BCA	BIN	BMA	BUN	BAL	BUT	BLA	BDA	BSA	BPA	CCO	CES	CSA	CSP	CCA	CRO	CAL	COF	CMU	CPA	CSI	CRA	COD	CIN	.H.
.I.	CEN	CGL	CPL	CED	CSA	CSP	CAN	CBA	CCH	CFR	CPU	CPA	CPU	CCA	CED	CGR	CTI	CCA	CIL	CSP	CAM	CUR	CED	CAL	CAU	.I.
.J.	CCC	CSE	CCR	CDE	CMO	CPU	CSA	CEL	CUL	CED	CRO	CPE	CCI	CIP	CSI	CSE	CAL	CBU	CAL	CAM	CQU	CGA	CVI	CEX	CTI	.J.
.K.	CCI	CCO	CCO	CMA	CSP	CCA	CDE	CAR	CEN	CIN	COF	CBU	CCA	CCA	CTA	CVE	CCO	CLA	CAU	CAU	CGR	CLI	CME	CPA	CRE	.K.
.L.	CSI	CTA	CFI	CPO	CDE	CLA	CLA	CTE	CUV	CNU	CAR	CBE	CCA	CCO	CLI	CLA	CAC	CNI	CVE	CAU	CAM	CPA	CES	COF	CCA	.L.
.M.	COL	CTE	CSA	CVA	COL	CAV	CCH	CCO	CCO	CFE	CHE	CMA	CTI	CAB	CMA	CSA	CBR	CJU	CLA	CPA	CSP	CTI	CAN	CME	CSA	.M.
.N.	CFI	CMA	CMI	CMO	CPE	CCY	CLO	CZE	CTE	CPE	COB	CCI	CFL	CMA	CNA	CWI	CCA	CSC	CDA	CPL	CTR	CES	CRO	CBE	DFR	.N.
.O.	DGL	DAE	DCA	DSP	DEL	DMA	DIN	DAN	DAR	DRE	DAL	DGR	DPU	DDE	DEX	DIB	DAL	DBU	DCO	DCO	DFL	DMA	DOP	DRO	DCU	.O.
.P.	DEB	DKA	DVI	DOD	DOP	DUN	DHE	DMA	DWI	DZI	DCO	EEL	ECR	ECR	EPY	ECA	EPH	EGU	EDU	ECA	ECO	EIN	ECA	ECI	ECO	.P.
.Q.	EGL	EJU	EGE	ESI	EAR	ECH	ECU	ELE	ETE	ETR	EOP	EGL	EJA	EPO	EVE	EFO	ECO	ESP	EUT	EUN	EAN	ELO	FES	FTA	FGR	.Q.
.R.	FSY	FSE	FLI	FAR	FID	FLO	FOV	FPR	FRU	FRU	FTI	FCA	FEL	FVO	FGL	FIN	FVU	FSP	FAN	FCH	FVE	FVI	FAF	FEL	FFO	.R.
.S.	GDU	GLA	GLI	GMA	GMU	GTI	GPR	GBA	GTI	GLU	GUR	GJA	GMA	GGL	GLE	GAN	GAR	GBA	GHE	GHI	GAS	GAB	GTI	HCO	HAN	.S.
.T.	HAN	HTU	HZE	HAL	HBE	HBR	HCA	HSA	HBE	HJA	HMU	HBU	HDI	HVU	HCO	HLU	HAN	HNI	HHI	HRU	HSP	HOF	IAM	IPA	IVE	.T.
.U.	IAR	IHI	ISP	ITI	IED	IHE	IAQ	IBA	IER	JOF	JCH	JAI	JCI	JHI	JNI	JRE	JMA	JCO	KGA	KGE	KSC	LPU	LSA	LVI	LSI	.U.
.V.	LIB	LHE	LKI	LOW	LGA	LDO	LHI	LSA	LNO	LLA	LSP	LST	LIN	LMI	LOL	LCU	LSA	LVI	LLA	LCU	LST	LST	LFE	LGO	LLE	.V.
.W.	LOF	LRI	LAL	LBA	LDO	LUS	LMI	LCH	LIN	LMU	LPE	LNI	LUR	LUT	LCO	LPE	LTE	LAC	LCY	LAL	LAN	LLU	LCH	LES	LPI	.W.
.X.	MSP	MLO	MHO	MON	MPU	MBA	MSY	MAM	MOF	MIN	MDI	MES	MGL	MBI	MZA	MAR	MVU	MCH	MVI	MAR	MFA	MLU	MOR	MPO	MSA	.X.
.Y.	MSC	MCA	MUS	MBI	MAL	MIN	MOF	MSU	MMI	MOF	MAR	MGE	MPI	MPU	MRO	MSP	MCH	MMI	MBA	MCH	MDE	MPE	MCI	MOL	MAL	.Y.
.Z.	MNI	MRU	MDE	MKO	MAC	MEN	MPA	MTE	MCA	MFR	MBA	MPE	MOD	MCO	NOF	NNU	NCA	NLA	NMU	NTA	NSA	NCO	OBA	OKI	OSA	.Z.
.a.	OJA	OEU	OVI	OJA	OFI	OCO	OVU	OSA	OGL	OSA	OHY	OMI	OTU	PER	PTU	PGU	PSP	PAN	PCO	PHE	PMA	PMI	POB	PRE	PVI	.a.
.b.	PBR	PSO	PFI	PCE	PED	PAR	PDI	PNO	PSC	PUR	PED	PLA	PLI	PMO	PQU	PSA	PCU	PYO	PHA	PPT	PAM	PCL	PPU	PMA	PIN	.b.
.c.	PPE	PFR	PAM	PCR	PAQ	PAR	PCA	PCO	PST	PAC	PAT	PCO	PLU	PVU	PPR	PDA	PTE	PAU	PAC	PEM	PSP	PIX	PPE	PAM	PDI	.c.
.d.	PRA	PAN	PED	PQU	PBE	PLO	PME	PNI	PRE	PVE	PSA	PES	PAM	PAN	PAR	PBU	PFE	PPR	PSE	PTR	PCA	PTU	PRU	PTR	POL	.d.
.e.	PAF	PCA	PSA	PVI	PAR	PAV	PCE	PCE	PDO	PDO	PDU	PLA	PMA	PPE	PSA	PSU	PCA	PFR	PGU	PGU	PMO	PTE	PER	PSA	PSO	.e.
.f.	PLO	PPH	PGR	PCO	PPY	QSU	RSA	RSE	ROD	ROF	RPA	RRH	RVE	RHI	RNI	RRU	RSA	RUV	RCO	RCO	ROF	RSP	RSP	RID	ROC	.f.
.g.	ROC	RGR	SOF	SCH	SOF	SSC	SCA	SCA	SMI	STR	SSE	SHO	SMO	SBA	SLO	SMO	SSC	SOL	SCA	SCE	SED	SAN	SAL	SIN	SRA	.g.
.h.	SBI	SEX	SIT	SSP	SOD	SGL	SCH	SAL	SAR	SAE	SAV	SFE	SGI	SHY	SIN	SIN	SKH	SLA	SMA	SME	SMU	SNI	SQU	STO	STU	.h.
.i.	SRO	SAV	SAL	SBI	SHA	SSU	SJU	SST	SAC	SOL	SCY	SMO	SPU	STU	SAI	SCR	SWR	SSI	SSE	SRE	SCO	STE	SVI	SGR	SHI	.i.
.j.	SKO	SSA	SNU	SAD	SER	SGU	SHU	SBE	STO	SCO	SPE	SDU	SAR	SCU	SJA	SMA	SSA	TIN	TVU	TKU	TKO	TOF	TOC	TPE	TCA	.j.
.k.	TVO	TCA	TTE	TPU	TDA	TBI	TCA	TGR	TSE	TVU	TAM	TPO	TBI	TNA	TAF	TAN	TCU	TAL	TAM	TFR	THI	THY	TIN	TNI	TPR	.k.
.l.	TRE	TRE	TSU	TVA	TVE	TWI	TSP	TFO	TTR	XTR	TAE	TCA	TCO	TDI	TDU	TMO	TPO	TSP	TTI	TTU	TTU	TCO	TRH	TTO	TTU	.l.
.m.	TFA	UTU	UGA	ULO	UDI	VAN	VAS	VCO	VMA	VVI	VLO	VMA	VPL	VCA	VAN	VZI	VAN	VAR	VBE	VDA	VER	VFA	VMO	VPA	VSA	.m.
.n.	VVI	VAC	VAN	VMU	VRA	VUM	VUN	VUN	VUN	VVE	VSE	VLA	VRO	VVI	VSU	XSA	YEL	ZMA	ZMA	ZOF	ZAQ	ZJU	ZMA	ZJA	ZMA	.n.
	1	2	3	4	5	6	7	8	9	10	11	12	13	14	15	16	17	18	19	20	21	22	23	24	25	

ALTERNATIVE CROP MATRIX

Table 15. Antitumor Agents

	1	2	3	4	5	6	7	8	9	10	11	12	13	14	15	16	17	18	19	20	21	22	23	24	25	
A	AES	AMO	ATH	AFA	AME	ANI	APY	ASE	ASE	ASA	ANA	ACA	AMA	ACH	ADI	AVA	ACY	AMA	AME	ASC	AFO	ALE	ASI	ACR	ADA	A
B	ADE	AEL	AIN	ARE	ARI	ASI	ASM	ASP	ATR	ACA	AGI	AST	ATE	AFO	AMO	AMO	APS	AAM	ACE	ACH	AFI	ASA	ASC	ATU	AMA	B
C	ABA	AFE	APE	AAR	APR	AOF	AVA	ACA	ACR	AHY	ALI	ATR	AVI	AAR	ABR	AOC	ACO	AGA	AGE	AHA	AGR	AAR	ACH	ADI	AMO	C
D	AMU	ARE	ASQ	AMO	ACE	AVU	ABU	AGR	AGR	AHY	ACO	ALA	ACA	API	AES	ARU	AMO	AXA	AEL	AAB	AAB	ACI	ADR	AMA	AVU	D
E	AAL	ACH	AIN	ASP	ADO	AGR	ATR	ADE	AOF	AOD	AGU	ACA	AHO	ABE	AAB	ABY	AFA	ASA	AST	AST	ABI	ACA	AAF	ACO	BMO	E
F	BRA	BRA	BGA	BFR	BAE	BSP	BNI	BVU	BBE	BAL	BES	BSY	BHI	BEX	BVU	BOR	BSA	BNI	BCE	BOF	BFL	BCA	BCA	BIN	BIN	F
G	BIS	BPE	BSA	BCU	BER	BGR	BBR	BDE	BEM	BMU	BRA	BCA	BCH	BJU	BNA	BNA	BNI	BOL	BOL	BOL	BOL	BOL	BOL	BPE	BRA	G
H	BHU	BCA	BIN	BMA	BUN	BAL	BUT	BLA	BDA	BSA	BPA	CCO	CES	CSA	CSP	CCA	CRO	CAL	COF	CMU	CPA	CSI	CRA	COD	CIN	H
I	CEN	CGL	CPL	CED	CSA	CSP	CAN	CBA	CCH	CFR	CPU	CPA	CPU	CCA	CED	CGR	CTI	CCA	CIL	CSP	CAM	CUR	CED	CAL	CAU	I
J	COC	CSE	CCR	CDE	CMO	CPU	CSA	CEL	CUL	CED	CRO	CPE	CCI	CIP	CSI	CSE	CAL	CBU	CAL	CAM	CQU	CGA	CVI	CEX	CTI	J
K	CCI	CCO	CCO	CMA	CSP	CCA	CDE	CAR	CEN	CIN	COF	CBU	CCA	CCA	CTA	CVE	CCO	CLA	CAU	CAU	CGR	CLI	CME	CPA	CRE	K
L	CSI	CTA	CFI	CPO	CDE	CLA	CLA	CTE	CUV	CNU	CAR	CBE	CCA	CCO	CLI	CLA	CAC	CNI	CVE	CAU	CAM	CPA	CES	COF	CCA	L
M	COL	CTE	CSA	CVA	COL	CAV	CCH	CCO	CCO	CFE	CHE	CMA	CTI	CAB	CMA	CSA	CBR	CJU	CLA	CPA	CSP	CTI	CAN	CME	CSA	M
N	CFI	CMA	CMI	CMO	CPE	CCY	CLO	CZE	CTE	CPE	COB	CCI	CFL	CMA	CNA	CWI	CCA	CSC	CDA	CPL	CTR	CES	CRO	CBE	DFR	N
O	DGL	DAE	DCA	DSP	DEL	DMA	DIN	DAN	DAR	DRE	DAL	DGR	DPU	DDE	DEX	DIB	DAL	DBU	DCO	DCO	DFL	DMA	DOP	DRO	DCU	O
P	DEB	DKA	DVI	DOD	DOP	DUN	DHE	DMA	DWI	DZI	DCO	EEL	ECR	ECR	EPY	ECA	EPH	EGU	EDU	ECA	ECO	EIN	ECA	ECI	ECO	P
Q	EGL	EJU	EGE	ESI	EAR	ECH	ECU	ELE	ETE	ETR	EOP	EGL	EJA	EPO	EVE	EFO	ECO	ESP	EUT	EUN	EAN	ELO	FES	FTA	FGR	Q
R	FSY	FSE	FLI	FAR	FID	FLO	FOV	FPR	FRU	FRU	FTI	FCA	FEL	FVO	FGL	FIN	FVU	FSP	FAN	FCH	FVE	FVI	FAF	FEL	FFO	R
S	GDU	GLA	GLI	GMA	GMU	GTI	GPR	GBA	GTI	GLU	GUR	GJA	GMA	GGL	GLE	GAN	GAR	GBA	GHE	GHI	GAS	GAB	GTI	HCO	HAN	S
T	HAN	HTU	HZE	HAL	HBE	HBR	HCA	HSA	HBE	HJA	HMU	HBU	HDI	HVU	HCO	HLU	HAN	HNI	HHI	HRU	HSP	HOF	IAM	IPA	IVE	T
U	IAR	IHI	ISP	ITI	IED	IHE	IAQ	IBA	IER	JOF	JCH	JAI	JCI	JHI	JNI	JRE	JMA	JCO	KGA	KGE	KSC	LPU	LSA	LVI	LSI	U
V	LIB	LHE	LKI	LOW	LGA	LDO	LHI	LSA	LNO	LLA	LSP	LST	LIN	LMI	LOL	LCU	LSA	LVI	LLA	LCU	LST	LST	LFE	LGO	LLF	V
W	LOF	LRI	LAL	LBA	LDO	LUS	LMI	LCH	LIN	LMU	LPE	LNI	LUR	LUT	LCO	LPE	LTE	LAC	LCY	LAL	LAN	LLU	LCH	LES	LPI	W
X	MSP	MLO	MHO	MON	MPU	MBA	MSY	MAM	MOF	MIN	MDI	MES	MGL	MBI	MZA	MAR	MVU	MCH	MVI	MAR	MFA	MLU	MOR	MPO	MSA	X
Y	MSC	MCA	MUS	MBI	MAL	MIN	MOF	MSU	MMI	MOF	MAR	MGE	MPI	MPU	MRO	MSP	MCH	MMI	MBA	MCH	MDE	MPE	MCI	MOL	MAL	Y
Z	MNI	MRU	MDE	MKO	MAC	MEN	MPA	MTE	MCA	MFR	MBA	MPE	MOD	MCO	NOF	NNU	NCA	NLA	NMU	NTA	NSA	NCO	OBA	OKI	OSA	Z
a	OJA	OEU	OVI	OJA	OFI	OCO	OVU	OSA	OGL	OSA	OHY	OMI	OTU	PER	PTU	PGU	PSP	PAN	PCO	PHE	PMA	PMI	POB	PRE	PVI	a
b	PBR	PSO	PFI	PCE	PED	PAR	PDI	PNO	PSC	PUR	PED	PLA	PLI	PMO	PQU	PSA	PCU	PYO	PHA	PPT	PAM	PCL	PPU	PMA	PIN	b
c	PPE	PFR	PAM	PCR	PAQ	PAR	PCA	PCO	PST	PAC	PAT	PCO	PLU	PVU	PPR	PDA	PTE	PAU	PAC	PEM	PSP	PIX	PPE	PAM	PDI	c
d	PRA	PAN	PED	PQU	PBE	PLO	PME	PNI	PRE	PVE	PSA	PES	PAM	PAN	PAR	PBU	PFE	PPR	PSE	PTR	PCA	PTU	PRU	PTR	POL	d
e	PAF	PCA	PSA	PVI	PAR	PAV	PCE	PCE	PDO	PDO	PDU	PLA	PMA	PPE	PSA	PSU	PCA	PFR	PGU	PGU	PMO	PTE	PER	PSA	PSO	e
f	PLO	PPH	PGR	PCO	PPY	QSU	RSA	RSE	ROD	ROF	RPA	RRH	RVE	RHI	RNI	RRU	RSA	RUV	RCO	RCO	ROF	RSP	RSP	RID	ROC	f
g	ROC	RGR	SOF	SCH	SOF	SSC	SCA	SCA	SMI	STR	SSE	SHO	SMO	SBA	SLO	SMO	SSC	SOL	SCA	SCE	SED	SAN	SAL	SIN	SRA	g
h	SBI	SEX	SIT	SSP	SOD	SGL	SCH	SAL	SAR	SAE	SAV	SFE	SGI	SHY	SIN	SIN	SKH	SLA	SMA	SME	SMU	SNI	SQU	STO	STU	h
i	SRO	SAV	SAL	SBI	SHA	SSU	SJU	SST	SAC	SOL	SCY	SMO	SPU	STU	SAI	SCR	SWR	SSI	SSE	SRE	SCO	STE	SVI	SGR	SHI	i
j	SKO	SSA	SNU	SAD	SER	SGU	SHU	SBE	STO	SCO	SPE	SDU	SAR	SCU	SJA	SMA	SSA	TIN	TVU	TKU	TKO	TOF	TOC	TPE	TCA	j
k	TVO	TCA	TTE	TPU	TDA	TBI	TCA	TGR	TSE	TVU	TAM	TPO	TBI	TNA	TAF	TAN	TCU	TAL	TAM	TFR	THI	THY	TIN	TNI	TPR	k
l	TRE	TRE	TSU	TVA	TVE	TWI	TSP	TFO	TTR	XTR	TAE	TCA	TCO	TDI	TDU	TMO	TPO	TSP	TTI	TTU	TTU	TCO	TRH	TTO	TTU	l
m	TFA	UTU	UGA	ULO	UDI	VAN	VAS	VCO	VMA	VVI	VLO	VMA	VPL	VCA	VAN	VZI	VAN	VAR	VBE	VDA	VER	VFA	VMO	VPA	VSA	m
n	VVI	VAC	VAN	VMU	VRA	VUM	VUN	VUN	VUN	VVE	VSE	VLA	VRO	VVI	VSU	XSA	YEL	ZMA	ZMA	ZOF	ZAQ	ZJU	ZMA	ZJA	ZMA	n

ALTERNATIVE CROP MATRIX

Table 16. Vermifuges

	1	2	3	4	5	6	7	8	9	10	11	12	13	14	15	16	17	18	19	20	21	22	23	24	25	
.A.	AES	AMO	ATH	AFA	AME	ANI	APY	ASE	ASE	ASA	ANA	ACA	AMA	ACH	ADI	AVA	ACY	AMA	AME	ASC	AFO	ALE	ASI	ACR	ADA	.A.
.B.	ADE	AEL	AIN	ARE	ARI	ASI	ASM	ASP	ATR	ACA	AGI	AST	ATE	AFO	AMO	AMO	APS	AAM	ACE	ACH	AFI	ASA	ASC	ATU	AMA	.B.
.C.	ABA	AFE	APE	AAR	APR	AOF	AVA	ACA	ACR	AHY	ALI	ATR	AVI	AAR	ABR	AOC	ACO	AGA	AGE	AHA	AGR	AAR	ACH	ADI	AMO	.C.
.D.	AMU	ARE	ASQ	AMO	ACE	AVU	ABU	AGR	AGR	AHY	ACO	ALA	ACA	API	AES	ARU	AMO	AXA	AEL	AAB	AAB	ACI	ADR	AMA	AVU	.D.
.E.	AAL	ACH	AIN	ASP	ADO	AGR	ATR	ADE	AOF	AOD	AGU	ACA	AHO	ABE	AAB	ABY	AFA	ASA	AST	AST	ABI	ACA	AAF	ACO	BMO	.E.
.F.	BRA	BRA	BGA	BFR	BAE	BSP	BNI	BVU	BBE	BAL	BES	BSY	BHI	BEX	EVU	BOR	BSA	BNI	BCE	BOF	BFL	BCA	BCA	BIN	BIN	.F.
.G.	BIS	BPE	BSA	BCU	BER	BGR	BBR	BDE	BEM	BMU	BRA	BCA	BCH	BJU	BNA	BNA	BNI	BOL	BOL	BOL	BOL	BOL	BOL	BPE	BRA	.G.
.H.	BHU	BCA	BIN	BMA	BUN	BAL	BUT	BLA	BDA	BSA	BPA	CCO	CES	CSA	CSP	CCA	CRO	CAL	COF	CMU	CPA	CSI	CRA	COD	CIN	.H.
.I.	CEN	CGL	CPL	CED	CSA	CSP	CAN	CBA	CCH	CFR	CPU	CPA	CPU	CCA	CED	CGR	CTI	CCA	CIL	CSP	CAM	CUR	CED	CAL	CAU	.I.
.J.	CCC	CSE	CCR	CDE	CMO	CPU	CSA	CEL	CUL	CED	CRO	CPE	CCI	CIP	CSI	CSE	CAL	CBU	CAL	CAM	CQU	CGA	CVI	CEX	CTI	.J.
.K.	CCI	CCO	CCO	CMA	CSP	CCA	CDE	CAR	CEN	CIN	COF	CBU	CCA	CCA	CTA	CVE	CCO	CLA	CAU	CAU	CGR	CLI	CME	CPA	CRE	.K.
.L.	CSI	CTA	CFI	CPO	CDE	CLA	CLA	CTE	CUV	CNU	CAR	CBE	CCA	CCO	CLI	CLA	CAC	CNI	CVE	CAU	CAM	CPA	CES	COF	CCA	.L.
.M.	COL	CTE	CSA	CVA	COL	CAV	CCH	CCO	CCO	CFE	CHE	CMA	CTI	CAB	CMA	CSA	CBR	CJU	CLA	CPA	CSP	CTI	CAN	CME	CSA	.M.
.N.	CFI	CMA	CMI	CMO	CPE	CCY	CLO	CZE	CTE	CPE	COB	CCI	CFL	CMA	CNA	CWI	CCA	CSC	CDA	CPL	CTR	CES	CRO	CBE	DFR	.N.
.O.	DGL	DAE	DCA	DSP	DEL	DMA	DIN	DAN	DAR	DRE	DAL	DGR	DPU	DDE	DEX	DIB	DAL	DBU	DCO	DCO	DFL	DMA	DOP	DRO	DCU	.O.
.P.	DEB	DKA	DVI	DOD	DOP	DUN	DHE	DMA	DWI	DZI	DCO	EEL	ECR	ECR	EPY	ECA	EPH	EGU	EDU	ECA	ECO	EIN	ECA	ECI	ECO	.P.
.Q.	EGL	EJU	EGE	ESI	EAR	ECH	ECU	ELE	ETE	ETR	EOP	EGL	EJA	EPO	EVE	EFO	ECO	ESP	EUT	EUN	EAN	ELO	FES	FTA	FGR	.Q.
.R.	FSY	FSE	FLI	FAR	FID	FLO	FOV	FPR	FRU	FRU	FTI	FCA	FEL	FVO	FGL	FIN	FVU	FSP	FAN	FCH	FVE	FVI	FAF	FEL	FFO	.R.
.S.	GDU	GLA	GLI	GMA	GMU	GTI	GPR	GBA	GTI	GLU	GUR	GJA	GMA	GGL	GLE	GAN	GAR	GBA	GHE	GHI	GAS	GAB	GTI	HCO	HAN	.S.
.T.	HAN	HTU	HZE	HAL	HBE	HBR	HCA	HSA	HBE	HJA	HMU	HBU	HDI	HVU	HCO	HLU	HAN	HNI	HHI	HRU	HSP	HOF	IAM	IPA	IVE	.T.
.U.	IAR	IHI	ISP	ITI	IED	IHE	IAQ	IBA	IER	JOF	JCH	JAI	JCI	JHI	JNI	JRE	JMA	JCO	KGA	KGE	KSC	LPU	LSA	LVI	LSI	.U.
.V.	LIB	LHE	LKI	LOW	LGA	LDO	LHI	LSA	LNO	LLA	LSP	LST	LIN	LMI	LOL	LCU	LSA	LVI	LLA	LCU	LST	LST	LFE	LGO	LLE	.V.
.W.	LOF	LRI	LAL	LBA	LDO	LUS	LMI	LCH	LIN	LMU	LPE	LNI	LUR	LUT	LCO	LPE	LTE	LAC	LCY	LAL	LAN	LLU	LCH	LES	LPI	.W.
.X.	MSP	MLO	MHO	MON	MPU	MBA	MSY	MAM	MOF	MIN	MDI	MES	MGL	MBI	MZA	MAR	MVU	MCH	MVI	MAR	MFA	MLU	MOR	MPO	MSA	.X.
.Y.	MSC	MCA	MUS	MBI	MAL	MIN	MOF	MSU	MMI	MOF	MAR	MGE	MPI	MPU	MRO	MSP	MCH	MMI	MBA	MCH	MDE	MPE	MCI	MOL	MAL	.Y.
.Z.	MNI	MRU	MDE	MKO	MAC	MEN	MPA	MTE	MCA	MFR	MBA	MPE	MOD	MCO	NOF	NNU	NCA	NLA	NMU	NTA	NSA	NCO	OBA	OKI	OSA	.Z.
.a.	OJA	OEU	OVI	OJA	OFI	OCO	OVU	OSA	OGL	OSA	OHY	OMI	OTU	PER	PTU	PGU	PSP	PAN	PCO	PHE	PMA	PMI	POB	PRE	PVI	.a.
.b.	PBR	PSO	PFI	PCE	PED	PAR	PDI	PNO	PSC	PUR	PED	PLA	PLI	PMO	PQU	PSA	PCU	PYO	PHA	PPT	PAM	PCL	PPU	PMA	PIN	.b.
.c.	PPE	PFR	PAM	PCR	PAQ	PAR	PCA	PCO	PST	PAC	PAT	PCO	PLU	PVU	PPR	PDA	PTE	PAU	PAC	PEM	PSP	PIX	PPE	PAM	PDI	.c.
.d.	PRA	PAN	PED	PQU	PBE	PLO	PME	PNI	PRE	PVE	PSA	PES	PAM	PAN	PAR	PBU	PFE	PPR	PSE	PTR	PCA	PTU	PRU	PTR	POL	.d.
.e.	PAF	PCA	PSA	PVI	PAR	PAV	PCE	PCE	PDO	PDO	PDU	PLA	PMA	PPE	PSA	PSU	PCA	PFR	PGU	PGU	PMO	PTE	PER	PSA	PSO	.e.
.f.	PLO	PPH	PGR	PCO	PPY	QSU	RSA	RSE	ROD	ROF	RPA	RRH	RVE	RHI	RNI	RRU	RSA	RUV	RCO	RCO	ROF	RSP	RSP	RID	ROC	.f.
.g.	ROC	RGR	SOF	SCH	SOF	SSC	SCA	SCA	SMI	STR	SSE	SHO	SMO	SBA	SLO	SMO	SSC	SOL	SCA	SCE	SED	SAN	SAL	SIN	SRA	.g.
.h.	SBI	SEX	SIT	SSP	SOD	SGL	SCH	SAL	SAR	SAE	SAV	SFE	SGI	SHY	SIN	SIN	SKH	SLA	SMA	SME	SMU	SNI	SQU	STO	STU	.h.
.i.	SRO	SAV	SAL	SBI	SHA	SSU	SJU	SST	SAC	SOL	SCY	SMO	SPU	STU	SAI	SCR	SWR	SSI	SSE	SRE	SCO	STE	SVI	SGR	SHI	.i.
.j.	SKO	SSA	SNU	SAD	SER	SGU	SHU	SBE	STO	SCO	SPE	SDU	SAR	SCU	SJA	SMA	SSA	TIN	TVU	TKU	TKO	TOF	TOC	TPE	TCA	.j.
.k.	TVO	TCA	TTE	TPU	TDA	TBI	TCA	TGR	TSE	TVU	TAM	TPO	TBI	TNA	TAF	TAN	TCU	TAL	TAM	TFR	THI	THY	TIN	TNI	TPR	.k.
.l.	TRE	TRE	TSU	TVA	TVE	TWI	TSP	TFO	TTR	XTR	TAE	TCA	TCO	TDI	TDU	TMO	TPO	TSP	TTI	TTU	TTU	TCO	TRH	TTO	TTU	.l.
.m.	TFA	UTU	UGA	ULO	UDI	VAN	VAS	VCO	VMA	VVI	VLO	VMA	VPL	VCA	VAN	VZI	VAN	VAR	VBE	VDA	VER	VFA	VMO	VPA	VSA	.m.
.n.	VVI	VAC	VAN	VMU	VRA	VUM	VUN	VUN	VUN	VVE	VSE	VLA	VRO	VVI	VSU	XSA	YEL	ZMA	ZMA	ZOF	ZAQ	ZJU	ZMA	ZJA	ZMA	.n.

ALTERNATIVE CROP MATRIX

Chapter Five

CONTACT ALLERGY FROM PLANTS

JOHN C. MITCHELL

Division of Dermatology
University of British Columbia
Vancouver, British Columbia

ADVERSE REACTIONS TO PLANTS

The causes of adverse human skin reactions to plants can be summarized as follows:-

1. Mechanical injury, e.g. from a spine.

2. Toxicological effects, e.g. alkaloidal poisoning from wood dust of *Gonioma* (Apocynaceae).

3. Pharmacological effects, e.g. histamine, acetylcholine liberation by stinging hair of nettle, *Urtica* (Urticaceae).

4. Contact dermatitis by irritancy; chemical, e.g. from *Euphorbia* (Euphorbiaceae), or mechanical and chemical, e.g. granuloma from *Opuntia* (Cactaceae).

5. Contact dermatitis by immunological effects; immediate hypersensitivity (endogenous or contact urticaria), e.g. from *Agave* (Agavaceae), or delayed hypersensitivity e.g. from *Toxicodendron* (Anacardiaceae).

Epiphytodermatitis: lichens and liverworts, e.g. *Frullania* (Jubilaceae), can cause dermatitis which has incorrectly been attributed to trees such as oak, *Quercus*, (Fagaceae) and 'cedar', *Thuja* (Cupressaceae), on which these epiphytic plants reside.

6. Phytophotodermatitis: evoked by linear furanocoumarins (psoralens) or some plants (Umbelliferae, Leguminosae, Rutaceae, etc.) and exposure of the skin to long-wave ultra-violet light (350 nm).

7. Pseudophytophotodermatitis: the skin lesions of phytophotodermatitis evoked by psoralen-containing plants can be mimicked by lesions produced by plant irritants with an added effect from sunburn (290-320 nm).

8. Pseudophytodermatitis: this term has been used for skin reactions such as those produced by the mite, *Carpoglyphus* on figs, *Ficus* (Moraceae), by *Schistoma* infestation in growers of rice, *Oryza* (Gramineae) and by contact with agricultural fungicides, pesticides, etc.

9. Parasitophytodermatitis, e.g. dermatitis produced by fungi parasitic on *Arundo* (Gramineae).

10. Parasitophytophotodermatitis, e.g. the effects of the fungus, *Sclerotinia*, parasitic on celery *Apium* (Umbelliferae), which, in the parasitised state, yields psoralens which can evoke photodermatitis.

11. Heterophytodermatitis: this term might be suitable for the results of presently ill-defined biological phenomena such as the skin effects of chemical compounds, derived from plants, which become incorporated into other plants and render the recipient plants injurious to the skin.

12. Plant-animal dermatitis: interactions of insects and plants can occur by which chemical compounds of plants, e.g. cucurbitacins and glucosinolates can act as insect 'hormones', and by which chemical compounds of insects influence the chemical constitution of plants. Such aspects of chemical ecology require investigation in respect to

possible effects on humans.

Contact Dermatitis

Two forms of contact dermatitis are recognized. Irritant contact dermatitis, from exposure of the skin to plants such as *Euphorbia* spp. and *Ranunculus* spp., affects all individuals if the amount and duration of the exposure is sufficient. Allergic contact dermatitis, from exposure of the skin to *Toxicodendron* spp., *Primula obconica*, etc., affects some individuals who acquire a specific altered capacity to react to the plants. Such a capacity is not present at birth but is acquired by postnatal exposure, is specific for the chemical compounds responsible, such as pentadecylcatechol (1) (*Toxicodendron*, Anacardiaceae), or the quinone, primin (2) (*Primula*, Primulaceae), depends

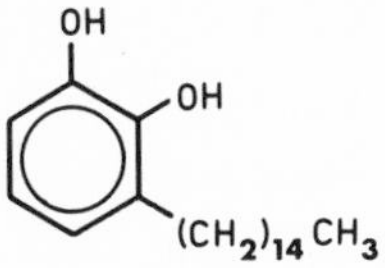

1

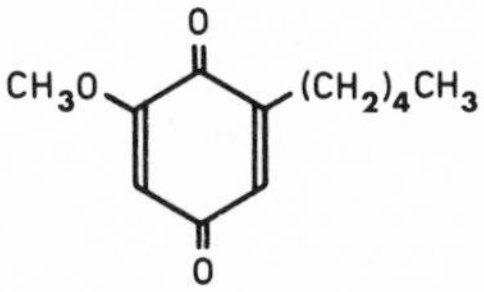

2

upon an altered immunological status of the individual, is mediated by sensitized lymphocytes rather than by antibody and is therefore designated 'cell mediated hypersensitivity', and is presented clinically as a skin reaction manifested by redness, vesiculation (small blisters within the epidermis) and itching. The pathogenetic mechanisms of allergic contact dermatitis are summarized in Figure 1.

Allergic Contact Dermatitis

Following initial contact with a chemical compound which is capable of producing 'cell-mediated hypersensitivity', a process of sensitization occurs in some individuals and, during the course of a minimum of six days, lymphocytes located in lymph glands become sensitized and are released into the blood stream. Following subsequent exposure to the specific sensitizer, circulating lymphocytes accumulate at the site and dermatitis becomes manifest one to two days later at the skin site of re-exposure.

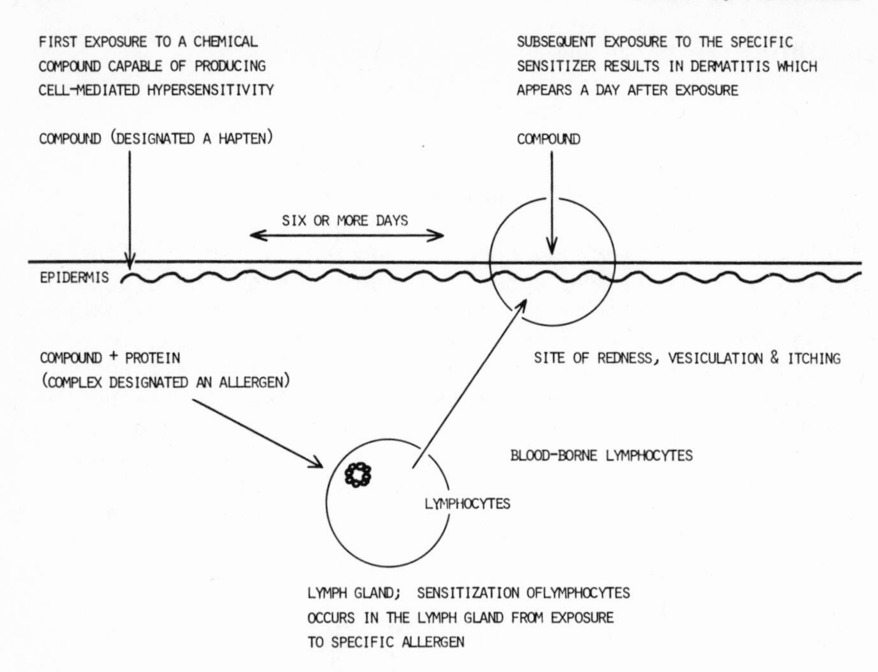

Figure 1. Pathogenetic mechanisms of allergic contact dermatitis.

Such delay in the appearance of dermatitis at the site of re-exposure is indicated by an alternative term for the reaction, namely 'delayed hypersensitivity'.

Delayed hypersensitivity may be detected by means of a 'patch test': a chemical compound is applied to the skin, by necessity in a non-irritant dose, and the skin site is examined 48 hours later; a positive patch test reaction, provided nonspecific irritancy is excluded, indicates a state of delayed hypersensitivity and is a deliberate elicitation, under controlled conditions, of allergic contact dermatitis as part of the clinical investigation of affected patients.

Atopy

The genetically determined state known as atopy, may

become manifest as hay-fever, allergic asthma and atopic dermatitis. 'Immediate hypersensitivity' is a feature of atopy and may be detected by a 'scratch' test of the skin. A wheal reaction appears within a few minutes of re-exposure to some allergens and is mediated by circulating antibody (immunoglobulin).

Immunochemistry of Contact Dermatitis

Implicit in current theory of the pathogenesis of allergic contact dermatitis is the notion that small molecular weight chemical compounds, designated haptens, become bound to skin protein to form higher molecular weight complexes, designated allergens or antigens, which induce sensitization of lymphocytes. The means and the site of attachment of compounds derived from plants to skin proteins are thus of interest to dermatologists and the immunochemical requisites of some compounds have been clarified by phytochemical studies. Two such examples are as follows: The main functional group of sesquiterpene lactones, e.g. alantolactone (3), derived from Composite species such as *Inula* is an α-methylene group attached to the γ-lactone; reduction of this group renders the compounds inactive for contact allergy[19]. *d*-Usnic acid (4), from lichens, is a sensitizer whereas *l*-usnic acid is

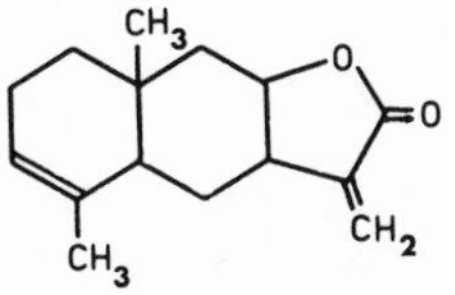

CH_3 O O CH_2 CH_3

3

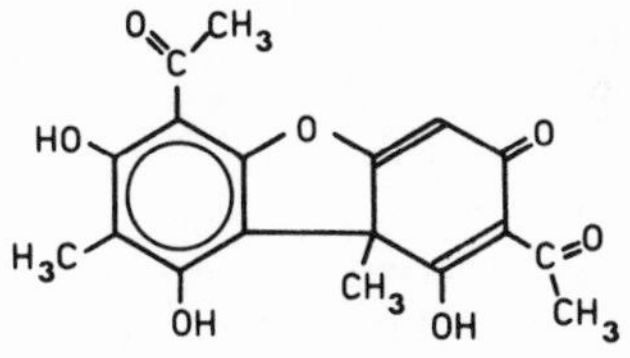

O CH_3 C HO O O H_3C C=O OH CH_3 OH CH_3

4

inactive[11,14]; the specificity can thus extend to stereochemical configurations and exemplifies the remarkable capacity of sensitized lymphocytes to distinguish between related compounds. Relatively few plant sensitizers have been identified, and include α-methylene-γ-butyrolactone (5, *Tulipa*, Liliaceae), δ-3-carene (6, *Pinus*, Pinaceae), allylisothiocyanate (7, Cruciferae spp.), 4-methoxydalbergione (8, *Dalbergia* spp., Leguminosae), oxyayanin (9, *Distemonanthus*, Leguminosae), anthothecol (10, *Khaya*,

Meliaceae), lapachol (11, *Tectona*, Verbenaceae), chlorophorin (12, *Chlorophora*, Moraceae), and thymoquinone (13, *Libocedrus*, Cupressaceae).

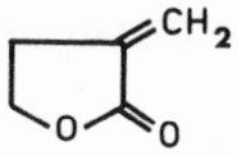

5

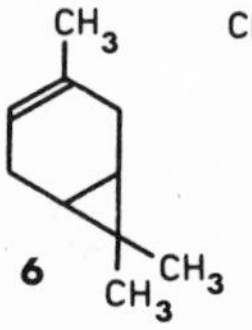

6

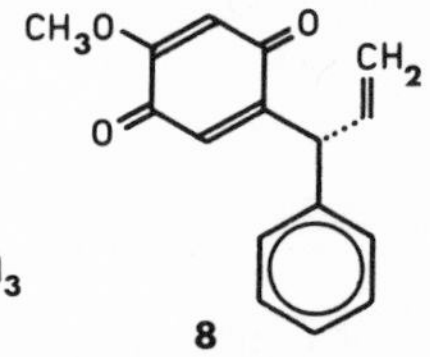

$CH_2=CH-CH_2-N=C=S$

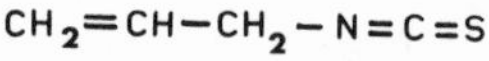

7

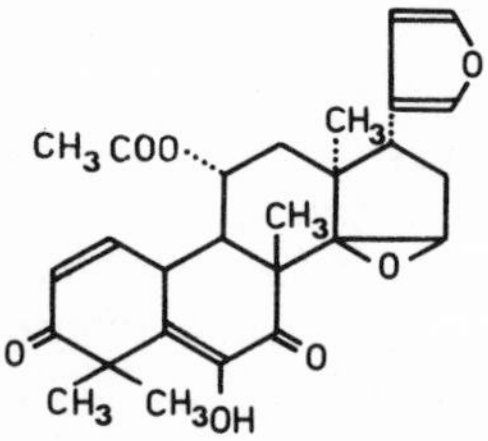

8

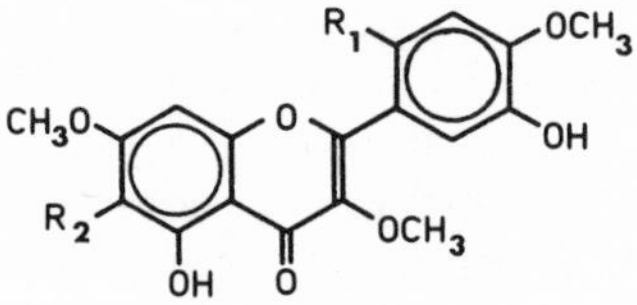

9 A R_1= OH; R_2= H
B R_1= H; R_2= OH

10

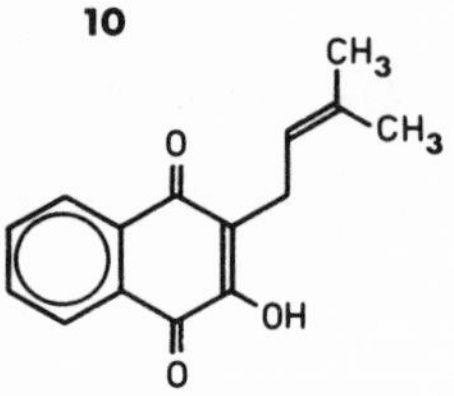

11

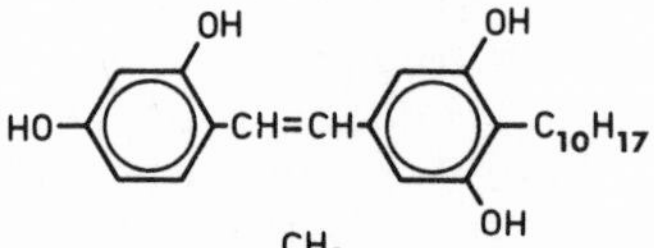

12

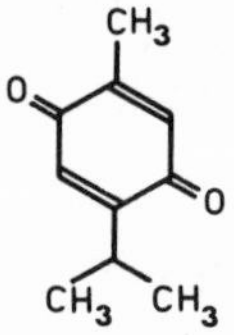

13

Some lactones have cytotoxic activity and it is of interest that the principal group responsible for tumoricidal effects is the methylene group attached to the lactone[9], which is the principal group responsible for allergenic effects[13-15].

In order to study the reactivity of a sesquiterpene lactone with some amino acids, we studied the capacity of alantolactone (3), derived from *Inula helenium*, to form adducts by Michael-type addition[4]. Alantolactone is capable of undergoing such addition with the sulfhydryl group of cysteine, with the imidazole group of histidine and with the ε-amino group of lysine but not with the guanido group of arginine, the hydroxyl group of serine or the thio-ether function of methionine. These amino acids were selected for study because of their nucleophilic side chain. Studies of reactivity with more complex protein models are in progress in the hope of finding out more information about the presumed carrier-protein of delayed hypersensitivity by this 'key and lock' approach. A study of the same problem from another direction involves a study of the localization of labeled compounds in subcellular fractions of skin cells and the capacity of the various fractions to elicit delayed hypersensitivity responses of specifically sensitized lymphocytes in vitro.

Cross-Sensitivity in Allergic Contact Dermatitis

An individual may become specifically sensitized to unrelated materials such as turpentine and nickel and is then said to exhibit 'multiple specific sensitivity'. Cross-sensitivity patterns are also observed; thus an individual may become sensitized to alantolactone and subsequently show positive patch test reactions to numerous related sesquiterpene lactones derived from various plants of the Composite family. When there is no evidence of prior exposure to such plants, 'cross-sensitivity' is implied.

STUDY OF THE BIOLOGY AND ECOLOGY OF A DERMATOLOGICAL PROBLEM USING PHYTOCHEMICAL INFORMATION. ALLERGIC CONTACT DERMATITIS FROM COMPOSITAE SPECIES AND THEIR SESQUITERPENE LACTONES

At the botanical level, apparently anomalous cross-sensitivity patterns have been observed. Thus our interest in plants of the family Compositae was aroused by the finding that one of some fifty forest-workers in British Columbia who had allergic contact dermatitis from the liverwort *Frullania* (Jubilaceae)[12-19] was also contact-sensitive to *Chrysanthemum* x *morifolium* (Compositae).[19] The patient actually found this out himself when he was obliged to retire from forest-work because of severe recurrent occupational dermatitis and took up growing chrysanthemums as a retirement hobby. Cross-sensitivity between a bryophyte and a botanically unrelated Composite, which is considered to be higher on the evolutionary scale, was found to be explicable in phytochemical terms, viz. the presence of frullanolide, a sesquiterpene lactone in *Frullania*[8-20] and related lactones in *Chrysanthemum* spp.[3,17,18]

The skeletal structure of five classes of sesquiterpene lactones are shown below (14, eremophilanolide; 15, germacranolide; 16, eudesmanolide=santanolide; 17, guaianolide; 18, pseudoguaianolide=ambrosanolide). Cross-sensitivity

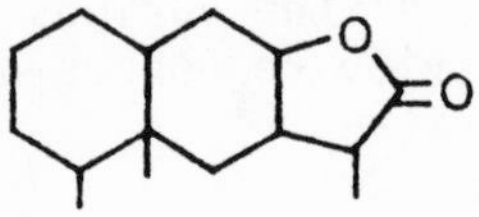

14

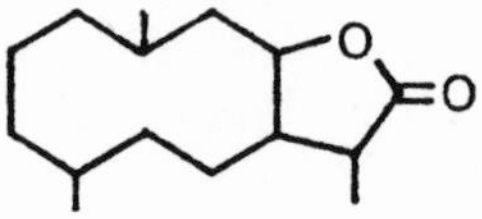

15

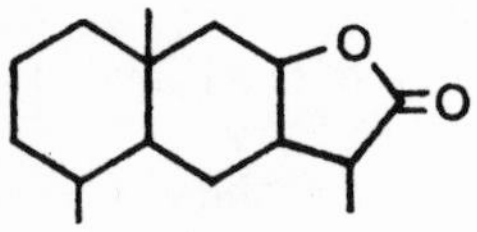

16

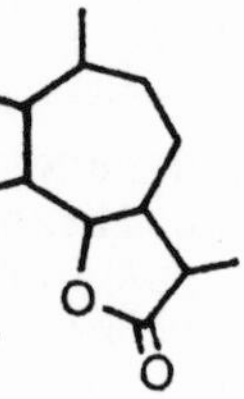

17

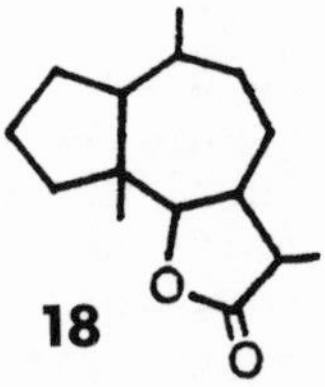

18

between members of more than one class can occur.[16] Plants of the Composite family which have been reported to cause allergic contact dermatitis are listed in Table 1.[10] Sesquiterpene lactones have been demonstrated to be the allergens of some of the plants, e.g. *Ambrosia, Artemisia.* [16-18] Some of the plants listed have not yet been examined for lactones.

Cross-sensitivity patterns vary in various sensitized individuals and, in general, highly sensitized individuals show a wider spectrum of cross-sensitivity than weakly sensitized individuals. One such study is summarized in Table 2; Six patients who had allergic contact dermatitis from the liverwort, *Frullania*, showed positive patch test reactions to acetone extracts of some plants of the Compositae. Five of the six patients showed positive reactions to alantolactone. *Frullania* and most of the Composite species which were tested have been reported to yield sesquiterpene lactones. Patients 1 - 4, who were further tested, showed positive patch test reactions to other sesquiterpene lactones. This study and other reports

Table 1. Plants of the family Compositae which have been reported to cause allergic contact dermatitis.[10]

Achillea millefolium, *Ambrosia* spp., *Anthemis*, *A. nobilis*, *Arctotheca calendulaceum (Cryptostemma)*, *Arnica montana*, *Artemisia* spp., *Aster multiflorus*, *Buphthalmum salicifolium*, *Cassinea aculeata*, *Centaurea americana*, *Chrysantheum* spp., *Cichorium endivia*, *C. intybus*, *Cosmos bipinnatus*, *Cynara scolymus*, *C. carduncalus*, *Dahlia* sp., *Erigeron canadensis (Leptilon)*, *Eupatorium serotinum*, *Gaillardia* sp., *Galinsoja parviflora*, *Haplopappus ciliata (Prionopsis)*, *Helianthus annuus*, *Helenium autumnale*, *H. microcephalum*, *H. tenuifolium*, *Helipterum charsleyae*, *Heterotheca subaxillaris (Chrysopsis)*, *Hieracium auricula*, *Humea elegans*, *Inula brittanica*, *I. graveolens*, *Iva angustifolia*, *I. microcephala*, *I. xanthifolia*, *Lactuca sativa*, *L. sativa* var. *longifolia*, *Matricaria chamomilla*, *Olearia* spp., *Oxytenia acerosa*, *Parthenium argentatum*, *P. hysterophorus*, *Pyrethrum* spp. *(Chrysanthemum)*, *Rudbeckia hirta*, *Rutidosis helichrysoides*, *Saussurea lappa*, *Solidago serotina*, *S. virga aurea* (sic), *Tagetes* spp., *Tanacetum vulgare*, *Taraxacum officinale*, *Telekia* sp. *(Buphthalmum)*, *Vernonia baldwinii*, *Xanthium californicum*, *X. canadense*, *X. chinense*, *X. spinosum*, *X. strumarium*.

suggest that these lactones were responsible for the dermatitis in these patients.

Clinical Aspects of Allergic Contact Dermatitis from Compositae species

Clinical manifestations of contact allergy to Compositae plants arise from various sources of exposure-risk,

Table 2. Patch test reactions to acetone extracts of some plants of the family Compositae in six patients who had allergic contact dermatitis from *Frullania* (Jubilaceae). Patients 1 to 5 showed positive reactions to the sesquiterpene lactone, alantolactone.

Tribe	Species	Patch test results - 6 patients 1	2	3	4	5	6
A	*Chrysanthemem cinerariifolium* (Yields pyrethrum)	++	-	-	-	-	NT
A	*Tanacetum vulgare* (Tansy)	++	++	+++	+	+++	+
H	*Gaillardia pulchella* (Gaillardia)	+	+	+	+	-	-
H	*Franseria acanthicarpa* (False ragweed)	+	++	++	-	-	?
H	*Ambrosia eliator* (Short ragweed)	++	++	+	-	-	-
H	*Helenium autumnale* (Sneezeweed)	++	++	-	?	-	?
A	*Chrysanthemum* x *morifolium* Florists "Chrysanthemum"	++	++	+	-	-	-
H	*Parthenium hysterophorus* (Wild feverfew)	-	+++	+	-	-	NT
A	*Chrysanthemum parthenium* (Feverfew)	++	++	++	++	++	++
A	*Artemisia absinthium* (Wormwood)	-	+	+	-	-	-
A	*Chrysanthemum maximum* (Shasta daisy)	-	-	-	-	-	NT
A	*Achillea millefolium* (Yarrow)	++	++	-	NT	-	?
H	*Iva angustifolia* *I. ciliata* (mixed) (Marshelder)	++	+	++	-	+	-
A	*Anthemis cotula* (Dog fennel)	++	+	+	NT	-	+
H	*Rudbeckia hirta* (Black-eyed Susan)	-	-	-	NT	-	NT
H	*Xanthium strumarium* (Cocklebur)	++	++	+	-	-	?
	Frullania (Liverwort)	++	+	+	++	++	+
	from *Inula helenium* Alantolactone	++	++	+	++	+	-

Tribe A = Anthemidae H = Heliantheae

+ Positive; - Negative; NT Not Tested.

which may be arbitrarily divided as follows:-

1. Farming. 'Weed-dermatitis' can affect the exposed skin surfaces of farmers. In the North-American mid-west such dermatitis is caused by ragweed (*Ambrosia*) and, in India, by the adventive *Parthenium hysterophorus*, and arises from contact with 'oleoresin' on the surface of the leaves. Air-borne fragments of dried plants represent a source of exposure but the role of pollen (unless pollen carries extraneous oleo-resin) is probably minimal. It might be noted, in this context, that hay-fever from ragweed (*Ambrosia*) results from inhalation of the proteinaceous-fraction of the pollen and from immediate hypersensitivity reactions in the mucous membranes of the nasal passages. Both the chemistry and the immunology are distinctive for these two types of hypersensitivity to ragweed - by contact with 'oleoresin' containing lactones and by inhalation of pollen containing protein. A contrast will be noted between ragweed *Ambrosia*, in which oil passages open to the leaf surface and poison ivy-oak (*Toxicodendron*) in which it is necessary to crush the leaf, and *Primula obconica* in which it is necessary to break the fine hairs on the undersurface of the leaves to release the allergenic material. The clinical entity known as 'weed-dermatitis' is far more common in males than in females; this difference cannot be related entirely to the degree of exposure to plants and is unexplained.

2. Grain Handling. Wheat (*Triticum*) etc. may be contaminated with the oleoresin of ragweed (*Ambrosia*) and such contaminated grains may cause dermatitis in grain-handlers. The distribution of the dermatitis is on the exposed skin surfaces.

3. Horticultural. *Chrysanthemum* x *morifolium* is the commonest reported cause of dermatitis from horticultural plants of this family. The dermatitis tends to be seasonal in gardeners and perennial in florists owing to the practice of bringing the plants to maturity at any time of the year. Mature plants are more troublesome than young plants reflecting an increasing yield of secondary products of metabolism with increasing age and maturity of the plant. Cultivar-specificity is known to occur and is important in rehabilitation of affected patients in the trade who may be able to grow cultivars to which they are not reactive. The dermatitis affects the hands, wrists and the face as a

result of touching the face with the hands.

4. Salad-making. Chicory (*Cichorium*), endive (*Endivia*) and lettuce (*Lactuca*) can cause occupational dermatitis in salad-makers. It seems that prolonged trade exposure rather than occasional exposure such as occurs in housewives is necessary to produce sensitization or to evoke dermatitis in sensitized individuals. It is possible that this situation indicates that the lactones of these species are relatively weak sensitizers but, more likely, immature plants used for salads yield a relatively small amount of the lactones.

5. Perfumery. Costus root oil (*Saussurea lappa*) is a strong sensitizer. Some ten metric tons of the oil are marketed per year but the sources of exposure-risk to the oil in perfumes have not yet been clarified. The arts of aroma (perfume-flavoring) chemistry have been veiled in secrecy for decades if not for centuries. At least one other perfume material, Oil of Champaca (*Michelia champaca*, Magnoliaceae) yields sensitizing sesquiterpene lactones.

6. Medicaments. A major part of dermatological practice is the treatment of 'over-treatment dermatitis': both patient and physician aggravate skin diseases by the application of irritant and sensitizing topical medications. Skin which is the site of dermatitis is more likely to become sensitized to such applications than is intact skin. Amongst a host of sensitizing medicaments are chamomile (*Matricaria chamomilla, Anthemis nobilis*, Compositae) and laurel oil (*Laurus nobilis*, Lauraceae). These plants yield sesquiterpene lactones.

A Biological Model for Chemotaxonomy

Of chemotaxonomic interest is an attempt to set up a human biological test model in order to detect the presence of allergenic sesquiterpene lactones in plants and natural plant products. Thus, a panel of five patients who were contact-sensitive to Composite plants were tested for their sensitivity level to alantolactone (Table 3); the patients were then patch-tested to materials which might be expected, from phytochemical data, to yield lactones. Positive results were obtained with laurel oil (*Laurus nobilis*, Lauraceae), Champaca oil (*Michelia champaca*, Magnoliaceae), the leaves

Table 3. Sources of initial sensitization, sensitivity level to alantolactone, and results of patch testing for five individuals.

Case Number	Age Years	Sex	Source of Initial Sensitization	Sensitivity Level to Alantolactone	Patch test materials:*								
					Magnolia	*Liriodendron*	Costus Root Oil	Oil of Champaca	*Michelia*	Laurel Oil (N.Y.)	Laurel Oil (Danish)	Laurel Oil (German)	German Chamomile
1	48	M	*Frullania*	2 y	+	++	++	-	-	++	++	+++	++
2	64	M	*Frullania*	0.5 y	++	+	++	+	-	++	++	++	++
3	75	M	*Chrysanthemum* x *morifolium*	0.016 y	++	++	++	+	-	++	++	++	++
4	55	M	*Frullania*	0.08 y	++	±	++	++	+	++	++	+++	++
5	78	M	Costus Root Oil (*Saussurea*)	0.1% - 1.0% + petrolatum	+	++	++	+	-	-	+	++	-
Controls (5)					-	-	-	-	-	-	-	-	-

Magnolia grandiflora, leaf; *Liriodendron tulipifera*, leaf; Costus Root Oil, 1% petrolatum; Oil of Champaca, 5% petrolatum; *Michelia champaca*, leaf; Laurel Oil (N.Y.), 2% petrolatum; Laurel Oil (Danish), 5% petrolatum; Laurel Oil (German), 5% petrolatum; German Chamomile, 10% w/v ethanol.

of *Magnolia grandiflora* and of *Liriodendron tulipera* (Magnoliaceae), costus root oil (*Saussurea lappa*, Compositae) and chamomile (*Matricaria chamomilla*, Compositae) suggesting that these materials may contain lactones. These materials are used in perfumery, with the exception of *Liriodendron*, American Whitewood, which has been reported to cause dermatitis in wood-workers. It is necessary to patch test with non-irritant concentrations of plant extracts; such manipulation may of course dilute out relevant compounds to a sub-threshold level at which the human test system does not detect their presence. Perhaps, with increasing usage, the dermatologist may have something to offer the chemotaxonomist. Up to the present time, such collaboration has usually been in the opposite direction with phytochemists generously donating their knowledge and compounds for human investigation.

Taxonomy and Chemotaxonomy in the Investigation of Dermatitis from Plants

For a brief review of the effects of some plants and plant products on the skin, only families beginning with the letter A in Willis' Dictionary[22] will be considered and plants of these families which have been reported injurious to the skin will be listed and briefly annotated. Only the genus name will be recorded although more than one species of a genus may have been implicated. Many of the reports on which this review is based do not clearly distinguish between irritant and allergic effects. Nevertheless, some pattern emerges of the type of skin reactions produced by the plants in relation to their botanical classification and the chemical compounds which they yield. The family Agavaceae includes irritant plants of the genera: *Agave*, *Cordyline*, *Furcraea*, *Phormium*, *Polianthes*, *Sansevieria* and *Yucca*. Sharp crystals of calcium oxalate lying in bundles (raphides) in the plant tissues account for some of the irritant effects, together with unidentified constituents and possibly saponins. Of the family Aïzoaceae, *Galenia* is reported as irritant and *Mesembryanthemum* is probably allergenic. In the Alismataceae, *Alisma* and *Sagittaria* are reported as irritant. Of the family Alliaceae, *Agapanthus* is reported as irritant and *Allium* is both irritant and allergenic; the allergen of garlic is a water-soluble, heat-labile component. Of the family Alstroemeriaceae, *Alstroemeria* is reportedly allergenic and cross-sensitized with *Tulipa* (Liliaceae); α-methylene-γ butyrolactone (5), which is the allergen of tulip is also reported from *Alstroemeria* and is probably responsible for the cross-sensitivity between the plants. In the Altingiaceae, *Liquidambar* yields a fragrant balsam, storax or styrax, which is allergenic and cross-sensitizes with the fragrant resin, gum-benzoin, derived from *Styrax*, (Styraceae), and also cross-sensitizes with certain natural plant products of *Myroxylon*, (Leguminosae; Balsam of Peru), *Vanilla*, (Orchidaceae; vanilla), *Citrus*, (Rutaceae), *Cinnamomum*, (Lauraceae; cinnamon), *Eugenia*, (Myrtaceae; clove) and *Pinus* (Pinaceae); phenolic compounds are probably responsible, at least in part, for the cross-sensitivity patterns which are observed in sensitized individuals.[7] Sources of exposure to Balsam of Peru are summarized in Table 4.

Of the Amaryllidaceae, *Crinum*, *Haemanthus* and *Pancratium* are reported as irritant; *Hippeastrum* is possibly allergenic as well as irritant; *Narcissus* is irritant and allergenic and presents a problem mainly from heavy industrial, occupational exposure to the plant juices rather than from casual contact. There is an extensive literature concerning dermatitis from 'daffodils' in the trade but no systematic investigation of the allergen has been undertaken and its chemical nature remains unknown. In the family Annonaceae, *Annona*, *Asimina* and *Xylopia* are reported as irritant; *Oxandra* is possibly allergenic as well as irritant; *Cananga* yields ylang-ylang oil which is allergenic. Of the Apoxynaceae, *Ocokanthera*, *Allemanda*, *Aspidosperma*, *Cameraria*,

Table 4. Cross-reactions and sources of allergic contact dermatitis from Balsam of Peru and cross-reacting materials.

Cross reactions: cinnamon, clove, orange peel, benzoin, pine rosin, vanilla, essential oils: cinnamic aldehyde, cinnamic acid, cinnamic alcohol, cinnamates, benzoic acid, benzyl benzoate, benzyl alcohol, eugenol.

Sources of exposure: Therapeutic applications, suppositories, liquids in dentistry, cosmetics, hair tonics and lotions, perfumes, flavors, cough mixtures, throat lozenges, balsams of pine and spruce, pastries and cakes with essences, Danish pastries, caramels, cola-type drinks, toothpaste, liquers (Vermouth, Curacao), poplar buds (bee keepers).

Carissa, Catharanthus, Cerbera, Geissospermum, Gonioma, Macoubea, Nerium, Plumeria, Rauwolfia, Rhabdadenia, Strophanthus, Tabernaemontana, Thevetia, Vinca (Catharanthus) and *Zschokkea* are reportedly injurious; dermatitis can occur from exposure to the plants; irritant and allergic effects have not been clearly distinguished and most reports are concerned with alkaloidal poisoning in woodworkers who are exposed to wood dusts by skin contact or by inhalation. A yield of alkaloids is a feature of many members of this family. Of the Araceae, *Aglaonema, Alocasia, Amorphophallus, Anthurium, Arisaema, Arisarum, Arum, Caladium, Calla, Colocasia, Culcasia, Dieffenbachia, Dracunculus, Homalomena, Lysichitum, Monstera, Philodendron, Pinellia, Pleurospa (Montrichardia), Pothos, Rhaphidophora, Scindapsus, Spathiphyllum, Symplocarpus, Syngonium, Typhonium, Xanthosoma* and *Zantedeschia* are all reported to be irritant. As in the case of the Agavaceae, sharp-pointed crystals of calcium oxalate contained in the plant tissues of araceous plants are responsible, at least in part, for irritation of the mucous membranes, but other, unidentified, irritant constituents are probably also responsible. This family contains members which are of historical interest and appear in the early texts of Materia Medica of the Western World, notably that of Dioscorides (First Century A.D.); *Dracunculus* is thought to be the Drakontium of Hippocrates. At the present time *Pothos* is the commonest cause of dermatitis from plants on a year round basis in the Hawaiian Islands although the mango *Mangifera*, Anacardiaceae, predominates during its fruiting season[1]. *Arum maculatum* is the subject of a delightful little book by Prime[21]; information in the book might well be of interest to chemotaxonomists of this family.

Of the Araliaceae, *Aralia* and *Polyscias (Nothopanax)* are reported as irritant; *Hedera* is quite clearly allergenic but the cross-sensitizing allergen(s) of English Ivy and Algerian Ivy have not been chemically defined. In the Araucariaceae, *Araucaria* is reported as irritant, as is *Sabal* of the Arecaceae (Palmae). Of the Aristolochiaceae, *Aristolochia* is reported to be irritant and *Asarum* is possibly irritant and/or allergenic. In the Asclepiadaceae, *Araujia, Asclepias (Asclepiodora), Calotropis, Metaplexis, Sarcostemma* and *Tylophora* are all reported as irritant; from plants of the last-named genus are derived skin-vesicant alkaloids, tylophorine, etc., which are chemically related to the irritant alkaloid cryptopleurine derived from *Cryptocarya* (Lauraceae). *Sarcostemma* is colloquially named Caustic Vine which indicates lay knowledge of its marked irritant effect.

Anacardiaceous plants have received attention as the commonest causes of plant dermatitis in North America and the responsible allergens and the immunochemical requisites of some species have been precisely defined.[2] Cross-sensitivity between plants and natural plant products of this family is observed and is explicable in biochemical terms. Species of twenty-one genera of the Anacardiaceae are reported to be allergenic: *Anacardium, Astronium, Campnosperma, Comocladia, Gluta, Heeria, Holigarna, Lithraea, Loxopterygium, Mangifera, Melanochyla, Melanorrhoea, Metopium, Pentaspadon, Rhus, Semecarpus, Schinopsis, Schinus, Smodingium, Spondias,* and *Toxicodendron*; the plant juices released by injury to the plant tissues are also irritant, but the allergenic activity is the major factor in producing clinical dermatitis. The position of a twenty-second genus *Swintonia* is ambiguous having been reported as being injurious and non-injurious by equally eminent authorities! The actual sources of exposure to the plant products at a clinical level can be rather subtle; for example, some American soldiers sensitive to poison ivy developed dermatitis from handling varnished Japanese rifles; the lacquer from the tree *Rhus vernicifera* cross-sensitizes with the poison ivy-oak plants, *Rhus (Toxicodendron)*. Cross-sensitivity has been observed from a taxonomically remote species *Ginkgo* (Ginkgoaceae) owing to a yield of related chemical compounds. Some cross-sensitivity patterns, as they have been reported at a clinical level, are shown in Table 5.[6]

This brief review of a limited number of plant families exemplifies the usefulness of biochemical systematics for clinical dermatology and also points out the need for further studies in this area including precise clinical reports with obligation to identify the plants responsible for dermatitis and for a distinction between irritant and allergic aspects of pathogenesis.

We have not considered the treatment of plant dermatitis. After many conflicting reports, a recent carefully controlled study has confirmed that hyposensitization (but not desensitization) of poison oak-sensitive patients can be achieved by administering purified urushiol by mouth. A large dose such as 300 mg during a period of 6 months

Table 5. Allergic contact dermatitis, reported in various clinical settings, due to cross-sensitizing compounds of plants of the families Anacardiaceae and Ginkgoaceae.

	Botanical species	Principal allergen[5,6,23]
Contact with poison ivy-oak plants	*Toxicodendron* spp.	Pentadecylcatechol. The side-chain of poison ivy has 17 carbons: of poison oak, 15 carbons.
Eating mango fruits (contact with the skin of the fruit).	*Mangifera indica*	?
Handling cashew nut shells; Handling letters contaminated with cashew nut shell oil	*Anacardium occidentale*	Anacardic acid, cardol
Cutting Renghas timber	*Gluta* spp., etc.	Glutarenghol
Contact of the fore-arms with bar top varnished with Japanese lacquer	*Rhus vernicifera*	Urushiol
Marking ink used for clothing	*Semecarpus* spp.	Laccol, thitsiol
Walking on squashed fruits of Maidenhair tree.	*Ginkgo biloba*	Ginkgolic acid

has been shown to produce beneficial effects; however, such treatment can produce untoward effects and should be carried out only under dermatological supervision in selected patients.[5]

REFERENCES

1. Arnold, H.L., Jr. (1972) Personal communication.
2. Baer, H., C.R. Dawson & A.P. Kurtz. 1968. Delayed contact sensitivity to catechols IV. Stereochemical conformation of the antigenic determinant. J. Immunol. 101: 1243-1247.
3. Bleumink, E., J.P. Nater & J.C. Mitchell. 1973. Contact dermatitis to Chrysanthemums. Arch. Derm. 108: 220-222.
4. Dupuis, G., G.H.N. Towers & J.C. Mitchell. Reaction of Alantolactone, and allergenic sesquiterpene lactone, with some amino acids. Resultant loss of immunologic reactivity. Can. J. Biochem. (in press).
5. Epstein, W.L., H. Baer, C.R. Dawson & R.G. Khurana. 1974. Poison oak hyposensitisation. Evaluation of purified urushiol. Arch. Derm. 109: 356-360.
6. Goldstein, N. 1968. The Ubiquitous Urushiols - Contact Dermatitis from Mango, Poison Ivy, and other 'poison' plants. Cutis 4: 679-685.
7. Hjorth, N. 1961. Eczematous allergy to Balsams. Munksgaard, International Booksellers and Publishers Ltd. Copenhagen.
8. Knoche, H., G. Ourisson, G.W. Perold, J. Foussereau & J. Maleville. 1969. Allergenic component of a liverwort: a sesquiterpene lactone. Science. 166: 239-240.
9. Kupchan, S.M. 1970. Recent advances in the chemistry of tumor inhibitors of plant origin. Trans. N.Y. Acad. Sci. 32: 85-106.
10. Mitchell, J.C. 1970. Allergic contact dermatitis from Compositae. Trans. St. John's Hosp. Derm. Soc. 55: 174-183.
11. _____ & S. Armitage. 1965. Dermatitis venenata from lichens. Biology of lichens related to criteria for diagnosis of contact dermatitis in forest workers. Arch. Environ. Health 11: 701-709.
12. _____ & M. Chan-Yeung. 1974. Contact allergy from *Frullania* and respiratory allergy from *Thuja*. Can. Med. Ass. J. 110: 653-657.
13. _____ & G. Dupuis. 1971. Allergic contact dermatitis from sesquiterpenoids of the Compositae family of plants. Brit. J. Derm. 84: 139-150.
14. _____ & S. Shibata. 1969. Immunologic activity of some substances derived from lichenized fungi. J. Invest. Dermat. 52: 517-520.

15. _____, G. Dupuis & T.A. Geissman. 1972. Allergic contact dermatitis from sesquiterpenoids of plants. Additional allergenic sesquiterpene lactones and immunological specificity of Compositae, liverworts and lichens Brit. J. Derm. 87: 235-240.
16. _____, G. Dupuis & G.H.N. Towers. 1971. Allergic contact dermatitis from ragweeds (*Ambrosia*). Arch Derm. 104: 73-76.
17. _____, G. Dupuis & G.H.N. Towers. 1972. Allergic contact dermatitis from pyrethrum (*Chrysanthemum* spp.). Brit. J. Derm. 86: 568-573.
18. _____, G. Dupuis, T.A. Geissman & G.H.N. Towers. 1971. Allergic contact dermatitis caused by *Artemisia* and *Chrysanthemum* species. J. Invest. Dermat. 56: 98-101.
19. _____, B. Fritig, B. Singh, & G.H.N. Towers. 1970. Allergic contact dermatitis from *Frullania* and Compositae. The role of sesquiterpene lactones. J. Invest. Derm. 54: 233-239.
20. Perold, G.W., J.C. Muller & G. Ourisson. 1972. Structure d'une lactone allergisante: le frullanolide - I. Tetrahedron 28: 5797-5803.
21. Prime, C.T. 1960. "Lords and Ladies." Collins - The New Naturalist. London.
22. Shaw, H.K.A. 1973. "Willis' Dictionary of the Flowering Plants and Ferns." 7th ed. Cambridge University Press.
23. Whalley, W.B. 1959. The toxicity of plant phenolics. In "The Pharmacology of Plant Phenolics" (J.W. Fairbairn, ed.) Academic Press, London.

Chapter Six

TERATOGENIC CONSTITUENTS OF POTATOES

JOSEPH KUĆ

Department of Plant Pathology
University of Kentucky
Lexington, Kentucky

INTRODUCTION

Solanum tuberosum L., the potato, ranks third in world production behind wheat and rice. The per capita consumption in the United States is estimated at 55 kg (120 lbs) per year. The potato also makes an important contribution to the diet in northern, central and eastern Europe, and portions of South America, and it provides a source of high quality protein in addition to carbohydrate. Any question concerning the safety of the potato as a food, therefore, presents a serious threat to the health of a large segment of the world's population. The potato and a serious disease of potato, late blight, incited by *Phytophthora infestans*, have already had a pronounced influence on history. In the middle of the nineteenth century the Irish had become highly dependent on the potato as a food, and in 1845 and 1846 the crop was almost totally destroyed by the disease. Ireland lost almost a third of its population between 1845 and 1860 as a direct result of the outbreak of late blight. A million people died from starvation or from disease following malnutrition. A million and a half more emigrated.

The family Solanaceae includes plants that are important food crops, e.g., potato, tomato, eggplant and pepper. It also includes some that are highly poisonous, e.g., bitter nightshade, henbane and tobacco. Even potato and tomato may be classified as poisonous since the foliage of both and other parts of the potato can be toxic to some mammals including man. Numerous cases of potato poisoning have been recorded involving humans that consumed underdeveloped tubers, green tubers, blossoms, tuber peel and tuber sprouts[29]. This toxicity in potato has been attributed in large part to the two major steroid glycoalkaloids of the solanine group, α-solanine and α-chaconine. However, other glycoalkaloids may individually or collectively be even more toxic[29]. Foliage and sliced tubers of the variety Kennebec contain α- and β-solamarine in quantities equal to or greater than α-solanine and α-chaconine[26]. The report by Renwick[21] suggested that consumption of blighted or otherwise stressed tubers may be harmful to human health for reasons other than the gross toxicity attributed to the glycoalkaloids. The report implied that compounds in the affected tubers were directly related to the incidence of two congenital malformations, anencephaly and spina bifida cystica (collectively termed ASB). The suggestions made in the report have become known as the "Renwick Hypothesis." The extent of the malformations in the United States alone, ca 10,000 cases, required a thorough examination of the hypothesis.

DISEASE RESISTANCE AND STRESS RESPONSE

Many phenolics, phenolic derivatives and terpenoids are normal constituents of potato tubers, including: esculetin, caffeic acid, α-chaconine, chlorogenic acid, coumarin, 3,4-dihydroxyphenylalanine, ferulic acid, glucose esters of caffeic and *p*-coumaric acids, *o*- and *p*-hydroxybenzoic acids, *p*-hydroxycinnamic acid, norepinephrine, protocatechuic acid, scopolin, scopoletin, sinapic acid, α-solanine, syringic acid, tyramine, umbelliferone, and vanillic acid[1,2,3,5,6,13,30,31]. Most of the phenolics and the steroid glycoalkaloids, α-solanine and α-chaconine, are localized in the potato peel with relatively low concentrations in the underlying tissue. The presence of these compounds, in peel, some of which inhibit the growth of fungi and bacteria, may serve as a passive mechanism for disease resistance in the tuber. Recently attention has been drawn to the profound shifts in metabolism that occur in infected tubers and tubers placed under stress. Injury to a tuber results in the rapid accumulation of some phenolics and terpenoids including chlorogenic and caffeic acids, α-solanine and α-chaconine around the site of injury. Peroxidase, phenoloxidase and phenylalanine deaminase also increase, and the activity of the two oxidases results in the oxidation of *o*-phenols to quinones. The quinones are highly reactive and rapidly form addition products with sulfhydryl, phenolic and susceptible amino compounds and also polymerize to form water insoluble melanin-like pigments. The *o*-quinones in an aqueous environment are short lived but their extreme reactivity makes them unselective toxicants to both host and infectious agents. Their rapid polymerization is a detoxification mechanism. Characteristic browning of the wounded surface follows and layers of suberized cells form at the injured site. This essentially "seals off" the potato from the environment with a physical and chemical barrier. Once suberized, the metabolism in affected tissue returns to normal, but the high levels of phenolics and steroid glyco-alkaloids persist during the life of the tuber as they do in peel. The wound response is clearly evident in sliced tubers 6-12 hr after slicing. Tubers exposed to sunlight during the growing season also accumulate α-solanine, α-chaconine, and chlorogenic and caffeic acids and the accumulation extends into the tuber beneath the peel[23,24,25]. Recently α-solamarine and β-solamarine were also reported to accumulate in foliage and freshly cut tuber slices of the variety Kennebec[26]. Infection with a number of fungi, including the late blight fungus, *P. infestans*, and the common soft rot bacterium, *Erwinia carotovora*, causes the increased synthesis of phenolics present in healthy tubers. In addition, at least 16 terpenoids undetected in healthy tubers also accumulate around sites of infection. These include rishitin[7,15,35], rishitinol[8], lubimin[17], and phytuberin[13,32]. These terpenoids are synthesized by the host and their accumulation can be induced by constituents of *P. infestans*[33,34]. The accumulation of these phenolic and terpenoid stress metabolites and their role in disease resistance has been reviewed[4,9,10,11,12,13,28]. The question that "Renwick's Hypothesis" raised is whether host metabolites or fungal products in blighted potatoes are responsible for

ASB, and this question has recently been expanded to include potatoes infected with other organisms or otherwise stressed[13-22].

TERATOGENICITY

The "Renwick Hypothesis"[21] is based on the consideration of three major aspects: A) the occurrence of solanidine glycoalkaloids in potato, B) the production of phytoalexins by blighted potatoes, and C) epidemiological evidence. The latter aspect includes a consideration of the geographical distribution of ASB, potato blight, and consumption of blighted potatoes; and the roles of occupational class, diet of the mother, season of the year, urbanization, maternal age and parity, recurrence risks in siblings, twin data, rare, secular trends, sex ratios, and softness of water. Having concluded that "The potato is clearly an intermediate agent between blight and ASB . . ." and "Even now, though the potato carries the mark of all these influences imprinted on its size, numbers, and chemical constitution . . .", Renwick considers in detail the route of access of the teratogen — whether percutaneous, by inhalation or oral. He concludes, "Absorption of teratogen is probably, but not certainly, from ingestion" and cautions expectant mothers to avoid blemished potatoes. An examination of each of the aspects of the hypothesis indicates Renwick's conclusions are thought provoking, subject to differing interpretations, and waiting for experimental verification. In aspect A, the occurrence of solanidine glycoalkaloids in potato, there is no evidence to my knowledge which has demonstrated a teratogenic effect of the solanidine glycoalkaloids in mammals[21-29]. Toxic effects have been recorded and these have been summarized in the introduction. An oral dose of 3 mg/kg is generally considered a toxic level for man. Based on our determinations of the steroid glycoalkaloid content of potatoes, and using the above figure for toxicity to man, toxic levels of steroid glycoalkaloids would be consumed by ingesting 14-20 g sprouts, 1 kg of the top millimetre of aged tuber slices, 5-7 kg of peeled tubers, 1-2 kg of peeled "greened" tubers, 500-600 g potato peel and 150-200 g of potato leaves (Table 1). Obviously some caution must be exercised in consuming parts of the potato plant. Chop suey prepared from potato rather than bean sprouts might be lethal, as might a stew of potato peels and foliage. These toxicities, however,

Table 1. Consumption of Potato Tissues Containing Levels of Steroid Glycoalkaloids Toxic to Man[a]

Tissue	Consumption Necessary for Toxicity
Sprouts	14-20 g
Leaves	150-200 g
Tuber peel	500-600 g
Top mm aged tuber slices	1 kg
Peeled "greened" tubers	1-2 kg
Peeled undamaged tubers	5-7 kg

[a]Based on levels of steroid glycoalkaloids reported in potato tissues (13), a toxic level of 3 mg/kg body weight, and a body weight of 154 lb.

have no scientific documentation to relate them to ASB. It is comforting to note that the steroid glycoalkaloids, α-solanine and α-chaconine, as well as the phytoalexins, rishitin and phytuberin, were not detected in a dehydrated potato mix obtained in a local supermarket (Kuć, unpublished data). It has also been demonstrated that the accumulation of α-solanine, α-chaconine, α- and β-solamarine in tubers is suppressed when rishitin and phytuberin accumulate[24,25]. Inoculation of tubers with an incompatible race of *P. infestans* or nonpathogens, as well as treatment with cell-free sonicates of fungi, induce rapid accumulation of rishitin and phytuberin with little accumulation of steroid glycoalkaloids. Inoculation with a compatible race of *P. infestans* also suppresses the accumulation of steroid glycoalkaloids, as compared to uninoculated slices, and little or no rishitin and phytuberin accumulate. Other unidentified terpenoids accumulate in the compatible reaction[32,33], as does the glycoside, scopolin[2,6]. Blighted potatoes, therefore, would not have high levels of solanidine glycoalkaloids. Analysis of 15 cultivars of potatoes in our laboratory have indicated there is no relationship between the resistance of tubers to late blight and the content of solanidine glycoalkaloids in foliage and peel or the rate and magnitude of their accumulation in response to injury. Sprouts of all varieties of potato tested were resistant to late blight and their content of solanidine glycoalkaloids was approximately fifty times that of potato peel. These data further weaken Renwick's speculations that solanidine glycosides play a part in resistance to late blight, that they would be high in surviving potatoes during blight years, and that they are therefore related to the incidence of ASB.

In aspect B, the suggestion is made that the accumulation of rishitin, phytuberin and other terpenoid phytoalexins in resistant potatoes exposed to blight is related to ASB. Thus, potatoes surviving during epidemics of blight would contain the highest levels of phytoalexins. Evidence in our laboratory indicates that the accumulation of rishitin and phytuberin and at least 14 other terpenoids varies with time after inoculation or treatment of tubers. Rishitin, phytuberin and derivatives of phytuberin do not appear to be stable in affected tissue (Table 2). Total accumulation increases up to 72-96 hr and then decreases. Thus 8 days after inoculation or treatment, little rishitin and phytuberin remain in the affected tuber. We have never observed the accumulation of rishitin or phytuberin in healthy tubers or untreated cut slices. Mechanical injury does not cause the accumulation of these phytoalexins, although conditions during incubation of treated tubers markedly influence the ratio of phytoalexins. In the presence of ethrel or ethylene, the accumulation of phytuberin and phytuberin derivatives is markedly increased and rishitin decreased, in potatoes inoculated with incompatible races of *P. infestans*, nonpathogens or treated with cell-free sonicates of these fungi[23]. Ethylene or ethrel do not cause rishitin or phytuberin accumulation without the presence of the infectious agents or their products.

The terpenoid phytoalexins in affected tubers are not dispersed throughout the tuber but are localized in 1-2 mm of cells around the affected areas. Therefore, the amount of these phytoalexins ingested by the consumption of even an obviously blighted or infected tuber would be small. In a survey of four supermarkets in West Lafayette, Indiana during a period of 6 months, I have never detected blighted potatoes being sold. Unfortunately, I have noticed on three occasions that green tubers were being sold. In one instance the green tubers were small and were offered on

Table 2. Rishitin and Phytuberin in the top two mm of Kennebec potato slices inoculated with an incompatible race of *Phytophthora infestans* (Race 4)

Time after	µg/g f. wt.	
Inoc. (hr.)	Rishitin	Phytuberin
24	5	-
36	15	1
48	40	3
60	65	5
72	80	18
96	96	20
120	40	14
144	15	11
168	7	7
192	2	5

special sale. The manager of the super market was apparently unaware of the possibility of toxicity from accumulated steroid glycoalkaloids. A program to educate consumers and the food industry in the potential danger from consumption of green tubers seems in order. Regulations exist against selling blemished potatoes, but are there regulations prohibiting the sale of green potatoes?

Though Renwick proposed the possible relation of potato phytoalexins to ASB in 1972, no published data is available to support the suggestion. The only data relating to the consumption of blighted potatoes with observed defects are those reported by Poswillo *et al.*[18], using marmosets as the test animal. Potatoes of the variety Kerr's Pink, infected with *P. infestans*, were cubed and boiled for 30 min. The cooked potato and water mixture was homogenized, freeze-dried and fed together with banana, bread, egg and vitamin supplement. Skins, sprouts and portions of potatoes affected by brown discoloration and softening were all included in the preparation. A control diet did not contain the "blighted" potato concentrate. Examination of 11 fetuses in the control group revealed no gross abnormalities. Four of 11 fetuses examined in the group fed the potato concentrate showed cranial osseous defects. The most severe defects were found in marmosets which had been conceived after the longest period of pre-conception exposure to the potato concentrate. Rats fed the potato concentrate had normal offspring. The experimental design was such that the defects could have been due to a number of factors including: the consumption of healthy tubers, sprouts, *P. infestans*, secondary microbial invaders and/or affected tubers.

Subsequently, Poswillo *et al.*[19] reported that feeding a concentrate from blemished tubers rejected by commercial graders and tubers infected by *E. carotovora* did not result in defective offspring in marmosets. The potato concentrate from tubers infected with *E. carotovora* was high in rishitin and phytuberin and low in total glycoalkaloids, whereas the rejected graded tubers contained little or no rishitin and phytuberin and total glycoalkaloid content ranged from one third to approximately three times that of healthy tubers. The concentrate from the "blighted" Kerr's Pink potatoes which produced defects contained little or no rishitin and phytuberin but had approximately twice the concentration of total glycoalkaloids found in healthy tubers. The data suggest that neither rishitin, phytuberin nor total glycoalkaloids were responsible for the cranial defects reported with marmosets.

In more recent data Stallknecht[27] saw no evidence of teratogenicity from feeding blighted potato material to marmosets, rats, rabbits, hamsters and mice. No evidence for teratogenicity has also been reported in preliminary data for marmosets and Rhesus monkeys fed blighted potatoes at the Food Research Institute, Madison, Wisconsin (personal communication). However, J.M. Wilson, Dept. Food Sciences, Maine Agricultural Experiment Station, has reported (personal communication) that α-solanine caused defects in the chick embryo test.

Aspect C, epidemiological data related to ASB, has been reviewed[13,16,19,20,22,29]. This chapter will not consider this aspect except to mention that exceptions to Renwick's interpretations have been presented. Renwick's contention that 95% of ASB occurrences could be prevented by avoidance of the potato during early pregnancy also requires further critical experimentation, since it was reported that a mother of two previous ASB children produced an anencephalic fetus while on an extended potato-avoidance diet[14].

SUMMARY

There is no experimental evidence to relate α-solanine, α-chaconine, rishitin, and phytuberin to ASB. The steroid glycoalkaloids are toxic to man and this observation has been recorded for many years. Only one test has indicated that blighted potatoes may be teratogenic, and this work was done with marmosets. The cranial osseous defects reported with the animals do not appear identical to ASB (Swinyard in ref. 16). The possibility persists, nevertheless, that teratogenic substances exist in infected or damaged potatoes. Feeding tests of blighted and damaged potatoes are being conducted with chick embryos, rats, marmosets and Rhesus monkeys. Feeding infected or damaged potato tissue to a broad spectrum of animals offers the best approach to resolving the "great potato debate." If teratogenicity is indicated, feeding of isolated purified components would be warranted.

The rapid withdrawal of a recently released potato variety "Lenape" because of its high level of steroid glycoalkaloids has focussed attention on the need for a testing program for all new crop varieties. The problem is especially pertinent when mutants or wild inedible species are used in developing varieties.

Acknowledgement

Journal Paper no. 74-11-134 of the Kentucky Agricultural Experimental Station, Lexington, Kentucky 40506. The author's work reported in this paper has been supported in part by a grant from the Herman Frasch Foundation and Cooperative State Research Service, USDA, Research Agreement 316-15-51.

REFERENCES

1. Allen, E., & J. Kuć, 1968. α-Solanine and α-chaconine as fungitoxic compounds in extracts of Irish potato

tubers. Phytopathology 58: 776-781.
2. Clarke, D. 1973. The accumulation of scopolin in potato tuber tissue in response to infection. Physiol. Plant Pathology 3: 347-358.
3. Corner, J., & J. Harborne, 1960. Cinnamic acid derivatives of potato berries. Chem and Ind. 76.
4. Cruickshank, I., D. Biggs, & D. Perrin, 1971. Phytoalexins as determinants of disease reaction in plants. J. Indian Bot. Soc. 50A: 1-11.
5. Harborne, J. & J. Corner, 1960. The metabolism of cinnamic acids and coumarins in higher plants. Biochem. J. 76: 53.
6. Hughes, J. & T. Swain. 1960. Scopolin production in potato tubers infected with *Phytophthora infestans*. Phytopathology 50: 398-400.
7. Katsui, N., A. Murai, M. Takasugi, K. Imaizumi, & T. Masamune, 1968. The structure of rishitin, a new antifungal compound from diseased potato tubers. Chem. Commun. 43-44.
8. Katsui, N., A. Matsunaga, K. Imaizumi, T. Masamune, & K. Tomijama, 1971. The structure and synthesis of rishitinol, a new sesquiterpene alcohol from diseased potato tubers. Tetrahedron Lett. 83-86.
9. Kosuge, T. 1969. The role of phenolics in host response to infection. Ann. Rev. Phytopathology 7: 195-222.
10. Kuć, J. 1966. Resistance of plants to infectious agents. Ann. Rev. Microbiol. 20: 337-370.
11. _____ .1971. Compounds accumulating in plants after infection. In "Microbial Toxins" (S. Ajl, G. Weinbaum, & S. Kadis, eds.) Vol. 8, pp. 211-247. Academic Press. New York.
12. _____. 1972. Phytoalexins. Ann. Rev. Phytopathology 10: 207-232.
13. _____. 1973. Stress metabolites in potato tubers. Teratology 8: 333-338.
14. Lorber, J., C. Stewart, & G. Ward, 1973. Alpha-ferroprotein in antenatal diagnosis of anencephaly and spina bifida. Lancet 1187.
15. Lyon, G.D. 1972. Occurrence of rishitin and phytuberin in potato tubers inoculated with *Erwinia carotovora* var. *atroseptica*. Physiol. Plant Path. 2: 411-416.
16. Medical World News 1973. Feb. 16, 29-40.
17. Metlitsky, L., & O. Ozeretskovskaya, 1970. Phytoncides and phytoalexins and their role in plant immunity. Mikol. Fitopatol. 4: 146-155.
18. Poswillo, D., D. Sopher, & S. Mitchell. 1972. Experimenta

induction of foetal malformation with "blighted" potato: a preliminary report. Nature 239: 462-464.
19. Poswillo, D., D. Sopher, S. Mitchell, D. Coxon, R. Price, & K. Price, 1973. Investigations into the teratogenic potential of imperfect potatoes. Teratology 8: 339-347.
20. Report of the Potato Marketing Board, 50 Hans Crescent, Knightsbridge, London, S.W. 1. 1972.
21. Renwick, J. 1972. Hypothesis: anencephaly and spina bifida are usually preventable by avoidance of a specific but unidentified substance present in certain potato tubers. Br. J. Prev. Soc. Med. 26: 67-88.
22. _____, 1973. Prevention of anencephaly and spina bifida in man. Teratology 8: 321-323.
23. Shih, M. 1973. The Accumulation of Isoprenoids and Phenols and its Control as Related to the Interaction of Potato (*Solanum Tuberosum* L.) with *Phytophthora infestans*. Ph.D. Thesis Purdue University, Lafayette, Indiana.
24. _____, & J. Kuć. 1973. Incorporation of ^{14}C from acetate and mevalonate into rishitin and steroid glycoalkaloids by potato tuber slices inoculated with *Phytophthora infestans*. Phytopathology 63: 826-829.
25. _____, J. Kuć, & E. Williams. 1973. Suppression of steroid glycoalkaloid accumulation as related to rishitin accumulation in potato tubers. Phytopathology 63: 821-826.
26. _____, & J. Kuć. 1974. α- and β-Solamarine in Kennebec *Solanum tuberosum* leaves and aged tuber slices. Phytochemistry 13: 997-1000.
27. Stallknecht, G. 1974. Role of potato stress response in human health, plant disease, and insect resistance. Report of Meeting Regional Project Participants and Observers, NE-94, Wyndmoor, Pa, June 4-5. pp. 3-4.
28. Stoessl, A. 1970. Antifungal compounds produced by higher plants. Rec. Adv. Phytochem. 3: 143-180.
29. Swinyard, C. & S. Chaube, 1973. Are potatoes teratogenic for experimental animals? Teratology 8: 349-357.
30. Undenfriend, S., W. Lovenberg, & A. Sjoerdsma, 1959. Physiologically active amines in common fruits and vegetables. Arch. Biochem. Biophys. 85: 487-490.
31. Van Sumere, C. 1960. In "Phenolics in Plants in Health and Disease" (J. Pridham, ed.) Pergamon Press, New York.
32. Varns, J. 1970. Biochemical Response and its Control in

the Irish Potato Tuber (*Solanum tuberosum* L.) - *Phytophthora infestans* interactions. Ph.D. Thesis, Purdue University, Lafayette, Indiana.

33. _____ & J. Kuć, 1971. Suppression of rishitin and phytuberin accumulation and hypersensitive response in potato by compatible races of *Phytophthora infestans* Phytopathology 61: 178-181.

34. _____, W. Currier, & J. Kuć, 1971. Specificity of rishitin and phytuberin accumulation by potato. Phytopathology 61: 968-971.

35. _____, J. Kuć, & E. Williams, 1971. Terpenoid accumulation as a biochemical response of the potato tuber to *Phytophthora infestans*. Phytopathology 61: 174-177.

Chapter Seven

PLANT NEUROTOXINS (LATHYROGENS AND CYANOGENS)

CHARLOTTE RESSLER*

Division of Protein Chemistry
Institute for Muscle Disease, and
Department of Biochemistry
Cornell University Medical College
New York, New York

INTRODUCTION

This paper deals with developments made in the area of toxins from lathyrus and vetch peas since our last review in 1964.[1] The active principles, which include β-aminopropionitrile (BAPN), β-cyanoalanine (β-CNala), 2,4-diaminobutyric acid (DAB), and 3-N-(oxalyl)-2,3-diaminopropionic acid (OX-DAP), were isolated in free or bound form in our or other laboratories over the past twenty years, in studies relating to the neurological disease lathyrism in man.

*Present address: Department of Pharmacology, Schools of Medicine & Dental Medicine, University of Connecticut Health Center, Farmington, Conn.

Their isolation, identification, and distribution have been treated adequately in separate reviews.[1,2] Studies bearing on their mechanism of action will be the main concern of this review. Also included are discussions of a new enzyme that participates in the biosynthesis of γ-cyanoaminobutyric acid (γ-CNabu), a related bacterial toxin, and the propertie of γ-thiocyanoaminobutyric acid (γ-SCNabu), a related toxin that may be a bacterial biosynthetic intermediate.

The etiology of neurolathyrism has long been unclear. Although this disease has been frequently associated with the consumption of large quantities of the seeds of *Lathyrus sativus, L. cicera,* or *L. clymenum,* attempts to induce the disease in experimental animals by dietary administration of these seeds have been uniformly unsuccessful. Moreover, the disease in man has sometimes been observed in the complete absence of *Lathyrus* in the diet, as in prisoner-of-war camps with poor nutrition. Cases of lathyrism have been reported when wheat or *L. sativus* was consumed and found to contain *Vicia sativa* seeds.[3,4] The latter plant is sometimes used as a cover crop, alternating with wheat and other cereals. Vetches also find use as roadside plantings to prevent erosion. The seeds can escape cultivation and appear as weeds. Contamination of cereal grains with vetch seeds may arise from such sources. Whethe established to be due to a toxic factor or to a nutritional deficiency, any proposed basis for lathyrism should be reconcilable with the various observations. For example, a relevant biochemical reaction compatible with all the observations could be one that is capable of being inhibited by one or more of the isolated toxins but also of being rendered defective by poor nutrition, where the absence of a required substance or a process involving inadequate synthesis of a required enzyme may be a contributing factor.

ROLE OF β-AMINOPROPIONITRILE

β-Aminopropionitrile, $H_2NCH_2CH_2C{\equiv}N$, the active principle of *Lathyrus odoratus* (sweet pea), has been well studie by pathologists, biochemists and others. In a variety of experimental animals such as the rat, chick, turkey, baboon, and tadpole, BAPN produces a defect in connective tissue that results in collagen and elastin with decreased tensile strength and collagen with greater than normal solubility. This can be manifested in abnormalities such as hernia, kyphoscoliosis and exostoses, aortic aneurysms and rupture, and displaced lenses. Medical interest in the sweet pea factor derives from gross and microscopic observations that the defects which this factor produces in experimental animals simulate many of those seen in certain heritable disorders of collagen. Among these are Marfan's syndrome[5] and that associated with iminodipeptiduria.[6] In both these anomalies the excretion of hydroxyproline is elevated as it can be in experimental lathyrism. Some manifestations of experimental lathyrism also occur in Paget's disease, Legg-Calve-Perthes disease, and in spontaneous ideopathic scoliosis, slipped epiphysis, and degenerative arthritis in man.

From comparative structural studies on normal and lathyritic collagen, biochemical evidence (including decreased aldehyde content, increased lysine content, and decreased content of cross-linking amino acids) has accumulated to confirm earlier deductions that lathyritic collagen is less cross-linked than normal.[7] Because of its action to prevent maturation of collagen, β-aminopropionitrile has been suggested as an anti-aging drug. Some evidence has been obtained that it will extend the average life-span of rats.[8]

Lysyl oxidase, a newly recognized enzyme present in connective tissue, catalyzes the oxidation of the epsilon amino group of some of the peptide-bound lysines and hydroxylysines to the corresponding aldehydes, i.e., residues of α-aminoadipic semialdehyde and α-aminohydroxyadipic semialdehyde - reaction 1.

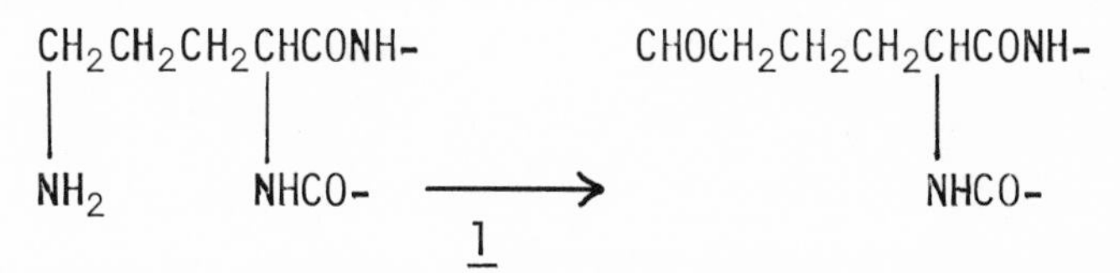

It has become clear that β-aminopropionitrile inhibits this enzyme. Thus, BAPN-treated animals have little lysyl oxidase activity. Inhibition of this enzyme by BAPN *in vitro* and, recently, binding of BAPN to the enzyme have been demonstrated.[9] Improvements in the isolation of lysyl oxidase in purified form have recently been accomplished.[10] Now that the enzyme affected by BAPN is recognizable, it will be of considerable interest to determine its activity in the suspect clinical collagen states in man.

ROLE OF 2,4-DIAMINOBUTYRIC ACID

2,4-Diaminobutyric acid, $NH_2CH_2CH_2CHNH_2COOH$, has had special interest as the first toxin, isolated from a species of *Lathyrus*, to produce neurological effects and to be recognized as a non-protein amino acid. It was possible for us to follow its isolation from *L. latifolius* by bioassay in feeding experiments and to account for the toxicity of the seed by the content of 2,4-DAB. A single meal of several grams of *L. sylvestris* W. containing this amino acid can be lethal to a rat. Instead of a neurotoxin that was highly active and present in small amount as had been sought previously[11], it became evident that the active factor had only low or moderate toxicity (LD_{50} 60 mg/100 g body weight) but it was present in high concentration. This has essentially been the pattern with the various toxic factors isolated since then from *Lathyrus* and vetch seeds (0.5 to 2.7 weight %). Moreover, being amino acid-like in nature, the toxic factors frequently form an important part of the amino acid profile of the seed, and can be seen readily and determined quantitatively by amino acid analysis (Fig. 1). DAB is present in appreciable amount in 13 of 49 species of *Lathyrus* and, usually in low concentration, in other members of the Leguminosae and Cruciferae.[2]

It has been suggested that the toxicity of DAB to the rat is due to chronic ammonia toxicity that results from inhibition of urea synthesis. This amino acid was shown to inhibit, in a competitive manner, ornithine carbamyltransferase which catalyzes the conversion of ornithine to citrulline, reaction 2. Both *Neurospora crassa* and rat liver served as sources of the enzyme.[12,13]

$$\begin{array}{c} CH_2NH_2 \\ | \\ CH_2 \\ | \\ CH_2 \\ | \\ CHNH_2 \\ | \\ COOH \end{array} + \begin{array}{c} O \\ \| \\ H_2N\text{-}C\text{-}OPO_3H \end{array} \xrightarrow[\underline{2}]{} \begin{array}{c} CH_2NHC\text{-}NH_2 \\ | \\ CH_2 \\ | \\ CH_2 \\ | \\ CHNH_2 \\ | \\ COOH \end{array}$$

ROLE OF 3-N-OXALYL-2,3-DIAMINOPROPIONIC ACID

3-N-Oxalyl-2,3-diaminopropionic acid,

$$\begin{array}{l} NHCH_2CHNH_2COOH \\ | \\ CO \\ | \\ COOH \end{array},$$

is the compound isolated from *Lathyrus sativus* and present in varying amount in 15 of 49 species of *Lathyrus* and 2 species of *Crotalaria*. Administered by ordinary routes, OX-DAP, like the seed from which it was isolated, was found to be either inactive or of low toxicity to experimental animals except in neonatal subjects. When given intrathecally at a dosage of 15 to 20 mg to five adult Macaca monkeys, however, it produced a flaccid paraplegia, in some cases a spastic paraplegia as in lathyrism in man. Histochemistry showed destruction of the nerve cells of the grey matter of the spinal cord accompanied by proliferation of microglial cells.[14] A comparison with the histopathology of lathyrism has not been available. Impermeability of the blood-brain barrier to the toxin when given by ordinary routes has been invoked to account for the insensitivity of adult animals.

When applied microelectrophoretically to spinal interneurons and Betz cells, OX-DAP is a potent excitant.[15] This activity may be related to convulsant activity[16], but its relation to the production of paralysis is not yet clear. Another study showed elevated levels of brain glutamine and suggested that OX-DAP, like DAB, leads to ammonia toxicity.[17] In resting yeast cells, OX-DAP appeared to block the uptake of glutamate and aspartate from the medium in a competitive manner, K_i=0.1mM. Glutamate and aspartate are putative neurotransmitters for mammals as they appear to be in lower forms of life. Extrapolation of the observations with yeast to the intact mammal led to the suggestion that OX-DAP may act to terminate synaptic activity by inhibiting the uptake of the dicarboxylic acids by nerve endings.[18]

ROLES OF β-CYANOALANINE AND ANALOGS

β-Cyanoalanine, $C{\equiv}NCH_2CHNH_2COOH$, is the toxin studied most extensively in our laboratory. This amino acid was synthesized originally to establish the structure of a transformation product of asparagine in synthetic peptide studies related to the posterior pituitary hormones, oxytocin and vasopressin.[19] N,N'-Dicyclohexylcarbodiimide and some other peptide-coupling agents were shown to effect an intramolecular carboxylate-assisted dehydration of the amide group of asparagine to the cyano group of β-cyanoalanine, reaction 3.[20] This type of coupling reaction is, therefore, now generally avoided when introducing asparagine into peptide linkage.

$$CONH_2CH_2CHNH_2COOH \xrightarrow{N=C(C_6H_{11})_2} C{\equiv}NCH_2CH_2CHNH_2COOH \qquad \underline{3}$$

The structural relation of β-cyanoalanine to α-aminopropionitrile, the sweet pea lathyrogen, and its neurotoxic properties prompted us to examine relevant *Lathyrus* and vetch species for the presence of β-cyanoalanine as the neurotoxin long sought in lathyrism. Among these, *Vicia sativa* (common vetch) had special interest for the neurotoxic effects it produces in ducks and monkeys which consume this seed. The availability of synthetic β-cyanoalanine and its ready detection by its green ninhydrin color facilitated its identification in seeds of *Vicia sativa.*[21]

The predominant form of β-cyanoalanine in *Vicia sativa* seeds and seedlings is γ-glutamyl-β-cyanoalanine (Fig. 1).[22,23] This dipeptide was first encountered as a major common product of β-cyanoalanine and cyanide assimilation in this species.[24,25] At that time cyanide had recently been found to enter bound aspartic acid in an insect[26] and asparagine in several plants.[27,28] Moreover, the synthesis of asparagine from aspartic acid in plants had not yet been firmly established. Studies with a variety of species of *Lathyrus* and *Vicia*, some of them available from our lathyrism studies, allowed our laboratory to establish in 1963 the role of β-cyanoalanine as the intermediate in the incorporation of of cyanide into asparagine, reactions 4 and 5.[1,24,25] This pathway was confirmed independently by Tschiersch in 1964,[29] and it has been extended by a number of investigators to a wide variety of plants belonging, in addition to the Leguminosae, to the Gramineae, Cucurbitacae, Linaceae, and Berberidaceae families, and to some bacteria, as well.

$$\overset{\circ}{C}{\equiv}\overset{\bullet}{N}^- \xrightarrow{\underline{4}} \overset{\circ}{C}{\equiv}\overset{\bullet}{N}CH_2CHNH_2COOH \xrightarrow{\underline{5}} \overset{\circ}{C}O\overset{\bullet}{N}H_2CH_2CHNH_2COOH$$

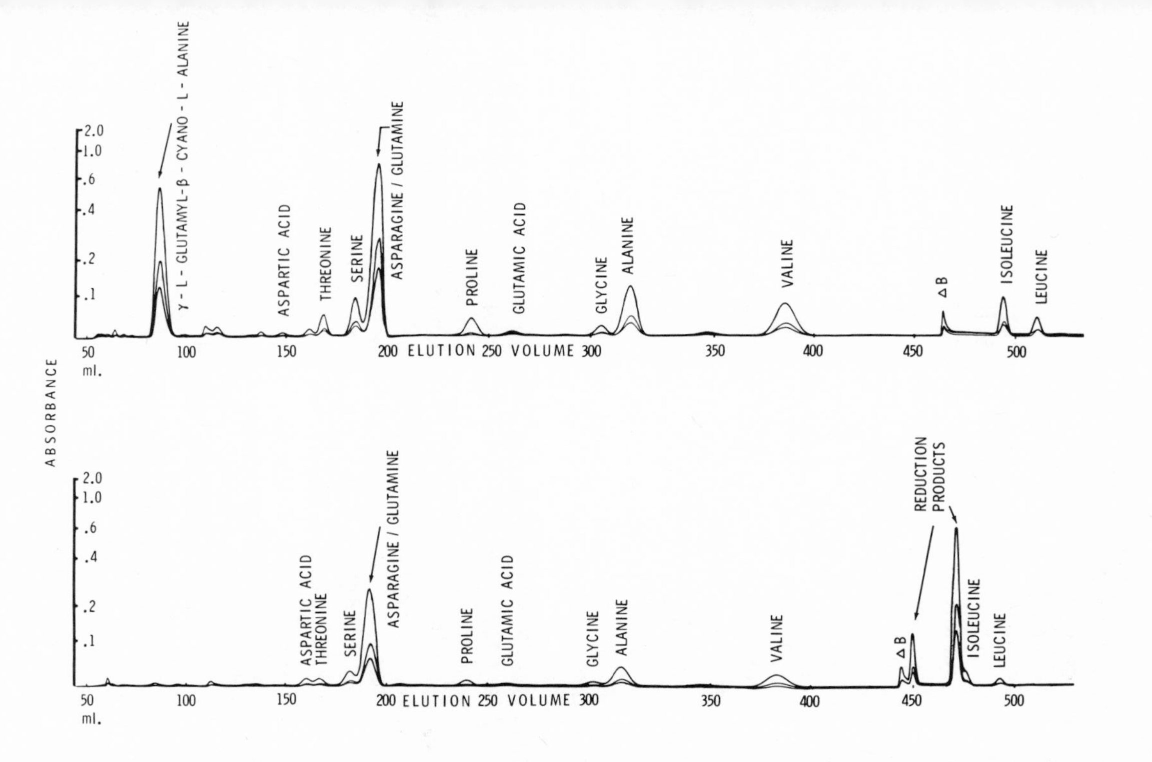

Figure 1. Amino acid analysis. The upper curve shows the chromatogram of a crude extract of tissues of 17-day-old seedlings of *Vicia sativa*; the lower curve, the chromatogram after treatment of the extracted material with sodium-ammonia-methanol which results in reduction of the γ-glutamyl-β-cyanoalanine (reprinted from reference 23).

Moreover, certain amino acids contribute their alpha carbon and alpha amino nitrogen as a unit to furnish the cyano group. Cyanogenesis and cyanide metabolism in plants and bacteria may therefore be considered to be special aspects of amino acid metabolism. Formation of β-cyanoalanine and of the asparagine derived from it could represent a mechanism for detoxifying cyanide. The detailed enzymatic reactions involved in each of these steps are being delineated in a number of laboratories. The subject has been reviewed elsewhere.[30]

β-Cyanoalanine was found in 16 of 48 *Vicia* species examined[1,2]. Although it has been observed principally in *Vicia*[23], in view of its apparent widespread role in cyanide metabolism in plants it would not be surprising to find it in other genera. Accumulation of β-cyanoalanine in bacteria has been observed in two strains of *Chromobacterium*[31,32,33].

β-Cyanoalanine produces, in the chick, a rigid opisthotonic, strychnine-like convulsion that can last 24 hours.[34] The effect in the rat is similar but of shorter duration. β-Cyanoalanine is moderately toxic to the chick and the rat. Its LD_{50} in the rat is 13.5 mg/100g and about half this in the chick. It is effective by all routes of administration including the diet. Administered β-cyanoalanine is found in the tissues of the rat and chick largely as the dipeptide, γ-glutamyl-β-cyanoalanine, and the glutathione analog, γ-glutamyl-β-cyanoalanylglycine.[35] Much of it is excreted unchanged as the free amino acid. From the results of pharmacological experiments with inhibitors it seems unlikely that β-cyanoalanine acts as a known type of central nervous system inhibitor.[36]

β-Cyanoalanine can enter into multiple biochemical reactions. It is a good competitive inhibitor *in vitro* of rat liver cystathionase, K_i=8μM. It inhibits reaction 7 of the transsulfuration process, i.e., the formation of cysteine from cystathionine derived from homocysteine (methionine), reactions 6 and 7.[37] It likewise inhibits spinach β-cystathionase, K_i=40μM, which catalyzes reaction 8.[38]

$$HSCH_2CH_2CH(NH_2)COOH + HOCH_2CH(NH_2)COOH \xrightarrow{\underline{6}} HOOCCH(NH_2)CH_2CH_2\text{-}S\text{-}CH_2CH(NH_2)COOH$$

$$\xrightarrow{\underline{7}} HSCH_2CH(NH_2)COOH + CH_3CH_2C(=O)COOH + NH_3$$

$$\xrightarrow{\underline{8}} HSCH_2CH_2CH(NH_2)COOH + CH_3C(=O)COOH + NH_3$$

β-Cyanoalanine also inhibits mammalian cystathionase in vivo. Except in vitamin B_6 deficiency in rats and children, certain neuroblastomas with high cystathionine synthetase activities (reaction 6), and the hereditary condition cystathionemia, excretion of cystathionine is seldom encountered. The rat receiving β-cyanoalanine for one month excretes cystathionine in an amount corresponding to 1.2 g/24 h[34] as compared with 0.6 g/24 h for a 40 kg child with cystathionemia.[39] It has been suggested that it may be possible to utilize β-cyanoalanine's property of inhibiting cystathionase to create an experimental model to evaluate the consequences of impaired transsulfuration as occurs in the genetic conditions cystathionemia and homocystinemia.

Other actions of β-cyanoalanine include its competitive inhibition of bacterial aspartate decarboxylase, reaction 9, K_i=0.3 mM.[40]

$$\overset{*}{C}OOHCH_2CHNH_2COOH \xrightarrow{\underline{9}} CH_3CHNH_2COOH + \overset{*}{C}O_2$$

β-Cyanoalanine is a substrate for guinea pig asparaginase and *Escherichia coli* asparaginase, reaction 10, K_m=12mM.[41,42,43]

$$C\ NCH_2CHNH_2COOH \xrightarrow{\underline{10}} COOHCH_2CHNH_2COOH + NH_3$$

β-Cyanoalanine inhibits *E. coli* glutaminase, reaction 11,

in a competitive manner, K_i=1.2 mM (O. Abe & C. Ressler, unpublished).

$$CONH_2CH_2CH_2CHNH_2COOH \longrightarrow COOHCH_2CH_2CHNH_2COOH + NH_3$$

$\underline{11}$

γ-Cyano-β-aminobutyric acid, $C{\equiv}NCH_2CH_2CHNH_2COOH$, like its homolog β-cyanoalanine, first was available synthetically.[19] It then became recognized as a natural product when it was found to accumulate as the chief product of cyanide assimilation in *Chromobacterium* D341.[33] This strain vigorously produces free HCN, and its incorporation of cyanide into γ-cyano-β-aminobutyric acid is therefore a likely physiological mechanism. This strain can also accumulate β-cyanoalanine. γ-Cyanoaminobutyric acid and β-cyanoalanine have several similar actions. It is a convulsant in the rat, and it inhibits *E. coli* glutaminase, reaction 11, K_i=2.6mM (O. Abe & C. Ressler, unpublished). γ-Cyanoaminobutyric acid effectively replaces glutamine in the glutamine-glyoxylate transaminase system from rat liver.[44]

γ-Cyano-γ-aminobutyric acid, $C{\equiv}NCHNH_2CH_2CH_2COOH$, has been reported to be the product of cyanide assimilation in an unidentified psychrophilic basidiomycete. A cycle was proposed for its formation from succinic semialdehyde and cyanide.[45] In structure-activity studies on osteolathyrogens and neurolathyrogens, our laboratory had the occasion to synthesize this cyanoamino acid.[46] We found it to differ in infrared spectrum and melting point from these properties reported for the natural product. Separate confirmatory evidence for the structure proposed for the natural product and for the cycle in which it participates is therefore awaited with interest, especially in view of its structural isomerism with γ-cyano-α-aminobutyric acid.

β-Cyano-β-alanine, $C{\equiv}NCHNH_2CH_2COOH$ (an isomer of β-cyanoalanine), and α-cyanoglycine, $C{\equiv}NCHNH_2COOH$, its lower homolog, were also synthesized. Enzymatic deacylation of N-acetyl cyanoglycine with hog kidney acylase provided the latter cyanoamino acid.[46] A residue of α-cyanoglycine has been postulated to be a central intermediate unit in a scheme for the prebiotic synthesis of proteins that does not require the intervening formation of free amino acids. This scheme has received support from finding several α-amino acids after treatment of a synthetic poly-α-cyanoglycine with HCN followed by hydrolysis.[47]

All three α-cyanoamino-acids inhibit baterial glutamate decarboxylase, reaction 12, the most effective being α-cyano-α-aminobutyric acid, $\overline{K_i}$=14mM.[48]

$$COOHCH_2CH_2CHNH_2\overset{*}{C}OOH \longrightarrow COOHCH_2CH_2CH_2NH_2 + \overset{*}{C}O_2$$

12

In man this enzyme is localized chiefly in nervous tissue, and since it is present in nerve endings it may serve a special function. Because of the tendency of α-cyano-α-amino acids to decompose in aqueous solution, we have turned to the simpler, more stable β-cyanopropionate. Like the α-cyanoamino acids, this compound inhibits the decarboxylase in a competitive manner. When preincubated with the enzyme, β-cyanopropionate changes to a noncompetitive inhibitor. Preliminary evidence indicates that it is bound to the enzyme in a linkage that involves participation of the pyridoxal 5'-phosphate cofactor (O. Abe & C. Ressler, unpublished). This compound may therefore serve as a probe for the active site of glutamate decarboxylase.

BIOSYNTHESIS OF NEUROLATHYROGENS

In parallel biosynthetic studies on the neurolathyrogens in progress in our laboratory, interest arose in the route by which γ-cyano-α-aminobutyric acid is formed. Results of isotope-labeling experiments with intact cells suggested a condensation of cyanide with a 4-carbon unit derivable from aspartic acid.[33] An active soluble preparation was then isolated from the cells for which the disulfide homocystine served as the unique cosubstrate of cyanide. Thiocyanate was formed as the co-product of γ-cyanoaminobutyric acid. This received support from their stoichiometry and common pH-rate-dependency. Although homocysteine participates in transmethylation and transsulfuration reactions, homocystine had not been previously known to enter synthetic reactions. Consideration of possible mechanisms for the enzymatic utilization of homocystine led to the proposal of the following sequence of reactions[49]:

$$(SCH_2CH_2CHNH_2COOH)_2 \xrightarrow[\underline{13}]{CN^-}$$

$$HOOCCHNH_2CH_2CH_2SH \quad + \quad HOOCCHNH_2CH_2CH_2SCN$$

$$SCN^- + HOOCCHNH_2CH_2CH_2CN \xleftarrow[\underline{14}]{CN^-,\ \text{enzyme}}$$

Nonenzymatic cyanolysis yields an organic thiocyanate, γ-thiocyanoaminobutyric acid, which serves as the enzyme substrate, reactions 13 and 14. It is interesting that cyanide participates in both steps. γ-Thiocyanoaminobutyric acid was synthesized by an independent route involving alkylation of homocysteine with cyanogen bromide. It served both *in vitro* and with intact cells as an excellent co-substrate of cyanide for the synthesis of γ-cyanoaminobutyric acid and thiocyanate, thereby confirming reaction 14. The enzyme which catalyzes reaction 14 has a molecular weight around 130,000 and a high pH-rate-optimum near 10. It is activated by simple thiols and it utilizes pyridoxal 5'-phosphate as co-factor. Enzymatic cleavage of an organic thiocyanate had not been encountered before this. Additionally, substitution of a thiocyano residue by a cyano residue represents an interesting new example of a B_6-enzyme-catalyzed displacement reaction. The requirement for pyridoxal phosphate and the simplicity with which thiocyanate is determined as its red ferric complex have incidentally led us to develop a convenient assay for this co-factor that depends upon reactivation of the inactive apoenzyme with pyridoxal 5'-phosphate (O. Abe & C. Ressler, unpublished).

Inorganic and organic thiocyanates in plants are thought to arise from the decomposition of glucosinolates[50]

$$R\text{-}C\begin{cases}SC_6H_{11}O_5\\ NOSO_3^-\end{cases}.$$

Although γ-thiocyanoaminobutyric acid has not yet been detected in nature, these findings raise the question whether cyanolysis of disulfides might offer an alternative route to natural organic and inorganic thiocyanates.

In a survey undertaken in our laboratory to determine the distribution of the new synthase enzyme, yeast was found to be capable of synthesizing γ-cyanoaminobutyric acid in this way. In the course of examination of rat liver, control experiments revealed that γ-thiocyanoamino-butyric acid could be utilized in the absence of cyanide. Cystathionase was known to be able to cleave carbon-sulfur bonds, and it was readily established that this enzyme in rat liver was liberating thiocyanate from γ-thiocyanoamino-butyrate (O. Abe & C. Ressler, unpublished). Substrates of γ-cystathionase are grouped below with the bonds indicated that are susceptible to cleavage.

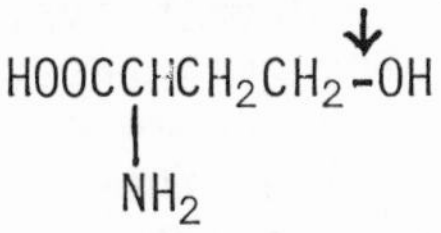

$$HOOCCH(NH_2)CH_2CH_2 \downarrow OH$$

Homoserine

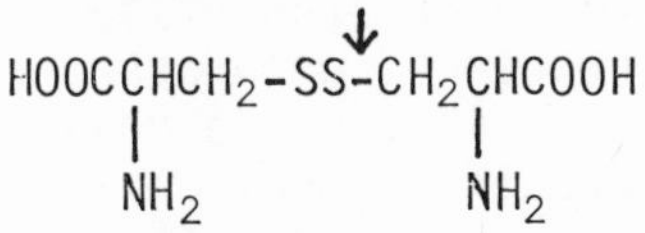

$$HOOCCH(NH_2)CH_2\text{-}SS \downarrow CH_2CH(NH_2)COOH$$

Cystine

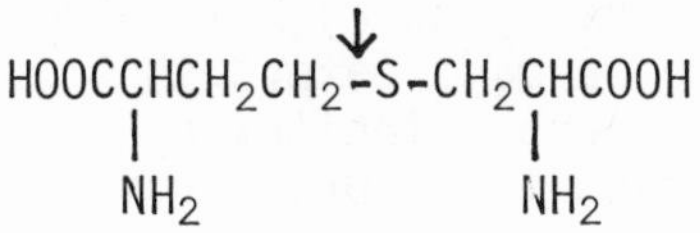

$$HOOCCH(NH_2)CH_2CH_2 \downarrow S\text{-}CH_2CH(NH_2)COOH$$

Cystathionine

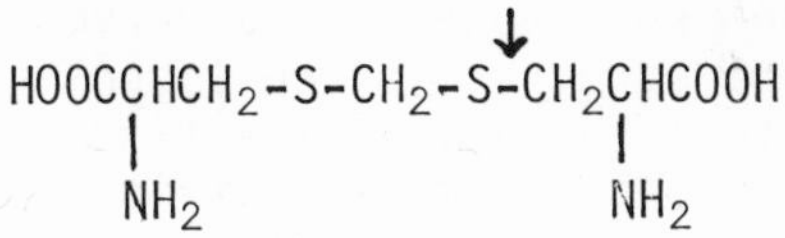

$$HOOCCH(NH_2)CH_2\text{-}S\text{-}CH_2\text{-}S \downarrow CH_2CH(NH_2)COOH$$

Djenkolic Acid

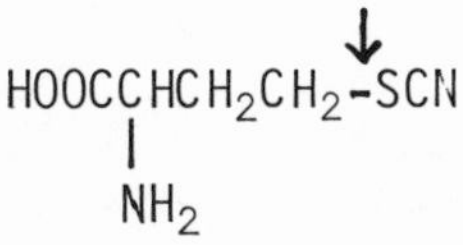

$$HOOCCH(NH_2)CH_2CH_2 \downarrow SCN$$

Thiocyanoaminobutyric Acid

γ-Thiocyanobutyrate was appreciably metabolized to thiocyanate by the intact rat, probably through a cystathionase-like mechanism. However, the thiocyanato-amino acid was far more toxic than administered thiocyanate. There was evidence that acute toxicity of γ-thiocyanobutyrate was accompanied by the formation of toxic amounts of cyanide. It is interesting to speculate that an amino acid transport process may serve to introduce γ-thiocyanoaminobutyrate into cells where it is cleaved enzymatically by cystathionase to thiocyanate. The introduction of thiocyanate into cells in this way may render it more susceptible to further metabolism than exogenous thiocyanate which is thought to be localized in extracellular fluid.

Investigations of plant seeds related historically or incidentally to the neurological disease lathyrism thus have led to the recognition of a new group of biologically active principles. These amino acid-like compounds of low molecular weight have structural features in common. They are only moderately toxic. Some are effective convulsants. Evidence has been accumulating that the principles can act as specific enzyme inhibitors, and, in some cases, as enzyme substrates. The enzymes affected include cystathionase, ornithine carbamyl transferase, a permease which mediates amino acid transport, asparaginase, glutaminase and glutamate decarboxylase as well as the ω-aminolysyl oxidase concerned with osteolathyrism. The relevance of enzyme inhibition to convulsion and to neurological disease remain as challenging questions. That these principles frequently can affect more than one biochemical reaction adds to the challenge. It may well be that clarification of this subject may have to await developments in the knowledge of the nervous system, just as new advances in the structure of collagen had to be made before the biochemical mechanism of action of β-aminopropionitrile was understood. Likewise it is hoped that these principles will be useful tools for studying the nervous system.

ACKNOWLEDGMENTS

This work was aided by U.S. Public Health Service Grant NS 04316 and by Muscular Dystrophy Associations of America.

REFERENCES

1. Ressler, C., 1964, Federation Proc. 23, 1350.
2. Bell, E.A. 1973. In "Toxicants Occurring Naturally in Foods" 2nd Ed., p. 153. Natl. Acad. Sci., Washington, D.C.
3. Anderson, L.A.P., A. Howard & J.L. Simonsen. 1925. Indian J. Med. Research, 12: 613.
4. Shah, S.R.A. 1939. Indian Med. Gas., 74: 385.
5. McKusick, V.A. 1972. "Heritable Disorders of Connective Tissue," 4th Ad., C.V. Mosby Co., St. Louis.
6. Goodman, S.I., C.C. Solomons, F. Muschenheim, G.A. McIntyre, B. Myles & D. O'Brien. 1968. Am. J. Med., 45: 152.
7. Page, R.C. & E.P. Benditt. 1967. Proc. Soc. Exper. Biol. Med. 124: 459.

8. LaBella, F.S. 1968. Gerontologist, 8: 13.
9. Narayanan, A.S., R.C. Siegel & G. R. Martin, 1972. Biochem. Biophys. Res. Commun., 46: 745.
10. _____, R.C. Siegel & G.R. Martin. 1974. Arch. Biochem. Biophys., 162: 331.
11. Schulert, A.R. & H.B. Lewis. 1952. Proc. Soc. Exper. Biol. Med., 81: 86.
12. O'Neal, R.M., C.-H. Chen, C.S. Reynolds, S.K. Meghal & R.E. Koeppe. 1968. Biochem. J., 106: 699.
13. Hermann, R.L., M.F. Lou & C.W. White. 1966. Biochim. Biophys. Acta, 121: 79.
14. Rao, S.L.N., P.S. Sarma, K.S. Mani, T.R.R. Rao & S. Sriramachari. 1967. Nature, 214: 610.
15. Watkins, J.C., D.R. Curtis & T.J. Biscoe. 1966. Nature. 211: 637.
16. McCrawford, J. 1963. Biochem. Pharmac., 12: 1443.
17. Cheema, P.S., K. Malathi, G. Padmanaban & P.S. Sarma. 1969. Biochem. J. 112: 29.
18. Mehta, T., A.-F. Hsu & B.E. Haskell. 1972. Biochem., 11: 4053.
19. Ressler C. & H. Ratzkin. 1961. J. Org. Chem., 26: 3356.
20. Kashelikar, D.V. & C. Ressler. 1964. J. Am. Chem. Soc., 86: 2467.
21. Ressler, C. 1962. J. Biol. Chem., 237: 733.
22. _____, S.N. Nigam, Y.-H. Giza & J. Nelson. 1963. J. Am. Chem. Soc., 85: 3311.
23. _____, S.N. Nigam & Y.-H. Giza. 1969. J. Am. Chem. Soc., 91: 2758.
24. _____, Y.-H. Giza & S.N. Nigam. 1963. J. Am. Chem. Soc., 85: 2874.
25. _____, Y.-H. Giza & S.N. Nigam. 1969. J. Am. Chem. Soc., 91: 2766.
26. Bond, E.J. 1961. Can. J. Biochem. Physiol., 39: 1793.
27. Blumenthal-Goldschmidt, S., G.W. Butler & E.E. Conn. 1963. Nature, 197: 718.
28. Tschiersch, B. 1963. Flora, 153: 115.
29. _____. 1964. Flora, 154: 445.
30. Ferris, J.P. 1970. In "Chemistry of the Cyano Group." (Z. Rappoport, ed.), p. 717. Interscience, New York.
31. Brysk, M.M., W.A. Corpe & L.V. Hankes. 1969. J. Bacteriol., 97: 322.
32. _____, C. Lauinger & C. Ressler. 1969. Biochim. Biophys. Acta, 184: 583.
33. _____ & C. Ressler. 1970. J. Biol. Chem., 245: 1156.
34. Ressler, C., J. Nelson & M. Pfeffer. 1967. Biochem. Pharmac., 16: 2309.

35. Sasaoka, K., C. Lauinger, S.N. Nigam & C. Ressler. 1968. Biochim. Biophys. Acta, 156: 128.
36. Feinstein, D.T. Curtis & C. Ressler. 1962. Federation Proc., 21: 651.
37. Pfeffer, M. & C. Ressler. 1967. Biochem. Pharmac., 16: 2299.
38. Giovanelli, J. & S.H. Mudd. 1971. Biochem. Biophys. Acta, 227: 654.
39. Perry, T.L., D.F. Hardwick, S. Hansen, D.L. Love & S. Israels. 1968. New Engl. J. Med., 278: 590.
40. Tate, S.S. & A. Meister. 1969. Biochem., 8: 1660.
41. Giza, Y.-H., H. Ratzkin & C. Ressler. 1963. Federation Proc., 22: 651.
42. Jackson, R.C., D.A. Cooney & R.E. Handschumaker. 1969. Federation Proc., 28: 601.
43. Lauinger, C. & C. Ressler, Biochim. Biophys. Acta, 198, 316 (1970).
44. Cooper, A.J.L. & A. Meister. 1973. In "The Enzymes of Glutamine Metabolism." (S. Prusiner & E.R. Stadtman, eds.) p. 207. Academic Press, New York.
45. Strobel, G.A. 1967. J. Biol. Chem., 242: 3265.
46. Ressler, C., G.R. Nagarajan, M. Kirisawa & D.V. Kashelikar. 1971. J. Org. Chem., 36: 3960.
47. Matthews, C.N. 1971. In "Chemical Evolution and Origin of Life," (R. Buvet & C. Ponnamperuma, eds.) pp. 231-235. North Holland.
48. Ressler, C. & T. Koga. 1971. Biochim. Biophys. Acta, 242: 473.
49. _____, O. Abe, Y. Kondo, B. Cottrell & K. Abe. 1973. Biochemistry, 12: 5369.
50. Van Etten, C.H. & T.A. Wolff. 1973. In "Toxicants Occurring Naturally in Foods." 2nd Ed., p. 210. Natl. Acad. Sci., Washington, D.C.

Chapter Eight

ADVANCES IN THE CHEMISTRY OF TUMOR-INHIBITORY NATURAL PRODUCTS

S. MORRIS KUPCHAN

Department of Chemistry
University of Virginia
Charlottesville, Virginia

INTRODUCTION

The discovery of the antileukemic activity of nitrogen mustard in the 1940's is considered by many to have been the start of modern cancer chemotherapy[2]. An enormous amount of work has followed on the synthesis and evaluation of biological alkylating agents[25-28]. Although the preparation of modifications of presently-known alkylating agents continues, some pessimism is evident among workers in the field. The extensive synthetic efforts to date have led only to minimal improvements over the prototype drugs. Evidently, the high degree of reactivity of simple alkylating agents leads to indiscriminate reactions with many cell constituents and, consequently, narrow therapeutic indices. There exists a need for new prototypes for the synthetic organic chemist to use in the design of selective alkylating agents.

Recent studies in the isolation and structural elucidation of tumor inhibitors derived from plants are yielding a fascinating array of novel types of growth-inhibitory compounds. Many of the new compounds possess structures and chemical properties which suggest that they may act by selective alkylation of growth-regulatory macromolecules. There appears to be reason for confidence that this approach will yield sorely-needed templates for the synthesis of potentially superior chemotherapeutic agents. As a corollary statement, the new growth inhibitors may provide powerful tools for the elucidation of new biochemical mechanisms of growth control which may be more amenable to selective regulation.

Studies of tumor inhibitors from plant sources are proceeding in many laboratories of wide geographic distribution. However, to limit the scope of the present discussion, the author will review only pertinent studies from his own laboratory.

The program started modestly, in 1959, with a screening study of crude extracts of a limited number of accessible plants for inhibitory activity against animal tumor systems. Some plants were procured by summer collections in Wisconsin, others by cooperative arrangements with botanists in India, Costa Rica, and other countries.

The results of testing of the first plant extracts prepared in this laboratory and elsewhere revealed that a small but significant number of the extracts showed reproducible tumor-inhibitory activity. Encouraged by these results, the National Cancer Institute (NCI) arranged with the U.S. Department of Agriculture to procure several thousand plant samples per year for evaluation. Shortly thereafter, the NCI arranged a contract with the Wisconsin Alumni Research Foundation, in Madison, to execute the initial extraction and screening studies. From that point onward, our program concentrated on the isolation and structural elucidation of new tumor inhibitors. To date, the active principles of more than one hundred plants have been isolated in our program, and the chemical studies of some of the most interesting compounds constitute the focus of this review.

One aspect of the approach of our program differs significantly from the classical, and most widely practiced, approach to the biological study of plant constituents. In the classical phytochemical approach, those compounds are studied which are most easily separated from a plant extract and most easily crystallized. In our program, however, the fractionation and isolation studies are guided at every stage by biological assays. The systematic fractionation, guided by biological assays, has made possible the isolation of important minor constituents which would most probably have been missed in the classical approach.

A consequence of our activity-directed fractionation and isolation procedures is the absence of any prejudicial selection of any particular class of compounds. Nevertheless, new and interesting relationships between chemical reactivity and growth-inhibitory activity are emerging. The focus of the remaining discussion will be the noteworthy chemical similarity of many of the tumor inhibitors which we have isolated, namely, the presence of highly electrophilic functions. The structures and reactivities of these compounds support the hypothesis that their tumor-inhibitory activity may be associated with selective alkylation of nucleophilic groups in enzymes which control cell division.

CHEMISTRY OF THE NEW TUMOR INHIBITORS

A systematic study of the cytotoxic principles of *Elephantopus elatus* Bertol. led to the isolation of two novel tumor-inhibitory germacranolide dilactones, elephantin and elephantopin. Although the compounds were concentrated and isolated solely on the basis of *in vitro* cytotoxicity, elephantopin has subsequently been found to show significant inhibitory activity against the P-388 lymphocytic leukemia in the mouse and the Walker carcinosarcoma-256 in the rat. (The KB cytotoxicity assay has been invaluable in many of our other isolation studies as well. Significant results were frequently obtained from the bioassay of 2 to 10 mg samples of materials which could be evaluated in animal tumor systems *in vivo* only at the cost of 0.5 to 1.0 g samples.) A combination of degradative, spectral, and X-ray crystallographic studies resulted in elucidation of the structures of elephantin and elephantopin[5,6].

Studies of *Vernonia hymenolepis* A. Rich led to the isolation and structural elucidation of two novel elemanolide dilactones, vernolepin and vernomenin. Vernolepin showed significant *in vitro* cytotoxicity and *in vivo* tumor-inhibitory activity against the Walker-256 carcinosarcoma

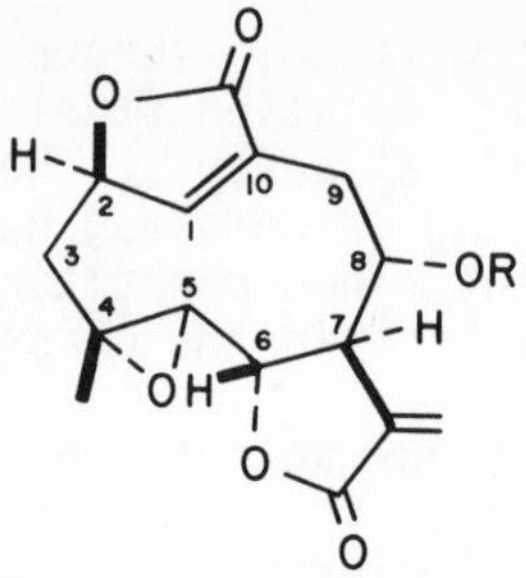

Elephantin : R : $(CH_3)_2C=CHCO-$

Elephantopin : R : $CH_2=C(CH_3)CO-$

in the rat. Elemental analysis and mass spectrometry indicated a $C_{15}H_{16}O_5$ molecular formula for vernolepin. Chemical and spectral evidence indicated the presence of two α,β-unsaturated lactone functions, a secondary alcohol, an additional double bond, and, therefore, a monocarbocyclic ring skeleton. The structure and stereochemistry for vernolepin, depicted at the upper left of Figure 1, were established by X-ray crystallographic analysis of the *p*-bromobenzene-sulfonate. Vernomenin showed chemical properties similar to vernolepin (R=H), and its structure was proved by (a) its conversion to the same methanol adduct as that obtained from vernolepin, and (b) a comparison of the NMR spectra of the respective acetate esters. In vernomenin acetate, the triplet centered at τ 4.78 could be assigned to the proton at acetate-bearing C-6, while the multiplet centered at τ 5.90 corresponded to the proton (spin-coupled to three protons) at lactone-bearing C-8. In contrast, the spectrum of vernolepin acetate showed a multiplet at τ 4.95, assigned to the proton at acetate-bearing C-8, while the lactone proton signal appeared as a triplet centered at τ 5.96, indicative of attachment to C-6[14-15].

In an attempt to elucidate a possible function of a series of new plant-derived tumor inhibitors, the collaboration of Professor Luis Sequeira was enlisted in an evaluation of the effects of these compounds on plant growth. Several of the sesquiterpene dilactones (specifically, elephantin, elephantopin, and vernolepin) were found to be strong inhibitors of extension growth of wheat coleoptile

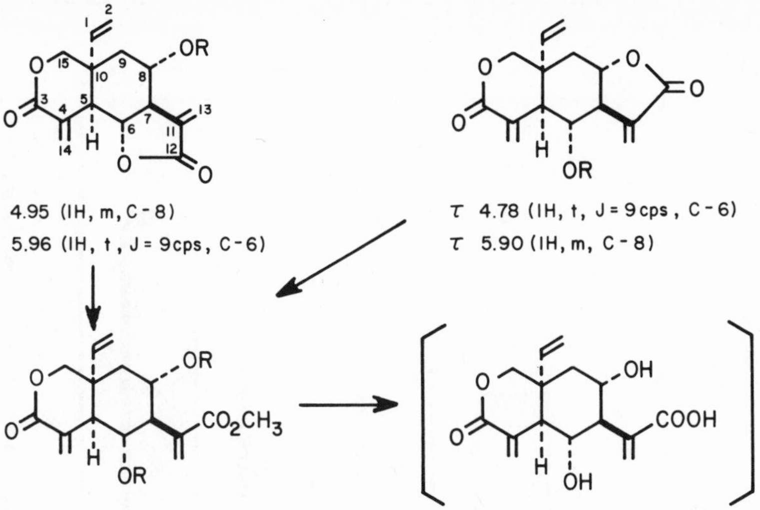

τ 4.95 (1H, m, C-8)
τ 5.96 (1H, t, J = 9cps, C-6)

τ 4.78 (1H, t, J = 9cps, C-6)
τ 5.90 (1H, m, C-8)

Figure 1. Structures of vernolepin and vernomenin.

sections[26]. Vernolepin in concentrations of 5-50 μg/ml inhibits extension growth from 20 to 80% as shown in Figure 2. If the inhibited sections are washed and subsequently treated with indole-3-acetic acid, the tissues respond to the auxin, but the degree of elongation is determined by the length of prior treatment with vernolepin. The fact that vernolepin's plant growth-inhibitory activity is reversible suggests that the compound may have a natural function in the regulation of plant growth.

Several recent observations have focused attention on the importance of the conjugated α-methylene lactone function for the biological activity of the sesquiterpene lactones. For instance, the plant-growth inhibitory effect of vernolepin is completely blocked by the addition of sulfhydryl compounds such as mercaptoethanol to the medium. Secondly, as shown in Figure 3, the cytotoxicity of vernolepin derivatives appears to be directly related to the presence of free conjugated α-methylene-γ-lactone functions. Thus, selective reduction of the ethylidene double bond does not appear to affect cytotoxicity. However, modification of the α-methylene-γ-lactone (by *trans*-esterification to the methanol adduct or by hydrogenation) results in a tenfold

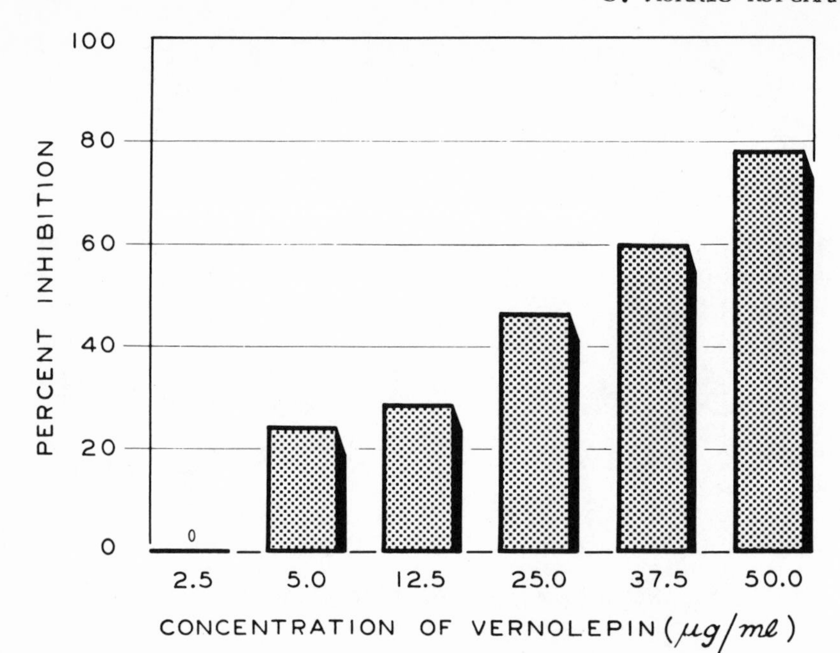

Figure 2. Wheat coleoptile bioassays.

VERNOLEPIN
ED_{50} 2.0 γ

DIHYDROVERNOLEPIN
2.0 γ

TETRAHYDROVERNOLEPIN
19.0 γ

HEXAHYDROVERNOLEPIN
> 100 γ

ACIDIC METHANOLYSIS PRODUCT
26 γ

Figure 3. Cytotoxicity of vernolepin derivatives.

diminution in cytotoxicity. Modification of both α-methylene lactone systems, as in hexahydrovernolepin, leads to a derivative which is essentially inactive.

We have studied the reactions of several conjugated α-methylene lactones with model biological nucleophiles. The rates of the reactions of vernolepin and elephantopin with cysteine at pH 7.4 were determined spectrophotometrically (Figures 4 & 6). The most reactive function in each case was the conjugated α-methylene-γ-lactone, and the second-order rate constants for the Michael-type addition of cysteine showed the same order of reactivity toward cysteine as iodoacetate, a commonly used sulfhydryl reagent. In contrast to the reactivity of the lactones toward sulfhydryl groups, their reaction with amino groups appeared to be very slow. When a solution equimolar in lysine and vernolepin at pH 7.4 was allowed to stand for 6 days at 25° C, 75% of the original lactone could be recovered. Similarly, guanine proved unreactive toward either vernolepin or elephantopin. The biscysteine adducts, as expected, were essentially inactive[10].

Structure-cytotoxicity relationships for elephantopin derivatives (Figure 5) paralleled those of vernolepin. Similarly, the reactivity of the functional groups toward cysteine (Figure 6) paralleled those of vernolepin. More recent studies of several dozen cytotoxic sesquiterpene lactones confirmed the requirement for an α-methylene-γ-lactone for cytotoxicity and pointed to the need for a second unsaturated carbonyl function for *in vivo* tumor-inhibitory activity[9].

The biomimetic addition of thiols to conjugated α-methylene lactones has been used to good advantage as a protecting reaction for the α-methylene groups. When vernolepin was treated with 1-propanethiol, the bis-1-propanethiol adduct was obtained in high yield (4, Figure 7)[11]. Methylation of the bis-adduct with methyl iodide followed by trituration of the sulfonium salt with sodium bicarbonate led to regeneration of vernolepin (1). To prepare dihydrovernolepin (6), the adduct 4 was hydrogenated with 10% palladium on carbon to 5, and the blocking groups were removed by the methyl iodide-sodium bicarbonate sequence.

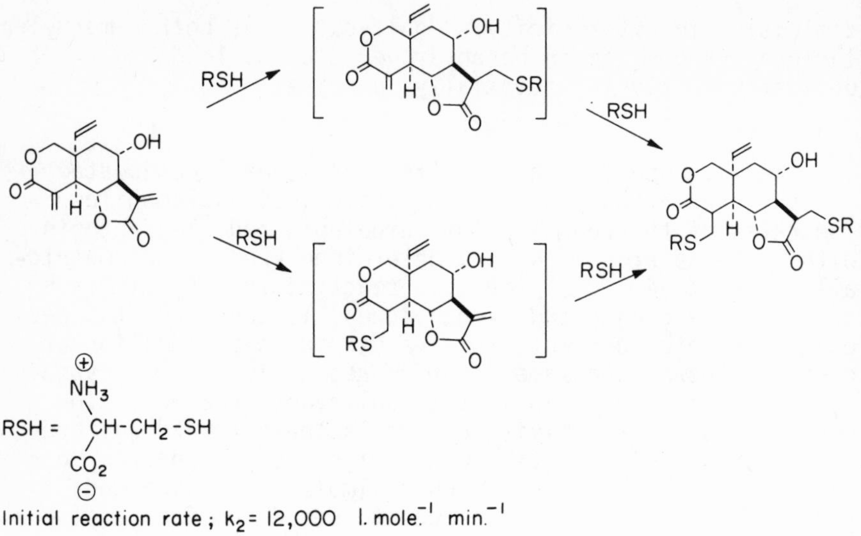

RSH = $NH_3^{\oplus}$-CH-CH_2-SH, $CO_2^{\ominus}$

Initial reaction rate; k_2 = 12,000 l. mole^{-1} min.$^{-1}$

Figure 4. Reaction of vernolepin with L-cysteine at 25°, pH 7.4.

ELEPHANTOPIN
ED_{50} 0.32γ

ELEPHANTOL
22.0γ

ELEPHANTOL ACETATE
1.80γ

TETRAHYDROELEPHANTOPIN
72γ
HEXAHYDROELEPHANTOPIN
>100γ

DIHYDROELEPHANTOL
>100

ELEPHANTOL METHACRYLATE
0.63γ

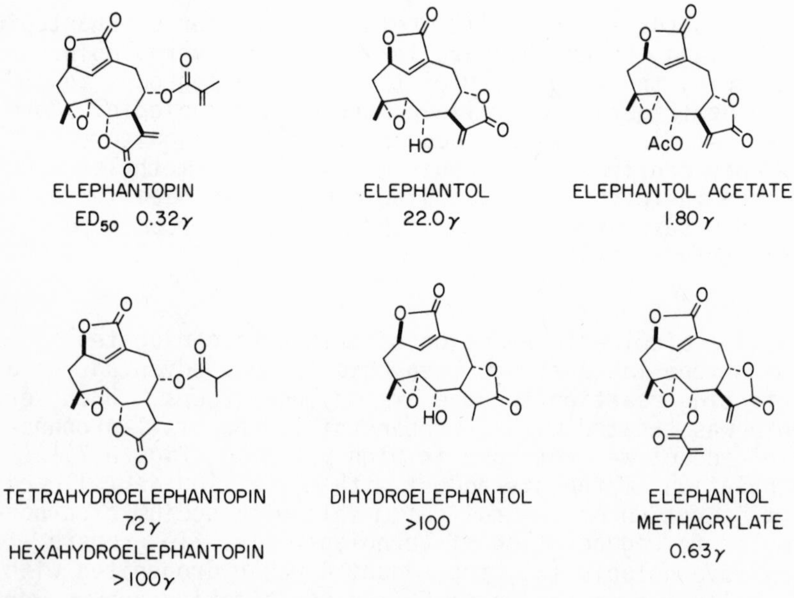

Figure 5. Cytotoxicity of elephantopin derivatives.

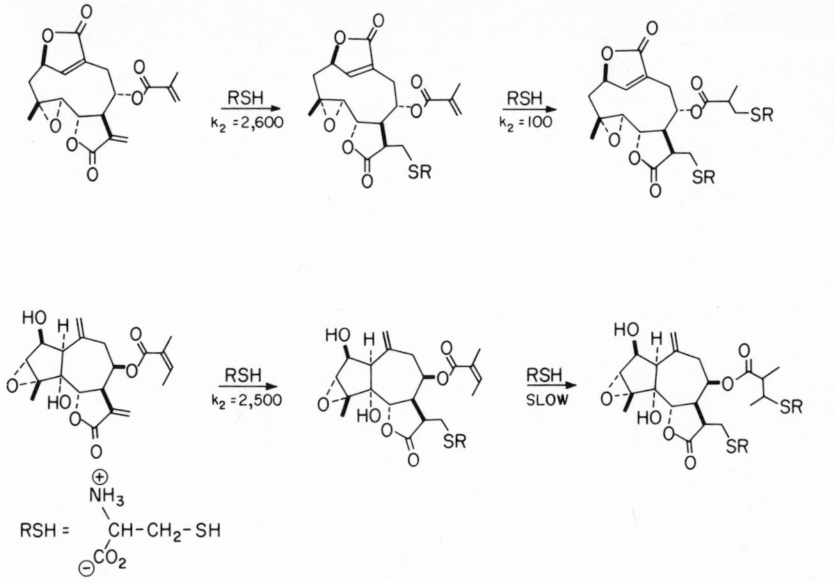

Figure 6. Reaction of elephantopin and derivatives with L-cysteine at 25° C, pH 7.4.

Similarly, the interrelation of elephantopin and deoxyelephantopin was accomplished *via* zinc-copper couple reductive elimination of the epoxide ring of the bis-thiol adduct 9 (Figure 8)[23]. Treatment with methyl iodide led to removal of the blocking groups and conversion to deoxyelephantopin (11).

The observed ease of the "retro-Michael" reaction of the methyl sulfonium salts of the thiol adducts of vernolepin and elephantopin may have significance for the selectivity reflected in the *in vivo* tumor-inhibitory activity of agents which contain conjugated electrophilic functionalities. If thioethers are methylated less efficiently in neoplastic tissues than in normal tissues, there could result a selective toxicity of tumor-inhibitory activity of agents such as α-methylene lactones toward neoplastic tissues[11]. Presumably, the greater effectiveness of thioether methylation would result in rapid detoxification of the alkylating agents in normal tissues.

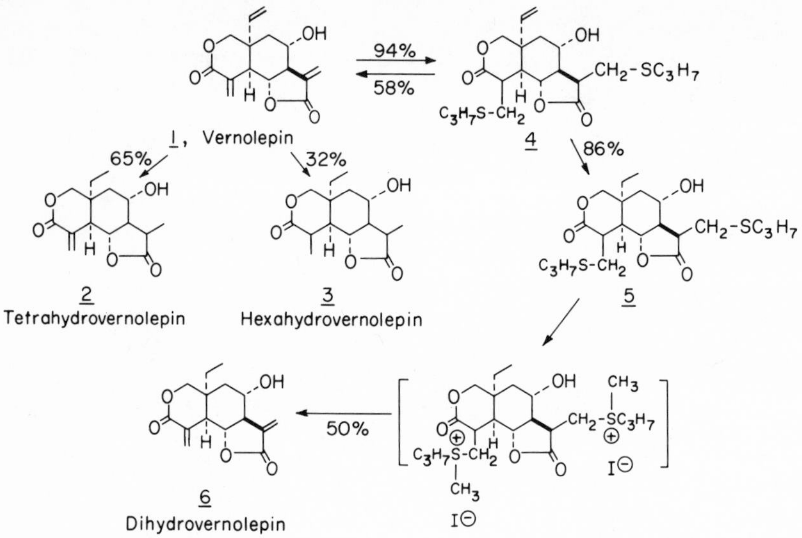

OH
94%
58%
OH
$CH_2-SC_3H_7$
H
H
$C_3H_7S-CH_2$
1, Vernolepin
4
65%
32%
86%
OH
OH
OH
H
H
$CH_2-SC_3H_7$
H
$C_3H_7S-CH_2$
2
Tetrahydrovernolepin
3
Hexahydrovernolepin
5
OH
OH
CH_3
50%
$CH_2-\overset{\oplus}{S}C_3H_7$
H
$C_3H_7\overset{\oplus}{S}-CH_2$
H
$I^\ominus$
CH_3
6
Dihydrovernolepin
$I^\ominus$

Figure 7. Reversible thiol addition in the synthesis of dihydrovernolepin.

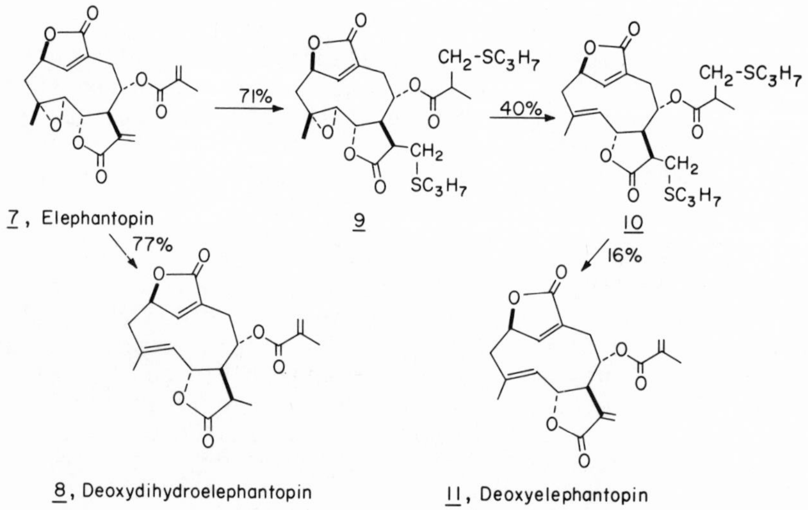

$CH_2-SC_3H_7$
$CH_2-SC_3H_7$
71%
40%
CH_2
CH_2
SC_3H_7
SC_3H_7
7, Elephantopin
9
10
77%
16%

8, Deoxydihydroelephantopin

11, Deoxyelephantopin

Figure 8. Reversible thiol addition in the interrelation of elephantopin with deoxyelephantopin.

Systematic studies of a tumor-inhibitory extract from *Taxodium distichum* Rich led to the isolation and structural elucidation of two novel diterpenoid quinone methides, taxodione and taxodone (1 and 5, Figure 9)[18,19]. The structures were deduced from their formulae and spectra, and confirmed by interrelation with a known diterpene. Both quinone methides were converted to 2, which was methylated to 3. Reduction of 3 with lithium aluminum hydride gave the 6β-alcohol, 4, which was dehydrated to 6. Catalytic hydrogenation of 6 gave 7, characterized by direct comparison with a sample prepared from sugiol (8). Among the six diterpenoid derivatives isolated from *Taxodium distichum*, only the quinone methide derivatives taxodione and taxodone showed significant inhibitory activity *in vivo* against the Walker carcinosarcoma-256 in the rat and *in vitro* against cells derived from human carcinoma of the nasopharynx (KB). This fact and the known sensitivity of quinone methides to nucleophilic

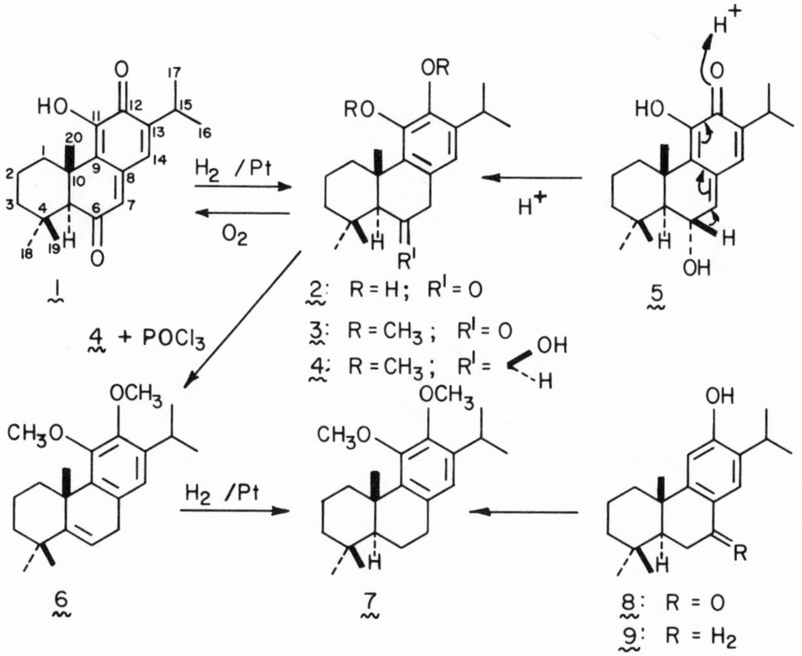

Figure 9. Taxodione (1) and taxodone (5), WM-inhibitory principles of *Taxodium distichum*.

attack suggested that taxodione and taxodone might exert their biological effect by interaction with a biological nucleophile at C-7.

Extracts of *Jatropha gossypiifolia* have been used to treat cancer in Costa Rica and elsewhere for many years. When an extract forwarded by Professor J.A. Saenz Renauld was tested by the National Cancer Institute, the extract was found to inhibit five standard tumor systems. Systematic fractionation led to the isolation of the novel antileukemic macrocyclic diterpenoid jatrophone (I), whose structure was elucidated with the help of X-ray crystallographic analysis of the dihydrobromide adduct II (Figure 10)[21]. To evaluate whether jatrophone's electrophilicity would be manifested in reactions with model biological nucleophiles, the compound was treated with *n*-propyl thiol. A well-characterized thiol mono-adduct (IIIa) was obtained[24].

Brucea antidysenterica Mill. is a simaroubaceous tree which is used in Ethiopia in the treatment of cancer. We found that an alcoholic extract of the plant showed significant inhibitory activity *in vitro* against cells derived

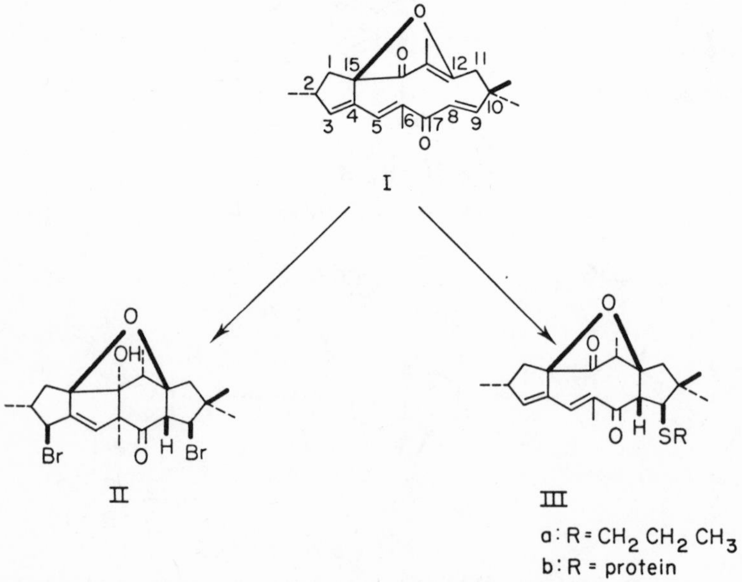

I

II

III

a: R = $CH_2CH_2CH_3$
b: R = protein

Figure 10. Products of nucleophilic addition to jatrophone.

from human cancer of the nasopharynx and against two standard animal tumor systems. Systematic fractionation led to the isolation and structural elucidation of two new bitter principles, bruceantin and bruceantarin[7]. Bruceantin shows potent antileukemic activity in the μg/kg

Bruceantin : R = OC
Bruceantarin : R = OCC_6H_5
Bruceine B : R = $OCCH_3$

dose range. In contrast, bruceantarin shows moderate activity and bruceine B shows only marginal antileukemic activity. In view of the noteworthy activity of bruceantarin, which lacks an α,β-unsaturated ester, the ester moiety may well serve as a carrier, rather than an alkylating group.

An ethanolic extract of *Acer negundo* L. was shown to possess significant inhibitory activity against the Sarcoma 180 and Walker intramuscular carcinosarcoma-256 tumor systems. Systematic fractionation led to the isolation of the active principles as the chromatographically homogeneous acids, saponin P and saponin Q[13]. Saponin P has shown a high therapeutic index in further biological studies, and the National Cancer Institute has procured a large collection of *Acer negundo* for extraction of saponin P in a quantity sufficient for preclinical toxicological studies and preliminary clinical trials. Our chemical studies of saponin P began with acid hydrolysis, which yielded glucose and arabinose, along with two aglycones, acerotin and acerocin. We have recently shown that the aglycones are diesters of the new acidic triterpene acerogenic acid. Each has a unique nonadienoate residue; one

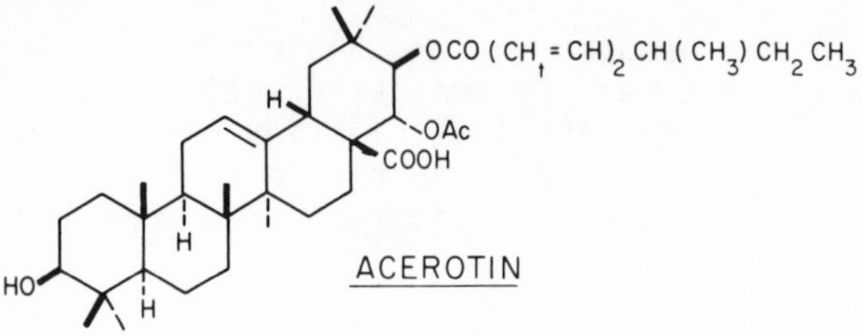

ACEROTIN

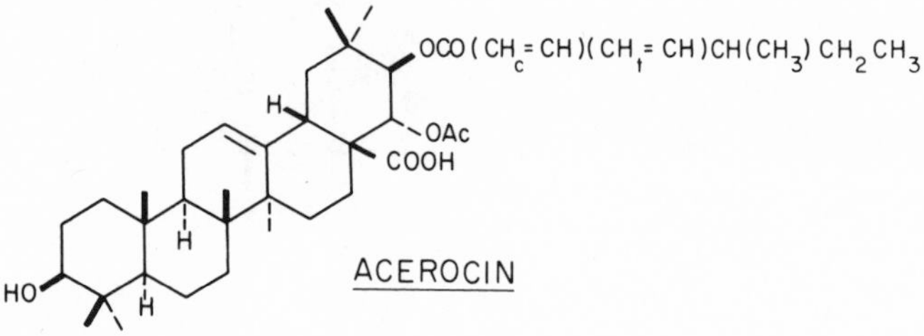

ACEROCIN

has *trans-trans*, and the other *cis-trans* stereochemistry[22]. As the presence of α,β-unsaturated carbonyl functions has been shown to be responsible for the tumor-inhibitory activity of other natural products, the electrophilic unsaturated esters in saponin P may contribute significantly toward making this material the most promising of the known tumor-inhibitory saponins.

An alcoholic extract of *Tripterygium wilfordii* Hook was found to show significant activity against L-1210 leukemia in mice. Systematic fractionation led to the isolation of triptolide and tripdiolide (1 & 2, Figure 11) as the antileukemic principles, and direct X-ray crystallographic analysis resulted in elucidation of their structures[8]. The diepoxide functionality has been shown to confer tumor-inhibitory activity upon certain classes of acyclic synthetic compounds[25], as well as the naturally occurring cyclohexane diepoxide, crotepoxide[12]. The α,β-unsaturated lactone function has been shown to be important for the tumor-inhibitory activity of several classes of terpenoids[4]. Consequently, the hypothesis was entertained that one or more of the four highly electrophilic functions of the antileukemic triptolides 1 and 2 may be involved in their mode of action. The NMR spectra of 1 and 2 display resonances attributable to strong hydrogen bonding between the 14-hydroxyl and the 9,11-epoxide groups[20]. The co-occurring keto-analog

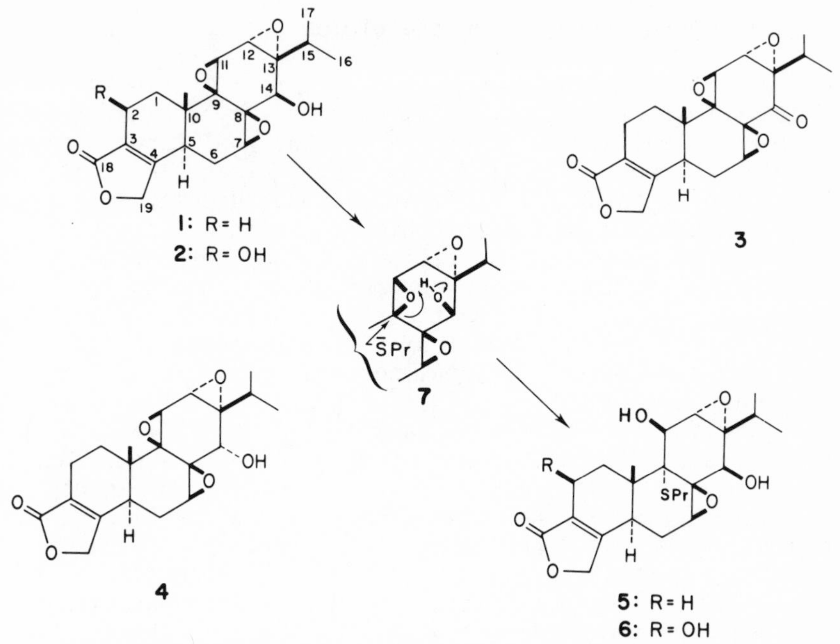

Figure 11. Structures of the triptolides and their derivatives.

triptonide (3), which differs from 1 solely at C-14, shows neither the resonances attributable to the hydrogen-bonded epoxide system nor antileukemic activity. In addition, subsequent testing of the minor variants 4, 5, and 6 revealed no antileukemic activity. Upon treatment with propanethiol, 1 and 2 underwent selective nucleophilic attack at C-9, to yield the adducts 5 and 6, respectively. In contrast, 4 was recovered largely unchanged after similar treatment. The increased reactivity of the antileukemic triptolides 1 and 2, relative to the inactive 14-epimeric derivative, 4, may be attributable to hydroxyl-group participation in the epoxide opening of the former compounds, as depicted in 7. The results agree with the hypothesis that the hydroxyl-assisted attack by nucleophiles on the 9,11-epoxide may mimic the mechanism by which the antileukemic triptolides exert their biological activity[20]. Conceivably, *in vivo*, one or more of the additional electrophilic functions may effect subsequent alkylation of other neighboring nucleophiles on the biological macromolecule.

Alcoholic extracts of *Maytenus serrata* (formerly *M. ovatus*) showed potent inhibitory activity against five standard animal tumor systems in the National Cancer Institute's screen. Fractionation of the alcoholic extract was guided by assay for activity against KB cells *in vitro* or P-388 leukemia in mice. A highly enriched concentrate was prepared by solvent partitions, column chromatography, acetylation, further column chromatography and repeated preparative thin layer chromatography. Maytansine (3-bromopropyl) ether was prepared from the concentrate and was characterized by X-ray crystallographic analysis as 3 in Figure 12 (see also Figure 13)[17]. Hydrolysis of this derivative yielded crystalline maytansine (1), which was used to seed the concentrate to yield additional maytansine. Maytansine is an exceptionally interesting antitumor agent, for it is active at the level of micrograms per kilogram of animal body weight. Furthermore, it exerts significant inhibitory activity against mouse P-388 lymphocytic leukemia over a 50- to 100-fold dosage range. Interest has been heightened by the finding that maytansine also shows significant inhibitory activity against L-1210 mouse leukemia, and the Lewis lung carcinoma and B-16 melanocarcinoma solid murine tumor systems, and the agent is under toxicological investigation in preparation for clinical trials.

Recent studies have led to the isolation and characterization of four new maytansinoids from *M. buchananii*[16]. Maytanvaline (4) is a highly-active antileukemic maytanside ester. Maysine (5), normaysine (6), and maysenine (7) lack antileukemic activity, and show only 1/10,000 the cytotoxicity of maytanside esters such as maytansine (1). The X-ray crystallographic structural elucidation of maytansine (3-bromopropyl) ether (3) revealed that the two longer sides of the 19-membered ring are roughly parallel and separated by about 5.4Å, so that there is a hole in the center of the ring (Figure 13)[1,17]. The two faces of the ring have a different character; the lower surface, opposite the ester residue, is predominantly hydrophobic, while the upper face is more hydrophilic. Furthermore, the ester residue is oriented in a manner which would sterically hinder the approach of reactants to the hydrophilic face. The ester function in the antileukemic maytansinoids may play a key role in the formation of highly selective molecular complexes with growth regulatory biological macro-

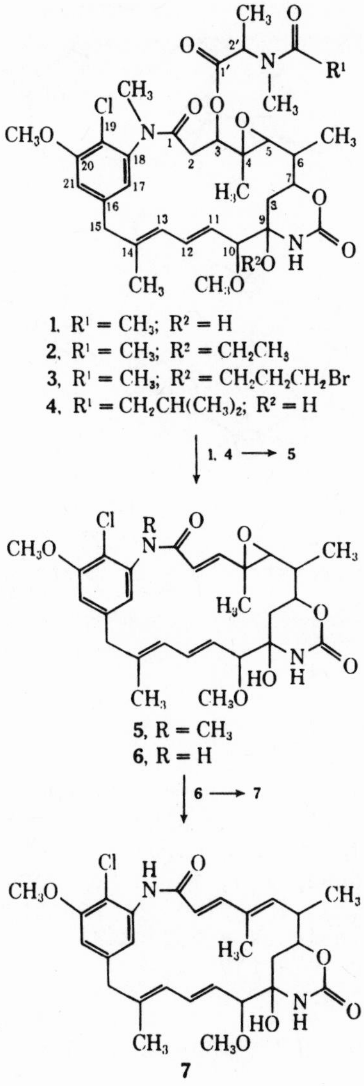

CH_3 O
O 2′ N R^1
1′
CH_3
Cl O O CH_3
CH_3O 19 N O
1 3 5 6 CH_3
20 18 2 4
21 17 H_3C 7
16 8 O
15 13 11 9
14 12 10 N O
R^2O H
CH_3 CH_3O

1, $R^1 = CH_3$; $R^2 = H$
2, $R^1 = CH_3$; $R^2 = CH_2CH_3$
3, $R^1 = CH_3$; $R^2 = CH_2CH_2CH_2Br$
4, $R^1 = CH_2CH(CH_3)_2$; $R^2 = H$

1, 4 ⟶ 5

Cl R O
CH_3O N O CH_3
H_3C
O
N O
HO H
CH_3 CH_3O

5, $R = CH_3$
6, $R = H$

6 ⟶ 7

Cl H O
CH_3O N CH_3
H_3C
O
N O
HO H
CH_3 CH_3O
7

Figure 12. Structures and interrelations of maytansinoids.

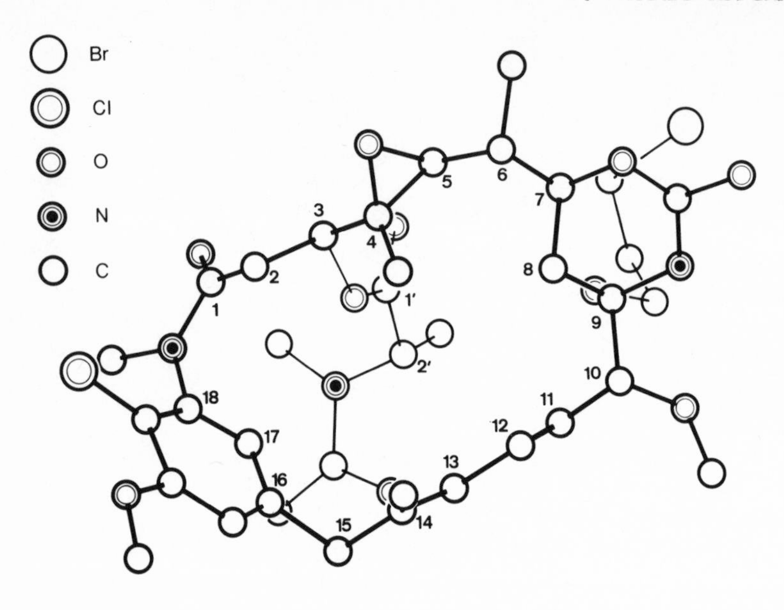

Figure 13. Molecular structure of maytansine (3-bromopropyl) ether.

molecules. Such molecular complex formation may be crucial for the subsequent selective alkylation of specific nucleophiles, for example, by the carbinolamide and epoxide functions. In this connection, it is noteworthy that maytansine ethyl ether (2), in which the reactive carbinolamide is no longer available as a potential alkylating function, shows no antileukemic activity.

BIOCHEMISTRY OF THE NEW TUMOR INHIBITORS

To test the possibility that the quinone methides, taxodone and taxodione, and α-methylene lactones, such as vernolepin, may react with sulfhydryl enzymes, their effects on phosphofructokinase were studied[3]. Phosphofructokinase contains 16 to 18 sulfhydryl groups per protomer of 93,200 daltons. The tumor inhibitors were found to inhibit phosphofructokinase, and evidence was presented that the inhibition results from their reaction with the sulfhydryl groups of the enzyme.

Taxodione was the most potent of the tumor inhibitors in diminishing phosphofructokinase activity. Addition of only 1.6 moles of taxodione per protomer of enzyme reduced activity by 50%. The other quinone methide, taxodone, inhibited 50% at a relative concentration of 32, and vernolepin and other sesquiterpene lactones inhibited 50% at relative concentrations of 1000 to 2000.

Similar results were obtained in a study of the action of vernolepin upon the enzyme glycogen synthase[27]. Vernolepin inactivates glycogen synthase. To measure the stoichiometry of the reaction with the enzyme protein, a procedure for tritiation of vernolepin was developed. Reaction with 3 moles of radioactive reagent per 90,000-dalton subunit caused virtually complete loss of activity. The concurrent disappearance of 3 titratable thiol groups (out of 6) indicated that thioether formation, i.e. alkylation of sulfhydryl groups, is the major mode of binding to the protein. The persistence of 1 to 2 residual thiol groups in the enzyme subunit, even after incubation with vernolepin in the presence of sodium dodecyl sulfate, pointed to the selectivity in the reaction with vernolepin.

In a recent study, jatrophone was shown to react with small molecular weight thiols as well as thiol groups on proteins such as bovine serum albumin and DNA-dependent RNA polymerase from *Escherichia coli*[24]. When bovine serum albumin was treated with jatrophone, all of the sulfhydryl groups in the albumin molecule had reacted after 15 hours. Gel filtration afforded a good separation between protein and excess jatrophone. The separated protein contained virtually no free sulfhydryl groups, and the UV spectrum was in agreement with a monoadduct product of type IIIb (Figure 10). When a portion of the jatrophone-treated protein was precipitated with trichloroacetic acid, no UV absorbing material was detected in the supernatant, indicating that the jatrophone was covalently linked to the albumin.

The reaction between jatrophone and protein sulfhydryl groups was further investigated with RNA polymerase from *E. coli*[24]. After 4 hours, approximately 50% of the activity had been lost, and approximately 8 sulfhydryl groups on the enzyme had reacted with jatrophone. Further incubation up to 22 hours resulted in almost complete loss of activity. At that point 11 sulfhydryl groups of the enzyme had reacted. When a sample of the polymerase which had been allowed to react with jatrophone for 4 hours was subjected to gel filtration, the spectrum of the treated polymerase was similar to that of jatrophone-treated bovine serum albumin. Evidently, in this case also, a monoadduct of type IIIb had been formed.

CONCLUSION

The activity-directed isolation of new tumor inhibitors of plant origin has yielded many novel compounds with significant growth-inhibitory properties. A significant proportion of the new compounds contain highly electrophilic functionalities, and chemical and biochemical studies are yielding a growing body of evidence to support the view that these compounds may act by selective alkylation of growth-regulatory biological macromolecules. The selectivity may result from many factors, among which are the transport of the tumor inhibitor into the cell and the chemical nature and steric environment of the specific nucleophile to be alkylated. Model studies support the hypothesis that the inhibition of tumor growth by the new agents may be attributable to selective alkylation of key enzymes which control cell division. The new tumor inhibitors may have a triple selectivity: (a) generally for thiols over other nucleophiles, (b) specifically for particular sulfhydryl enzymes, and (c) specifically for particular sulfhydryl groups within those enzymes. The definitive characterization of the nature of the nucleophiles and of the biological macromolecules which bear these nucleophiles number among the goals at the very frontier of research in cancer chemotherapy.

ACKNOWLEDGMENT

I would like to pay tribute to the skill and devotion of my collaborators and students whose names are given in the various references. I cannot adequately express my indebtedness to these colleagues.

REFERENCES

1. Bryan, R.F., C.J. Gilmore & R.C. Haltiwanger. 1973. J.Chem. Soc., Perkin II: 897.
2. Gilman, A., & F.S. Philips. 1946. Science, 103: 409.
3. Hanson, R.L., H.A. Lardy & S.M. Kupchan. 1970. Science, 168: 368.
4. Kupchan, S.M. 1970. Pure Appl. Chem. 21: 227.
5. _____, Y. Aynehchi, J.M. Cassady, A.T. McPhail, G.A. Sim, H.K. Schnoes & A.L. Burlingame. 1966. J. Am. Chem. Soc. 88: 3674.
6. _____, Y. Aynehchi, J.M. Cassady, H.K. Schnoes & A.L. Burlingame. 1969. J. Org. Chem. 34: 3867.
7. _____, R.W. Britton, M.F. Ziegler & C.W. Sigel. 1973. J. Org. Chem. 38: 178.
8. _____, W.A. Court, R.G. Dailey, Jr., C.J. Gilmore & R.F. Bryan. 1972. J. Am. Chem. Soc. 94: 7194.
9. _____, M.A. Eakin & A.M. Thomas. 1971. J. Med. Chem. 14: 1147.
10. _____, D.C. Fessler, M.A. Eakin & T.J. Giacobbe. 1970. Science. 168: 376.
11. _____, T.J. Giacobbe & I.S. Krull. 1970. Tetrahedron Lett. 2859.
12. _____, R.J. Hemingway, P. Coggan, A.T. McPhail & G.A. Sim. 1968. J. Am. Chem. Soc. 90: 2982.
13. _____, R.J. Hemingway, J.R. Knox, S.J. Barboutis, D. Werner & M.A. Barboutis. 1967. J. Pharm. Sci. 56: 603.
14. _____, R.J. Hemingway, D. Werner & A. Karim. 1969. J. Org. Chem. 34: 3903.
15. _____, R.J. Hemingway, D. Werner, A. Karim, A.T. McPhail & G.A. Sim. 1968. J. Am. Chem. Soc. 90: 3596.
16. _____, Y. Komoda, A.R. Branfman, R.G. Dailey, Jr. & V.A. Zimmerly. 1974. J. Am. Chem. Soc. 96: 3706.
17. _____, Y. Komoda, W.A. Court, G.J. Thomas, R.M. Smith, A. Karim, C.J. Gilmore, R.C. Haltiwanger & R.F. Bryan. 1972. J. Am. Chem. Soc. 94: 1354.
18. _____, A. Karim & C. Marcks. 1968. J. Am. Chem. Soc. 90: 5923.
19. _____, A. Karim & C. Marcks. 1969. J. Org. Chem. 34: 3912.
20. _____, & R.M. Schubert. 1974. Science. 185: 791.

21. _____, C.W. Sigel, M.J. Matz, J.A. Saenz Renauld, R.C. Haltiwanger & R.F. Bryan. 1970. J. Am. Chem. Soc. 92: 4476.
22. _____, M. Takasugi, R.M. Smith & P.S. Steyn. 1971. J. Org. Chem. 36: 1972.
23. Kurokawa, T., K. Nakanishi, W. Wu, H.Y. Hsu, M. Maruyama & S.M. Kupchan. 1970. Tetrahedron Lett., 2863.
24. Lillehaug, J.R., K. Kleppe, C.W. Sigel & S.M. Kupchan. 1973. Biochim. Biophys. Acta, 327: 92.
25. Ross, W.C.J. 1962. "Biological Alkylating Agents," Butterworths, London.
26. Sequeira, L., R.J. Hemingway & S.M. Kupchan. 1968. Science, 161: 789.
27. Smith, C.H., J. Larner, A.M. Thomas & S.M. Kupchan. 1972. Biochim. Biophys. Acta, 276: 94.
28. Wheeler, G.P. 1962. Cancer Res. 22: 651.

Chapter Nine

LABORATORY MODELS FOR THE BIOGENESIS OF INDOLE ALKALOIDS

A. IAN SCOTT

Sterling Chemistry Laboratory
Yale University
New Haven, Connecticut

INTRODUCTION

In an earlier discussion, we described some useful techniques for the study of indole alkaloid biosynthesis based mainly on relatively short term incubations (5 min - 7 days)[1]. The problems presented by strychnine biosynthesis in *Strychnos nux-vomica* called for the adoption of another variant in the methodology of reaching the site of biosynthesis in a higher plant, i.e. long-term incubation. Since this approach might be useful in other studies we begin our discussion with a description of these experiments. The delineation of the major pathway to the indole alkaloids having reached a fairly advanced stage, we have undertaken a systematic investigation of the biogenetic-type synthesis of all of the major classes of alkaloid, starting with and as far as possible retaining Nature's own intermediates throughout. The following account of our researches and complementary studies by other workers in the field indicates the state of the art of this endeavor and at the same time hopefully points the way to further biochemical experiments. We also comment on the relevance of these studies for chemotaxonomy and for the enzymology of alkaloid biosynthesis.

BIOSYNTHESIS OF STRYCHNINE IN *STRYCHNOS NUX-VOMICA*

In contrast to the modest but measurable incorporation of the larger intermediate substrates in *Catharanthus roseus*, species such as *Rauwolfia serpentina* and *Strychnos nux-vomica* have, from the outset, displayed most disappointing properties as far as precursor uptake has been concerned. Indeed if these had been the only plants available for study it might be suggested that our knowledge of the intricate mechanisms whereby mevalonate and tryptophan are co-metabolized to the complex indole alkaloids of the major structural types (for example, the *Strychnos* alkaloids) would still be in its infancy. Thus the administration of geissochizine (a known precursor of the "*Strychnos*" alkaloid akuammicine in *V. rosea*) to *S. nux-vomica* under "normal" feeding conditions (5 to 7 days) led to insignificant ($<2 \times 10^{-3}$ percent) incorporations into strychnine. More surprisingly, the Wieland-Gumlich aldehyde and diaboline, which seem almost certain candidates for intermediacy (near the end of the pathway), failed to incorporate over the same time scale. Reinvestigation of the problem involved the technique of replanting young seedlings after 5 day's incubation with 2-, 15-, 16-, and 17-labeled intermediates (^{3}H, aryl group). After a further 100 days, strychnine was isolated from each feeding, crystallized to constant radioactivity, and (in the case of positive incorporation) degraded to locate the position of the tritium label. Reference to Figure 1 shows that indeed geissoschizine fulfills the same pivotal role in *S. nux-vomica* as it does in *C. roseus* but that the result would have been in doubt without resource to the "prolonged contact" technique. Similarly the anticipated but hitherto unrealized conversion of Wieland-Gumlich aldehyde to strychnine (1.60 percent) could be demonstrated. The nonincorporation of diaboline after 70 days suggests strongly that the seventh ring of the complex strychnine molecule is built in

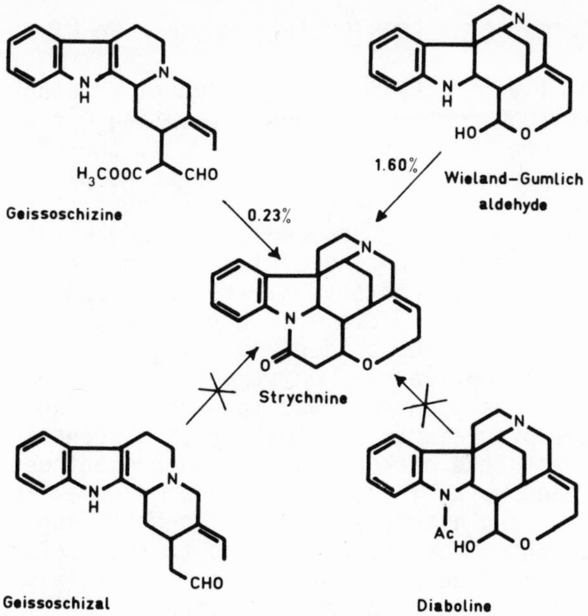

Figure 1. Interrelationships between strychnine and potential precursors in *Strychnos nux-vomica* revealed by prolonged feeding.

a clockwise carbon-to-carbon addition of the acetic acid unit, rather than from the nitrogen-carbon condensation which would have required diaboline as an intermediate. The nonincorporation of geissoschizal reveals that, by a hitherto undisclosed mechanism, a carbon atom is lost somewhere between geissoschizine and Wieland-Gumlich aldehyde but not by a simple decarbomethoxylation reaction.

These experiments do not begin to answer the question of genetic control and evolution in a plant such as *S. nux-vomica*, which, although producing quite massive concentrations of strychnine and brucine by a process now shown to emanate from geissoschizine, does not elaborate the variety of rearranged structures of the *Aspidosperma* and *Iboga* families engendered by *V. roseus* from the latter precursor, as discussed below.

BIOGENETIC-TYPE SYNTHESIS OF THE INDOLE ALKALOIDS

The family of alkaloids derived from tryptophan now boasts over 1000 members[2,3] thus accounting for almost 10% of known natural product structures[4]. It therefore seemed particularly appropriate to apply the principles of biogenetic analysis and biogenetic-type synthesis* to this group, many of whose members and indeed whole sub-classes are endowed with interesting biological and chemo-therapeutic properties. To this end several laboratories have, over the years, devoted considerable effort towards the laboratory construction of tryptophan-derived alkaloids based on authenticated or presumptive biochemistry. Much of the early work in this field has been reviewed in depth[3,6,7] and a seminal essay on the progress and philosophy of biogenetic-type synthesis[8] sets the background for the present description of more recent work, largely but by no means exclusively selected from the author's own published and unpublished experiments which have been directed towards solutions to the many and diverse problems in the intermediary metabolism of tryptophan in higher plants. Several detailed accounts of the nontryptophan-derived segment of the major classes of indole alkaloid have appeared[9,10,11,12] and we shall therefore only refer briefly to the major incorporation results in this area.

The Tryptophan - Corynanthe Sequence

The observation[10,13] that the *Corynanthe* alkaloids geissoschizine (1), corynantheine (2) and ajmalicine (3) are laid down as the earliest recognisible members of the indole family in *Catharanthus roseus* could be easily reconciled with the recognition of vincoside (4) as the "primordial" alkaloid of the series. Thus the structure and stereochemistry enjoyed by 4 represents the first encounter of tryptophan (or tryptamine) with the "C_{10}" unit contained

---

*For definitions of biogenetic-type synthesis see Van Tamelen[5]. The experiments described in this review fall into the class of "amphosynthesis" whereby the *in vitro* chemistry of a recognized of presumptive intermediate is used to carry out a partial synthesis of a complex natural product.

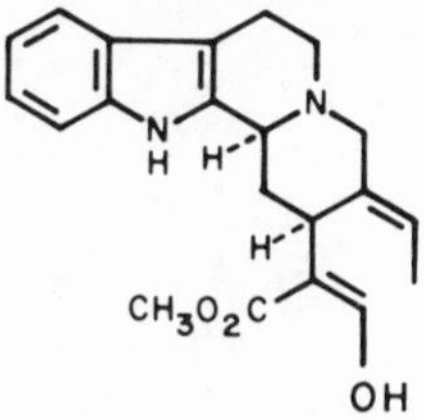

1, Geissoschizine

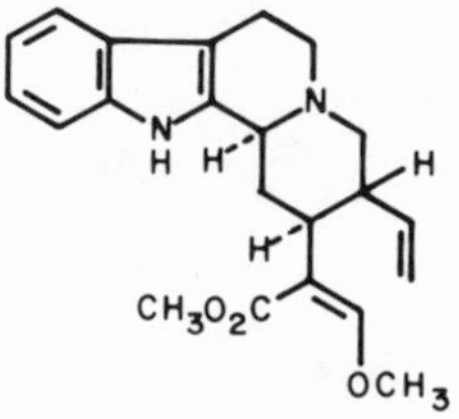

2, Corynantheine

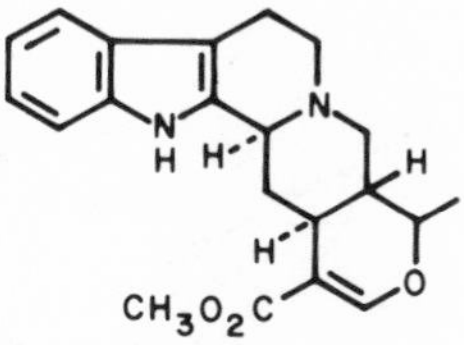

3, Ajmalicine

in the proven monoterpenoid precursor secologanin (5). In order to maintain a logical rather than a historical account of this field, a most important reaction, viz the condensation of tryptamine and secologanin to a separate mixture of vincoside (4) and the C-3 epimer isovincoside (6) described by Battersby, can be regarded as the first biogenetic-type transformation. Both 4 and 6 occur as natural products (e.g. in *C. roseus*) but only the 3βH-diastereisomer, vincoside, serves as an efficient precursor of the *Corynanthe*, *Aspidosperma* and *Iboga* alkaloids *in vivo*[11,14,15,16,17]. Although until recently of extreme scarcity, secologanin has now become available in considerable abundance, thus paving the way for further experiments designed to transform vincoside (4) and/or its isomer (6) into the *Corynanthe* and other alkaloids. Earlier work in this field had been performed with vincoside to give the C_3 epimer of vallesiochotamine (7) and vincoside lactam (8) by appropriate condensations as shown in Scheme 1[16,17]. Similarly the isolactam (9) and vallesiochotamine (10) could be prepared from 6.

The effect of removing the glucoside function with β-glucosidase at pH 5.2 has been re-examined. In the case of vincoside (3βH) (4) the resultant product was again

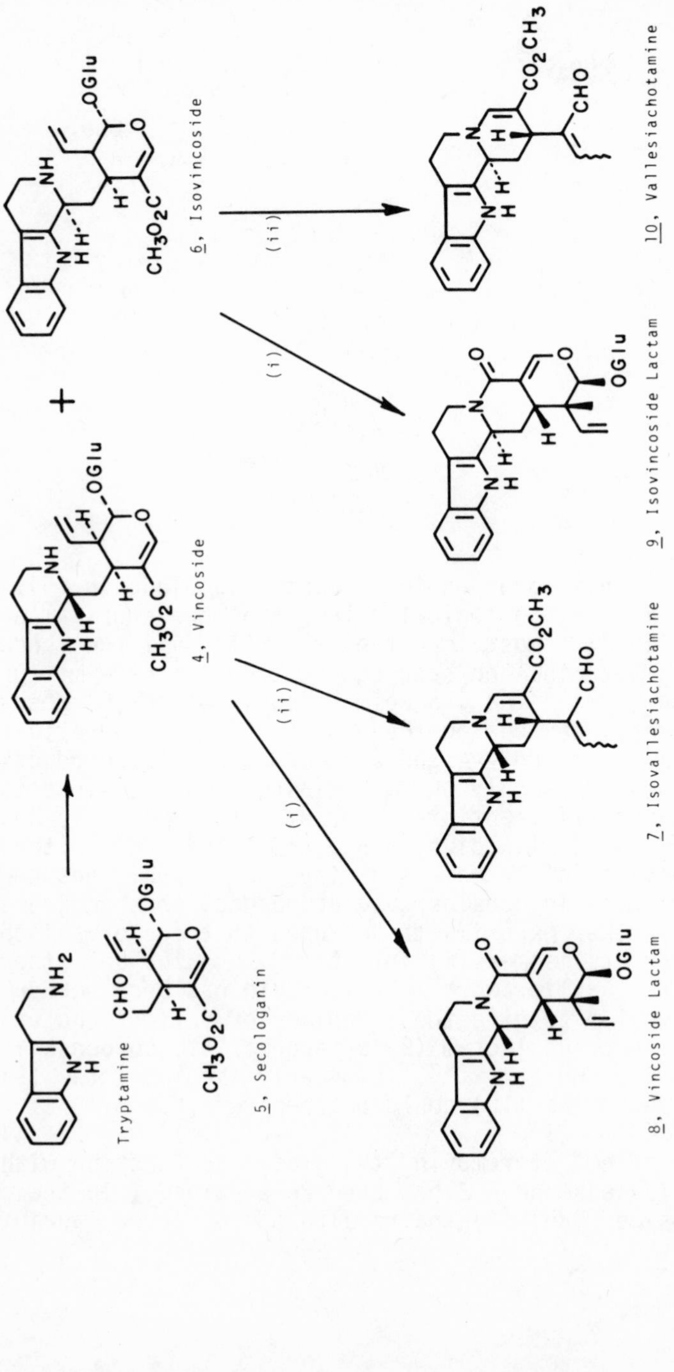

Tryptamine

5, Secologanin

4, Vincoside

6, Isovincoside

(i) (ii)

8, Vincoside Lactam

7, Isovallesiachotamine

9, Isovincoside Lactam

10, Vallesiachotamine

Scheme 1

(see Scheme 1) mainly the C-3 epimer of vallesiachotamine (7). However, as portrayed in Scheme 2, when the same reaction was carried out using isovincoside (3αH) (6) as substrate there was isolated in addition to vallesiachotamine (10) a small but reproducible yield (2-3%) of an alkaloid $C_{21}H_{22}O_3N_2$ which has been assigned the structure of the dieneamine (11) corresponding to a "dehydro geissoschizine" (A.I. Scott *et al.*, unpublished). The gross structure of 11 was readily demonstrated (Scheme 2) by catalytic hydrogenation to the tetrahydro compound (12) identical with the reduction product (12) of geissoschizine (1). The structure of 11 is of more than passing interest since this formula was proposed for a very active biosynthetic intermediate uncovered in previous work with short term incubations of *C. roseus*[18]. Comparison of the autoradiographs of 11 from these latter studies and the behavior of the new dehydro *Corynanthe* alkaloid (11) showed complete identity (A.I. Scott *et al.*, unpublished). Thus the *Corynanthe* series has been reached in this simple experiment which however does not simulate the unusual switch of the stereochemistry at C-3 implicit in the biochemical conversion vincoside (3βH) to geissoschizine (3αH). In a recent, related investigation, Brown and Chapple describe the preparation of 13 by an ingenious variant on this theme in which tryptophan is condensed with dihydrosecologanin and the subsequent condensation is controlled via N-benzylation (Scheme 3)[19]. Obviously, the next refinement in this series requires the exercise of control in the generation of the reactive species such as 6a and 6b (Scheme 2).

With the successful conversion of secologanin and tryptophan to the *Corynanthe* and vallesiachotamine series the stage now appears to be set for the logical progression to the more complex alkaloids, e.g. *Strychnos, Aspidosperma,* and *Iboga.*

From Corynanthe to Strychnos - The missing link

In spite of many interesting theories and even more ingenious experiments it has not yet proved possible to emulate Nature in passing between structural classes (A) and (B). The pattern of the "C_{10}" unit is not disturbed in this process, the new result being the migration of bond marked "0" from the α- to the β-position of the indole nucleus, followed by formation of the new bond (*) from the activated

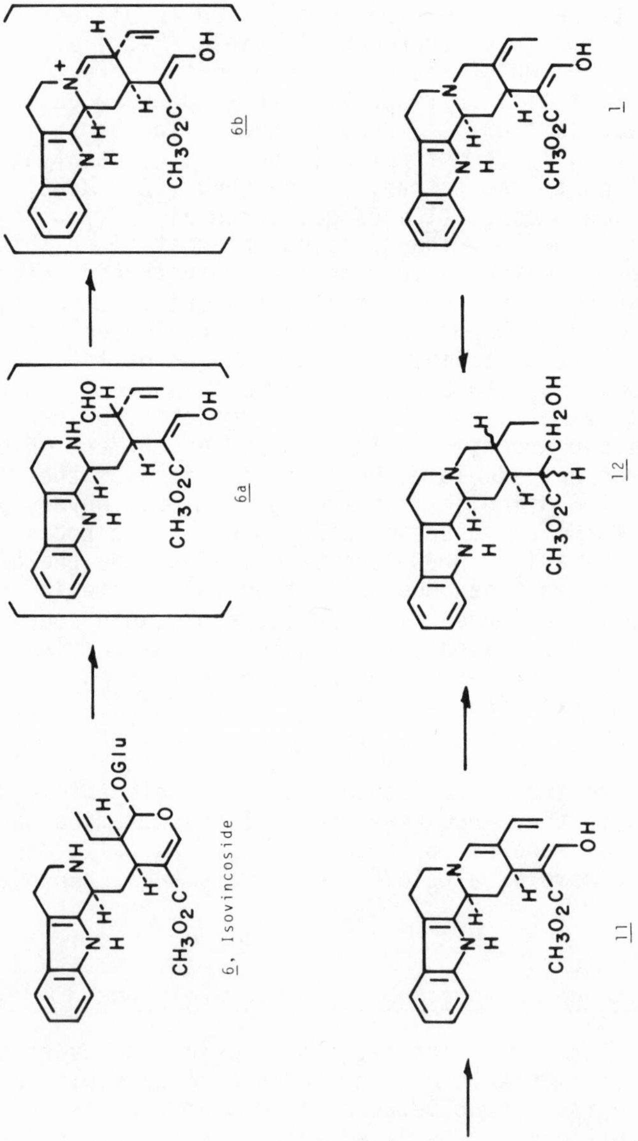

6, Isovincoside

6a

6b

11

12

1

Scheme 2

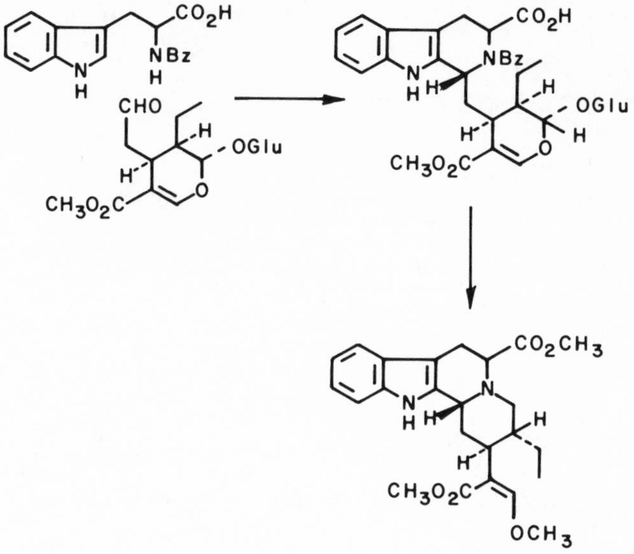

13

Scheme 3

C-16 position to the α-indole position. The sole working process for this theme is the special but very instructive case of Harley-Mason and Waterfield[20]. An earlier experi-

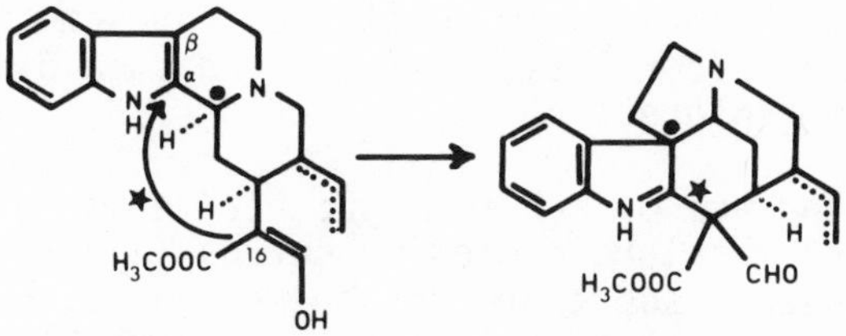

A B

ment with geissoschizine (refluxing acetic acid) in which conversion to type B was observed[21] suffers from extensive side reactions and variability in yield and obviously requires further definition. Other analogies involving β-oxidation (1 → 14) and oxindole chemistry (15 →16) have been proposed[10], there being preliminary feeding evidence

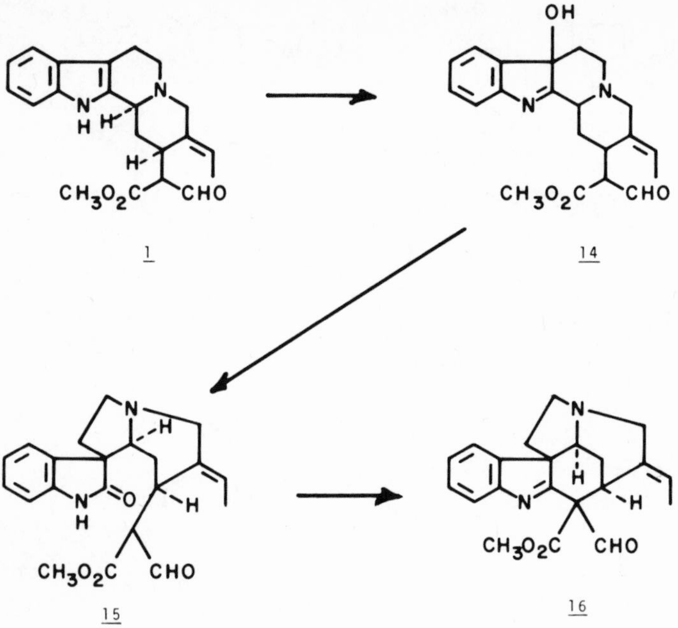

Scheme 4

to support the latter pathway *in vivo* (Scheme 4). However no satisfactory chemical operation on a *Corynanthe* alkaloid has yet been discovered to bridge this major gap.

At this juncture we should note that several *Corynanthe* alkaloids have been used as substrates for some interesting interconversions to related families and some of these branch-lines are now explored.

Oxindoles. The use of *t*-butyl hypochorite has provided entry into the β-chloroindolenine series (17) which in turn serves as a useful substrate for the oxindole, corynoxeine (19), and imino ether formation (18) (Scheme 5). The problems of the stereochemical control of these reactions and the equilibration at C-3 and C-7 have been reviewed in full detail[22].

Camptothecin (20). An ingenious partial synthesis (based on biogenetic analogy) of the camptothecin system has been described by Winterfeldt and his colleagues[23,24] as summarized in Scheme 6.

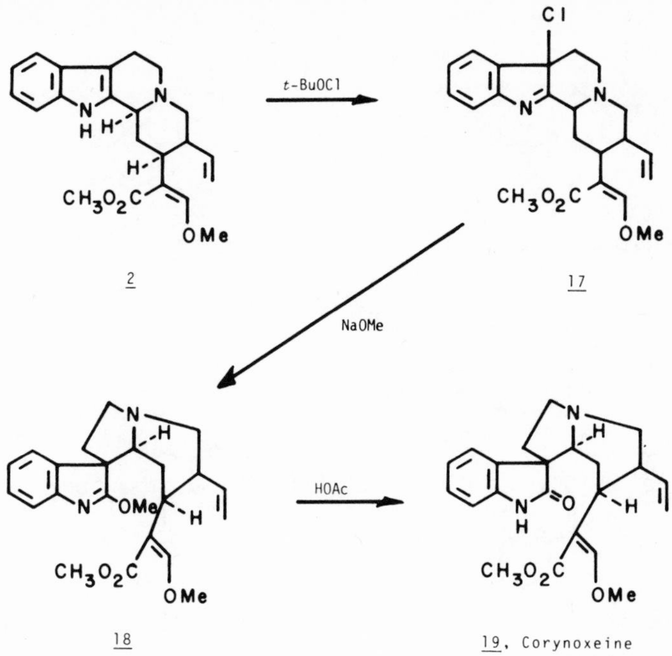

t-BuOCl

2 17

NaOMe

HOAc

18 19, Corynoxeine

Scheme 5

Ajmaline (22). A most plausible and interesting model for the formation of ajmaline from the aldehydocarboxylic acid (21) [Scheme 7] has been presented[25], in which generation of the immonium species (21a) leads to the desired bond formation (C-16 → C-8) and thence to the relay substance (21b). This is a particularly important model for the timing of the biochemical decarboxylation of the tryptophan moiety of the *Corynanthe* series.

Flavoperierine and Related Alkaloids. A simple model for the generation of the flavoperierine system (24,25) has been found in the reverse Mannich chemistry of the enamine (23) (Scheme 8) which can be formed by mercuric acetate oxidation of geissoschizine (1) (A.I. Scott *et al.*, unpublished results). Although the appropriate biochemical correlation has not been made, the above experiment suggests that loss of the "C_3" segment may well arise by such a process which occurs with facility *in vitro*.

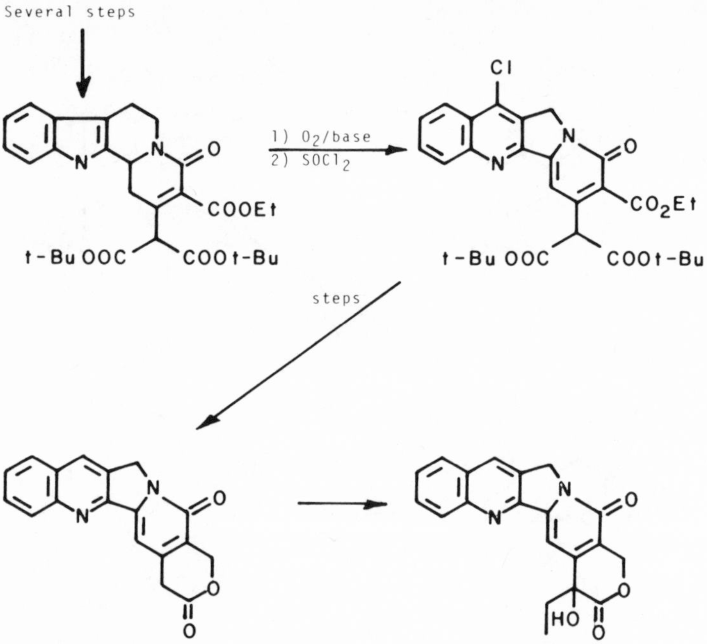

Several steps

1) O_2/base
2) $SOCl_2$

steps

20, Camptothecin

Scheme 6

Related reactions of the *Picralima* alkaloids. A close analogy for *Strychnos* alkaloid biosynthesis from a *Corynanthe* precursor concerns the manipulation[26-27] of the indolenine system of 26 (Scheme 9) in the presence of potassium *t*-butoxide in a remarkable rearrangement to norfluorocurarine (27). Successful conversion of a *Corynanthe* alkaloid (e.g. 1 to 26) would complete this intriguing system but, with the exception of some work by Dolby on models for echitamine synthesis, a satisfactory method for the vital bond-making process from C-16 has yet to be described. In fact all attempts to utilize the anionic or radical-forming propensity of C-16 have been uniformly unsuccessful, indicating the necessity for finding a new approach to the chemistry of this apparently reactive center.

With the key link (*Corynanthe* → *Strychnos*) still missing from the chain of events in the biogenetic synthesis of the major classes we now consider some of the successful

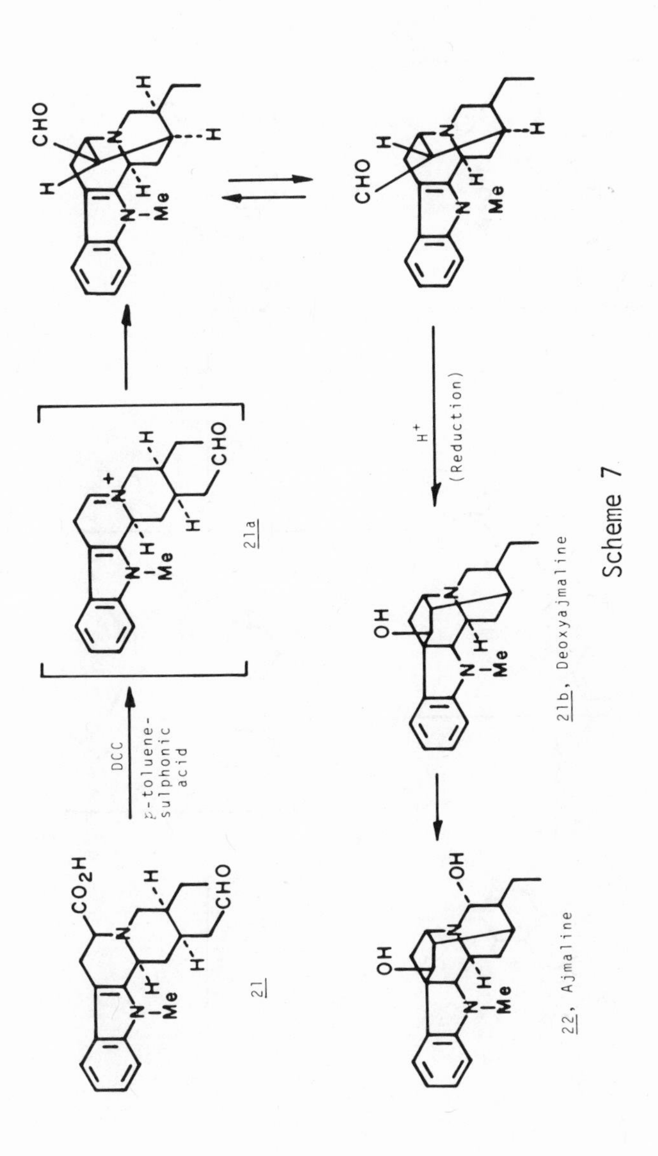
DCC
p-toluene-sulphonic acid
21
21a
H+
(Reduction)
21b, Deoxyajmaline
22, Ajmaline

Scheme 7

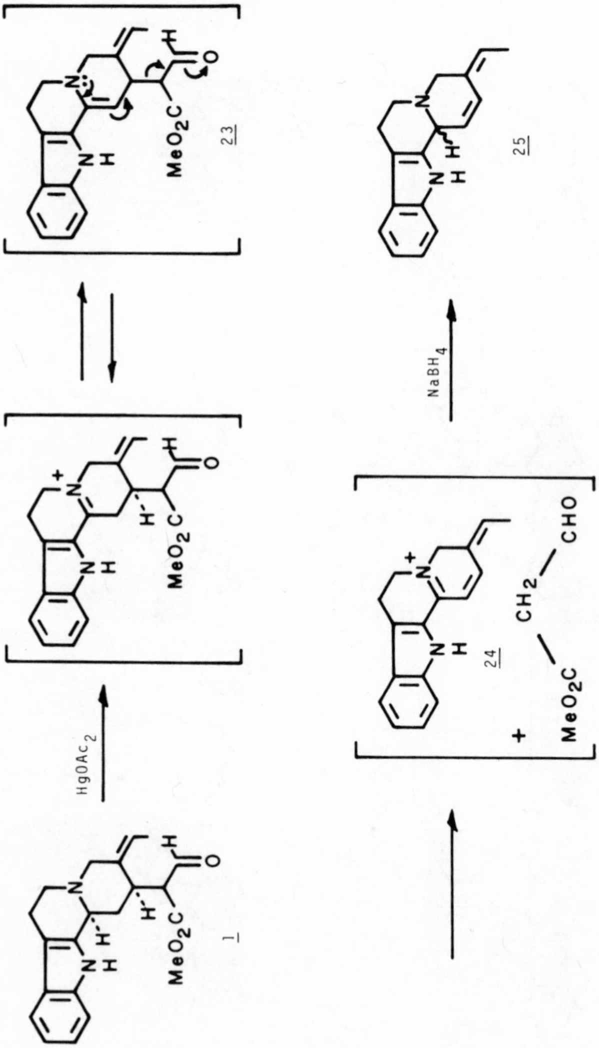

Scheme 8

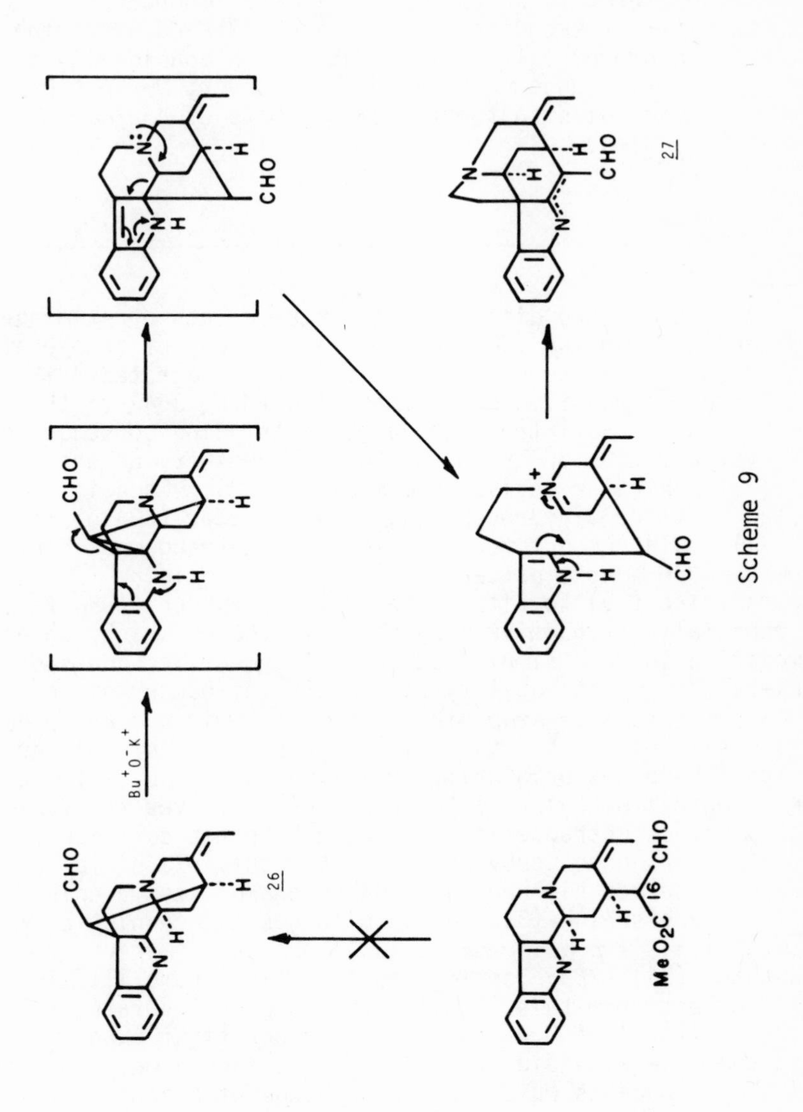

Scheme 9

conversions wrought upon the *Strychnos* and related systems in the quest for further analogies in the biochemical pathway. The overall scheme is shown again for the sake of clarity in Scheme 10 which emphasizes the importance of the acrylic ester or secodine intermediates (31 and 32). The latter have proved valuable touchstones in considering the deep-seated rearrangements which must convert *Corynanthe/ Strychnos* alkaloids to the typical members of *Aspidosperma* and *Iboga* series.

From Stemmadenine to Aspidosperma and Iboga Alkaloids - The "336" Concept

As soon as the discovery and intermediacy of stemmadenine (29) and tabersonine (33) in the main pathway of indole alkaloid biosynthesis were established it was suggested[10-13] that the formal dehydration of stemmadenine (M.W. 354) to the species with M.W. 336 introduced, as an intriguing consequence, the cycloaddition or extended Mannich chemistry of the hypothetical achiral reactive intermediate (32). Thus it was predicted that the dihydropyridine acrylic ester (32) or its isomer (31) could account for the skeletal rearrangements in going from the *Corynanthe* pattern (1) and preakuammicine (28) or stemmadenine (29) to either the *Aspidosperma* or *Iboga* families. An especially attractive facet of this theory which represents a modification and simplification of the earlier ideas of Wenkert[28-29] is its ability to explain the occurrence of pentacyclic *Aspidosperma* alkaloids in racemic and antipodal versions. Yet another interesting consequence of this speculation (which was originally conceived as a result of the mass spectral behavior of tabersonine!) involves the various recombination characteristics of the hypothetical intermediate (32) which not only serves as a highly plausible model for the observed biochemical conversions of stemmadenine (29) to tabersonine (33) but also forges a mechanistic link between (33) (*Aspidosperma* series) and the isomeric catharanthine (34) (*Iboga* series) as shown in Scheme 11. In its simplest form this isomerization process operates at the oxidation level of M.W. 336. In fact our first experiments[30] with the rare alkaloid stemmadenine provided a welcome model for these concepts which, as the subsequent biochemical experiments and the structures of new isolates of Apocynaceae were to show, constituted not only a working laboratory analogy for the labyrinth of inter-connecting pathways between families, but allowed the prediction of hitherto

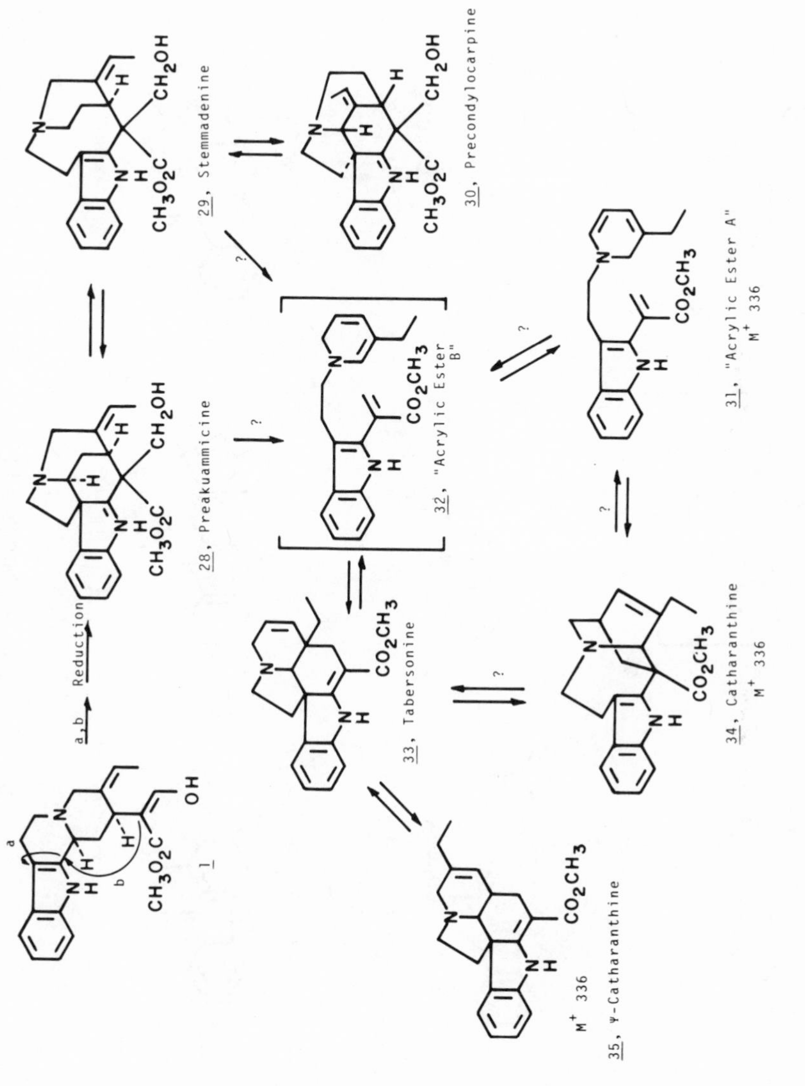

1

a,b → Reduction

28, Preakuammicine

29, Stemmadenine

30, Precondylocarpine

32, "Acrylic Ester B"

31, "Acrylic Ester A" M^+ 336

33, Tabersonine

34, Catharanthine M^+ 336

35, Ψ-Catharanthine M^+ 336

Scheme 10

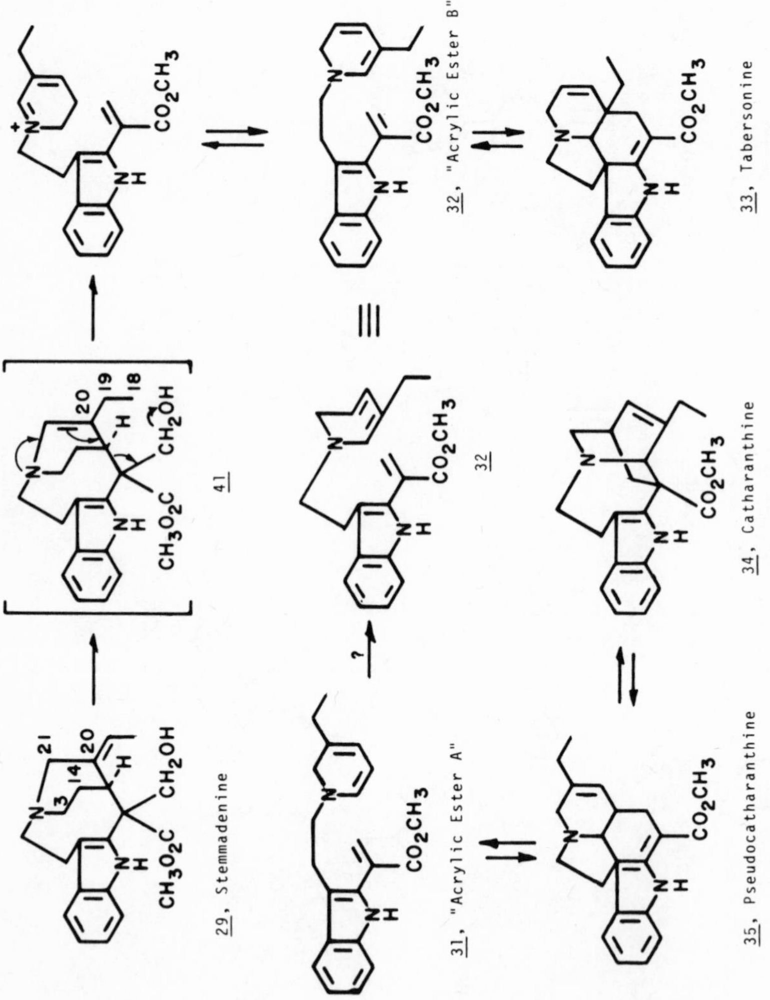

29, Stemmadenine

41

31, "Acrylic Ester A"

32

32, "Acrylic Ester B"

35, Pseudocatharanthine

34, Catharanthine

33, Tabersonine

Scheme 11

undiscovered structural types. The logical evolution of this area of bioorganic chemistry was however not to prove facile and trenchant criticism[31,32,33] of the first communication[30] on this topic required some reinvestigation in order to define extremely critical reaction conditions. Due to the scarcity of stemmadenine a series of expeditions to the state of Vera Cruz (Mexico) was made to identify and collect the fruits of *Stemmadenia Donelli-Smithii* in early November 1971[34]. Harvesting and extraction of this plant at other times of the year appear to be particularly unpropitious, the yield of 29 falling from 0.5% to less than 0.001%. With a fresh supply of the vital *Corynanthe-Strychnos* alkaloid in hand, the early experiments were examined with a most satisfactory outcome: the achievement of both regiospecific and stereospecific control in the transformation of stemmadenine to both *Aspidosperma* and *Iboga* alkaloids as described below. In addition to the *in vivo* conversion indicated it could be shown that platinum-catalyzed oxidation of 29 gave the pentacyclic system corresponding to oxygenation or dehydrogenation of the carbon adjacent to N (b) in stemmadenine. These reactions could be reversed by treatment of 28 and 30 with sodium borohydride (Scheme 12). Thus the facile redox equilibria depicted in Scheme 12 may indicate the branch point at which the stemmadenine molecule is drawn enzymatically into either the *Aspidosperma* or *Iboga* series. In this first set of preliminary experiments, stemmadenine was heated under vigorous reflux conditions in acetic acid over silica boiling stones at high (~200°) external bath temperatures. Great variability in composition and yield of the products was encountered from such reactions which were frequently maintained for 1-2 days with concomitant loss of solvent. However, it was shown by analytical and preparative tlc techniques that stemmadenine was converted first to the O-acetate (a reaction that can be effected very simply by reflux at 150° bath temperature). The latter undergoes a deep-seated rearrangement to give products of the M.W. 336 set, i.e. tabersonine (33) and the structural isomer of catharanthine, pseudocatharanthine (35). A small amount of catharanthine could also be detected. Thus although the reactions are extremely sensitive to concentration, surface and/or thermal conditions, the discovery of the formal dehydration products of stemmadenine in racemic form - as would be required by any mechanism involving 32 - lends credence to a most rewarding concept. As originally described[30], this biogenetic-type interconversion turned out

CH_3O_2C CH_2OH

29

Pt-O_2 / $NaBH_4$

CH_3O_2C CH_2OH

30, Precondylocarpine (major)

+

CH_3O_2C CH_2OH

28, Precondylocarpine (minor)

H_2-Pd

CH_3O_2C CH_2OH

38, 19,20-Dihydrostemmadenine (R=H)

Pt-O_2 / $NaBH_4$

CH_3O_2C CH_2OH

36, Dihydroprecondylocarpine (R=H) (minor)

+

CH_3O_2C CH_2OH

37, Dihydropreakuammicine (R=H) (major)

Scheme 12

to have severe limitations as a practical and reproducible route to the more complex alkaloids of the indole family. A complete account of this phase of the study is outside the scope of our review and the interested reader is referred elsewhere for a discussion of the problem and its eventual solution[30,32,33].

EVOLUTION OF A REGIO- AND STEREOSPECIFIC MODEL

In order to rationalize the promising but erratic results of the early (acetic acid) experiment the new source of stemmadenine was used to explore and define the conditions necessary, not only for the generation of the seco ester (32), but also its recombination to the known isomers of the "336" series in a more rigorous manner.

The Corynanthe-Aspidosperma Relationship

The reverse Mannich chemistry discussed above (Schemes 10, 11) for *Aspidosperma* biosynthesis via dehydrosecodine B (32) differs in regio-specificity from the *Iboga* model described below in that the formation of the secodine system B can take place via C-20, C-21 rather than C-3, C-14. The introduction of unsaturation at C-21 could, in principle, be realized by two methods. First it was hoped that the simple isomerization process (Scheme 11) previously adduced for reaction of stemmadenine (acetate) in hot acetic acid solution, might be capable of more rigorous control and thus lead only to the *Aspidosperma* framework by formation and recombination of dehydrosecodine B (32). A further spur towards a search for isomerization conditions was the discovery that, whereas stemmadenine (29) is hydrogenated unexceptionally to the 19, 20-dihydro derivative (38) in the presence of platinum catalyst (Scheme 12), the behavior of stemmadenine acetate (39) is quite different towards reduction. Thus, at atmospheric pressure, hydrogenation of 39 over platinum in ethanol solution leads to (±)-tetrahydrosecodine (40) (Scheme 13) in 75% yield. The facile cleavage of the 15, 16 sigma bond in 39, but not in 29 is suggestive that the primary acetoxyl at C-17 aids the irreversible loss of acetate from the $\Delta 20,21$ isomer (41) which is in equilibrium with the $\Delta 18,19$ exocyclic olefinic group of the starting material. This high yielding partial synthesis[36] of the naturally occurring member of the secodine family could be

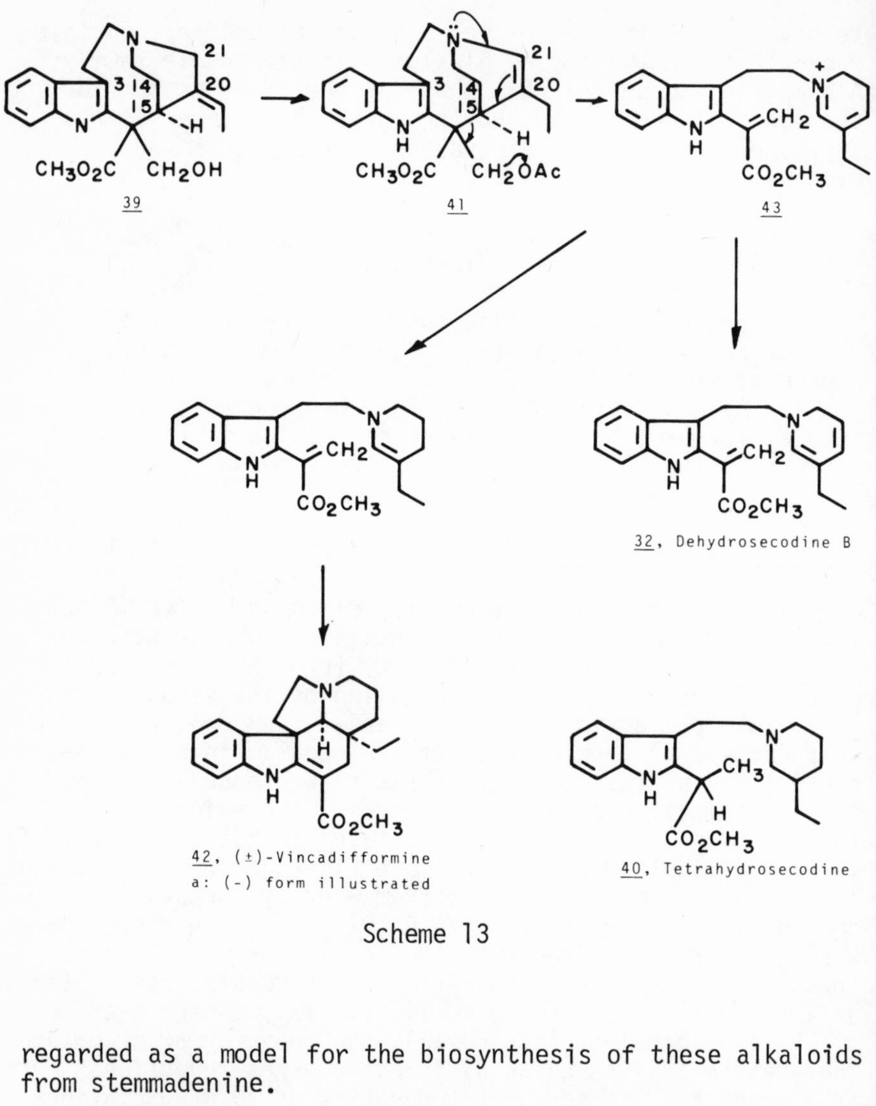

39

41

43

32, Dehydrosecodine B

42, (±)-Vincadifformine
a: (-) form illustrated

40, Tetrahydrosecodine

Scheme 13

regarded as a model for the biosynthesis of these alkaloids from stemmadenine.

Experiments designed to capture the fugitive dihydropyridine (32) in the cyclized form corresponding to the *Aspidosperma* alkaloids were carried out as follows. Firstly in a reaction designed to separate the *Aspidosperma* pathway of the "early" experiments, conversion of (39) to (±)-vincadifformine (42) was achieved in low yield (0.15 - 0.20%) by thermolysis at 150^{0} on a silica gel surface for 25 minutes

The analytical and preparative procedures used for this and the succeeding experiments left no doubt that tabersonine was not a product of the reaction. A plausible reaction pathway via 41 is shown in Scheme 13, which suggests that the intermediate immonium species (43) suffers reduction by disproportionative release of hydride to dehydrosecodine B (32) which then cyclizes to (±)-vincadifformine and that this sequence is favored over the more direct rearrangement (without reduction) to (±)-tabersonine. An alternative method was then studied, which takes into account the ready aerial oxidation (which must have occurred under the vigorous conditions of the "early" experiment) in acetic acid solution of 39 to the acetate (43) of the naturally occurring pentacyclic alkaloid, precondylocarpine (30)[37], a step which can be achieved more efficiently using platinum catalyzed oxidation (Scheme 14)[38]. Catalytic reduction of the ethylidene group of 43 affords the dihydroacetate (44) which is now formally capable of regiospecific collapse to dehydrosecodine B as shown in Scheme 14. When the acetate was subjected to thermolysis (150°, silica gel, 25 min) a separable mixture of (±)-tabersonine (33) (0.2%) and (±)-vincadifformine (42) (0.2%) was produced, while no trace of pseudocatharanthine (35) could be detected. It was concluded from these results that the generation of the acrylic ester B takes place without rearrangement to the A isomer according to Scheme 11. Although the yields in these models for "*Aspidosperma* synthetase" are extremely low, the reactions provide a working hypothesis for the complex series of rearrangements which accompany the biochemical conversion of the *Corynanthe* to the *Aspidosperma* and Secodine families. At the same time, it is believed that the multi-step reaction sequence which was concealed in the earlier model experiments can now be understood in terms of the operation of indiscriminate oxidative and reductive attack on stemmadenine acetate, the first formed product of the reaction between acetic acid and the original substrate, stemmadenine. Depending on the timing of the oxidative and reductive steps, the acetate is converted to both the preakuammicine and precondylocarpine skeletons which at the appropriate oxidation level collapse to *Iboga* and *Aspidosperma* alkaloids respectively.

The Corynanthe - Iboga Relationship[39]

The thermolysis of 19,20-dihydrostemmadenine was next

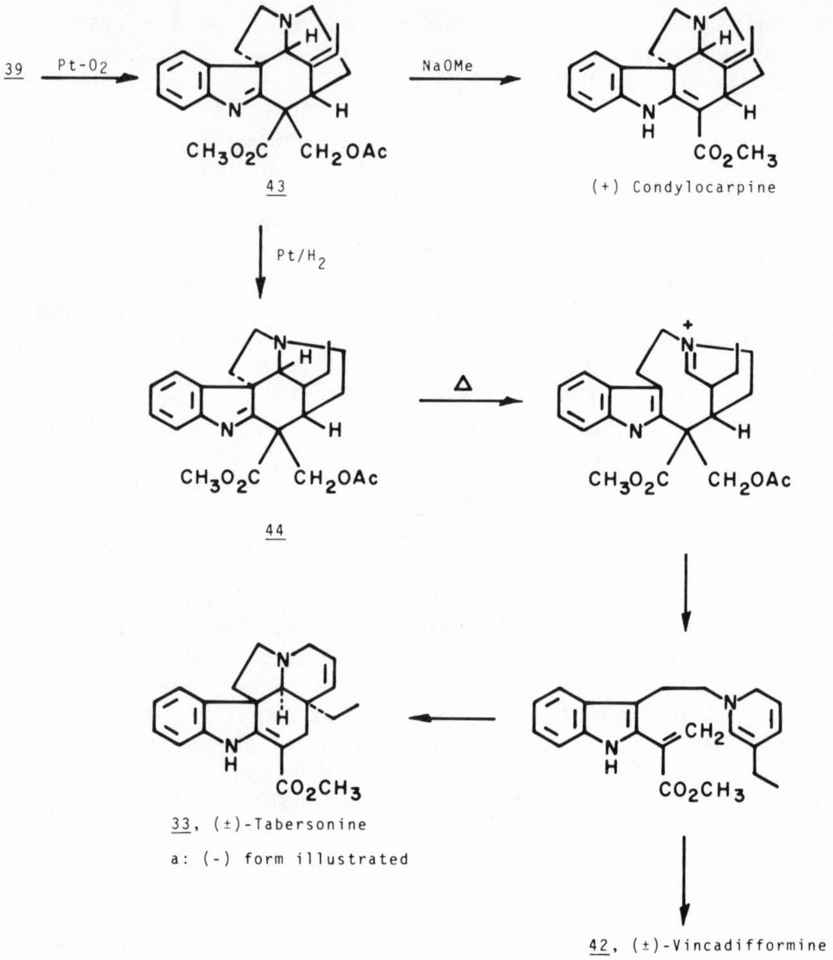

39
Pt-O2
NaOMe
CH3O2C CH2OAc
43
CO2CH3
(+) Condylocarpine
Pt/H2
Δ
CH3O2C CH2OAc
44
CH3O2C CH2OAc
CH2
CO2CH3
CO2CH3
33, (±)-Tabersonine
a: (-) form illustrated
42, (±)-Vincadifformine

Scheme 14

examined. This work took the reductive factor of the "early" experiment into account, for it is well known that systems such as 45, 46 readily disproportionate to the pyridinium salt (47) and the tetrahydropyridine (48). Allowance was also made for the regiospecific (aerial) oxidation of 38 at C-3 to dihydropreakuammicine (37). In order to complete the analogy with the earlier work using acetic acid (which has been shown to form the primary acetate),

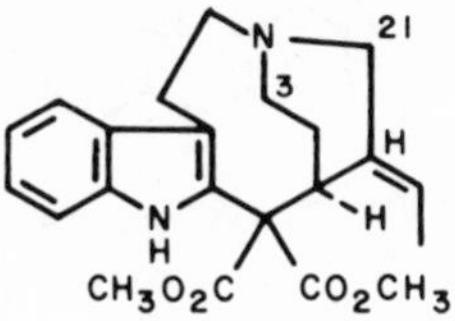

38

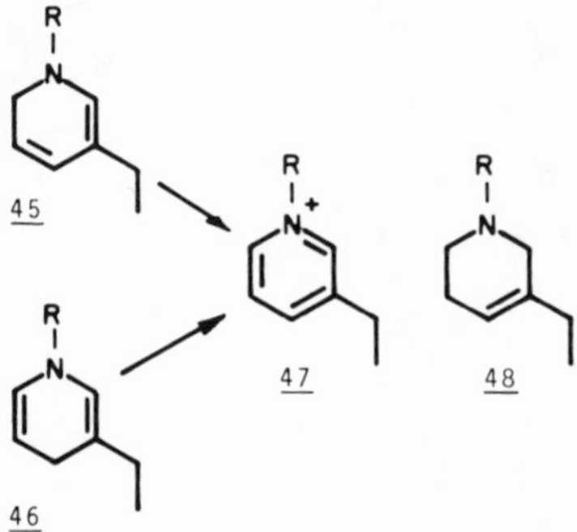

45

46

47

48

dihydrostemmadenine was converted to the acetate (50). When the latter compound was heated on a silica gel tlc plate at 150° for 45 min the resultant mixture afforded (±)-pseudocatharanthine (35) (1%) - and its dihydro derivative (51) (0.5%) (Scheme 15). No trace of tabersonine (33) was detectable in this experiment. In order to rationalize this result it is necessary to generate a double bond at C-3,C-14 (50a) so that the extended reverse Mannich chemistry shown in Scheme 16 can operate. The further reaction of the iminium salt (52) is seen as a rearrangement to 3,14-dehydrosecodine (31), which we have designated dehydrosecodine A (See Scheme 16).

Recombination of the achiral ester (31) affords (±)-pseudocatharanthine (35)[40] whereas conjugate reduction of 52 leads to the secodine (53) whose cyclization to (±)-dihydropseudocatharanthine (51) is unexceptional. A more efficient way of demonstrating this regio- and stereospecific rearrangement was found by preparing 19,20-dihydropreakuammicine acetate (49), a known autoxidation product of 50 by platinum/oxygen oxidation of 50. Thermolysis of 49 at 150° for 20 min afforded (±)-35 in yields which, although by no means optimized, average 5%. The known equilibrium between 35 and catharanthine (34) completes the partial synthesis of the latter member of the *Iboga* family in racemic form.

Further insight into the mechanism of this remarkable reaction was gained by methanolysis of the acetate (49) at room temperature for 4 hr, or at 80° for 15 min. The products of this reaction were now optically active and were separated to afford a dextrorotatory 15-methoxydihydropseudocatharanthine (54) and the levorotatory diastereomer

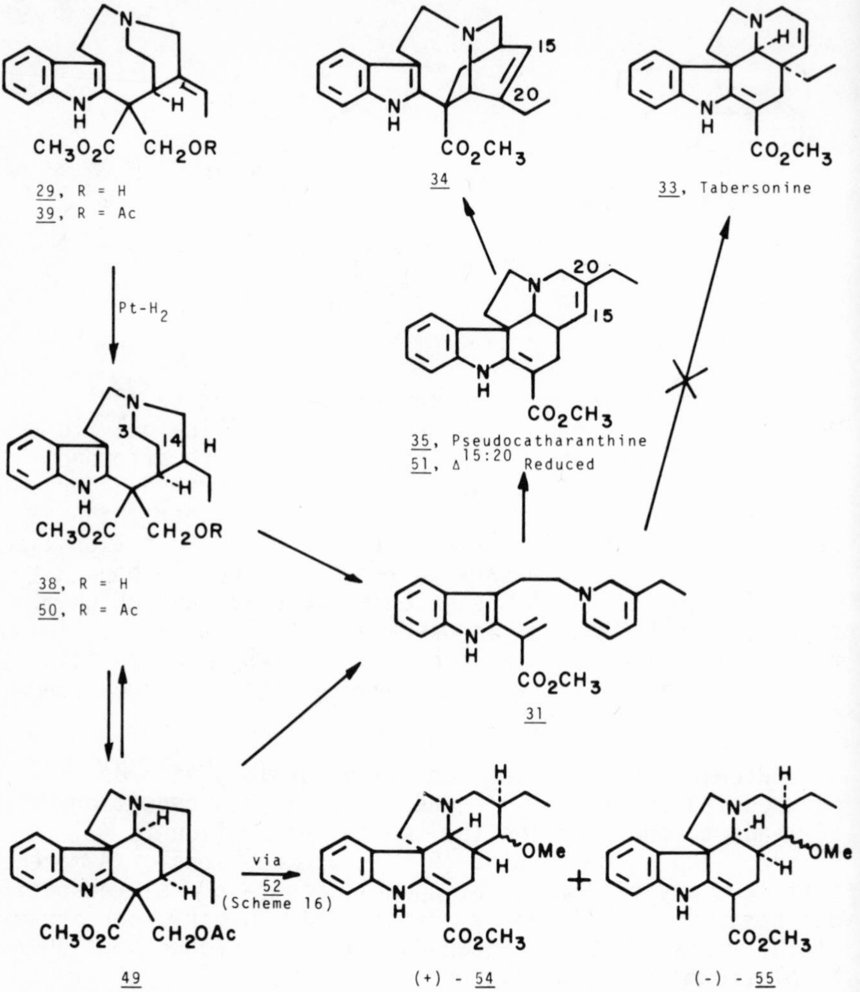

29, R = H
39, R = Ac

Pt-H_2

38, R = H
50, R = Ac

49

34

35, Pseudocatharanthine
51, $\Delta^{15:20}$ Reduced

33, Tabersonine

31

via 52 (Scheme 16)

(+) - 54

(-) - 55

Scheme 15

(55) in the ratio 9:1. The combined overall yield of this reaction from dihydrostemmadenine (38) was 3.5%. This result is in full accord with the postulate that the immonium species (52) is an intermediate in the rearrangement process and that conjugate addition of methanol affords the diastereomeric mixture 54 & 55 via 56. The stereospecific preference for the absolute configuration 54 over 55 may

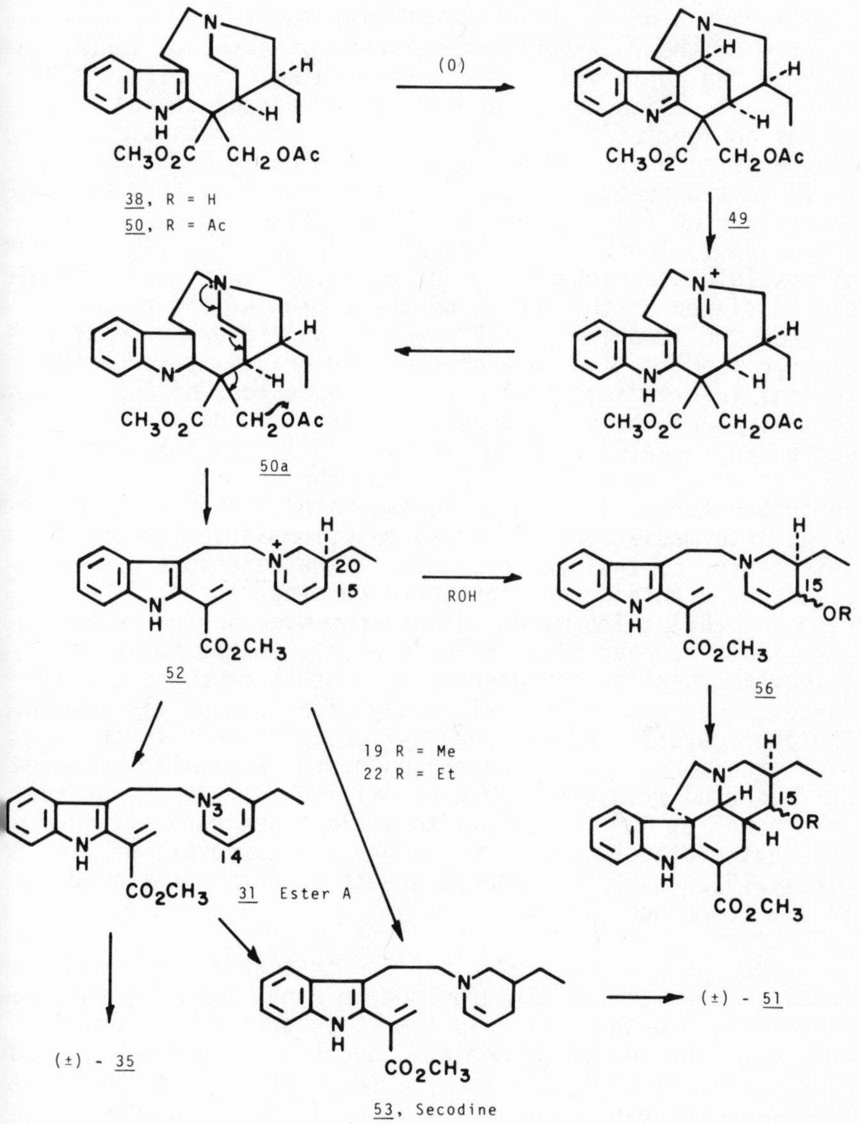

(O)

38, R = H
50, R = Ac

49

50a

ROH

52

56

19 R = Me
22 R = Et

31 Ester A

(±) - 35

(±) - 51

53, Secodine

Scheme 16

be dictated by the configuration of the methoxyl group at C-15 which controls the observed stereochemistry of the cyclization process. This particular model is also illustrative of the remarkable variation[41] in absolute configuration of the pentacyclic *Aspidosperma* alkaloids within the same plant, which could be mediated by conjugate addition of the appropriate prosthetic group of the synthesizing enzyme to such an immonium species. Another pertinent example is the co-occurrence of (-)-coronaridine (57) and (+)-catharanthine (34) within the same species (*Catharanthus roseus*)(A.I. Scott *et al.* unpublished results). This dualit of absolute stereochemistry for such complex alkaloids again can be viewed as the result of the timing of the reduction step of the immonium ion (52) which can either be reduced and cyclized to give (-)-coronaridine or pass through the achiral intermediate (31) with protropic loss of C-20 stereo chemistry and thence by enzymic control to the antipodal series represented by catharanthine (34). In this connection it is of interest to note that the *in vitro* conversion described above in which 52 is suggested as the first formed *chano* intermediate, the thermal reaction serves to epimerise this center so that the products of the reactions are (±) pseudocatharanthine and the corresponding racemic dihydro compound (51). The milder conditions used in the methanolys experiment however give products which, on the basis of molecular rotation differences with other memebers of this series, indicate an optical purity of approximately 70-80%. The configuration of the methoxyl group at C-15 in these diastereoisomers is at present unknown. A similar reaction was observed when dihydropreakuammicine acetate was heated in ethanol at 80° for 15 min to yield a separable mixture of the dextrorotatory and levorotatory diastereomers of 15-ethoxy-15,20-dihydropseudocatharanthine in the ratio 10:1, in 10% (combined) yield.

The above experiments indicate that an "*Iboga* synthetas model in the form of dehydrosecodine A has been demonstrated to operate and that not only the regio specific generation and recombination of 31 mediated by the collapse of reduced preakuammicine (49), but the stereospecificity of the recyclization process leads to the *Iboga* isomer, (±)-pseudocatharanthine, which in turn is convertible to (±)-catharanthine Although the yields in this reaction are not yet of preparative value, it is our view that the synthesis and reactivity of the dehydrosecodine system is worthy of more detailed study as a new method of preparing quite complex pentacycli alkaloids from simple starting materials, with full stereospecific control.

Further isomerizations of the "336 series. The Aspidosperma-Iboga-Secodine Relationship

In 1968, a report[30] on the *in vivo* and *in vitro* transformations of the *Aspidosperma* alkaloid tabersonine (33) to the *Iboga* alkaloid catharanthine included the proposal for an intermediate (31) in this reaction which has since been found in a variety of stabilized versions. In 1969, Smith and his co-workers concluded[31] that this reaction, which involves rupture of both the 7-21 and 17-20 bonds of 33, proceeds only as far as cleavage of the 7-21 bond (Scheme 17). The resultant immonium species (58) then rearranges (59) and cyclizes to allocatharanthine (60), an optically active isomer of catharanthine (34) and pseudocatharanthine (35), the latter two *Iboga* structures being interconvertible (Scheme 17). Scott & Cherry stated, however, that both ionic and thermal requirements must be met in carrying out the full transformation[43] which, by passing through 60 affords the racemic products described earlier. In a reconfirmation and simplification of the 1968 report, the separation of the ionic and thermal components has been described recently[42].

(-)-Tabersonine (33) was heated in acetic acid for 15 hr (external bath temperature 140°). The resultant mixture was separated and (+)-allocatharanthine (60) (11%) isolated. A solution of 60 was applied to a silica gel tlc plate, the solvent removed, and the plate heated at 150° for 30 minutes. Elution and chromatography on $AgNO_3$-impregnated silica gel plates afforded two major products. These were (±)-pseudocatharanthine (35) (4%) and optically pure (-)-tabersonine (33) (4%). The absence of any racemization of tabersonine indicates reversal of the formation of allocatharanthine without cleavage at C-17, C-20 i.e. by the pathway 60 → 59 → 58 → 33 (Scheme 18). Thus not only is the *Aspidosperma* framework stable towards retro-Diels-Alder reaction but, in confirmation of the complete specificity noted in the previous section for the reaction of dehydro secodines A and B, only the racemic product, 35, of cyclization of dehydrosecodine A (31) is observed, there being no evidence for equilibration with the B isomer (32) which would have yielded (±)-tabersonine.

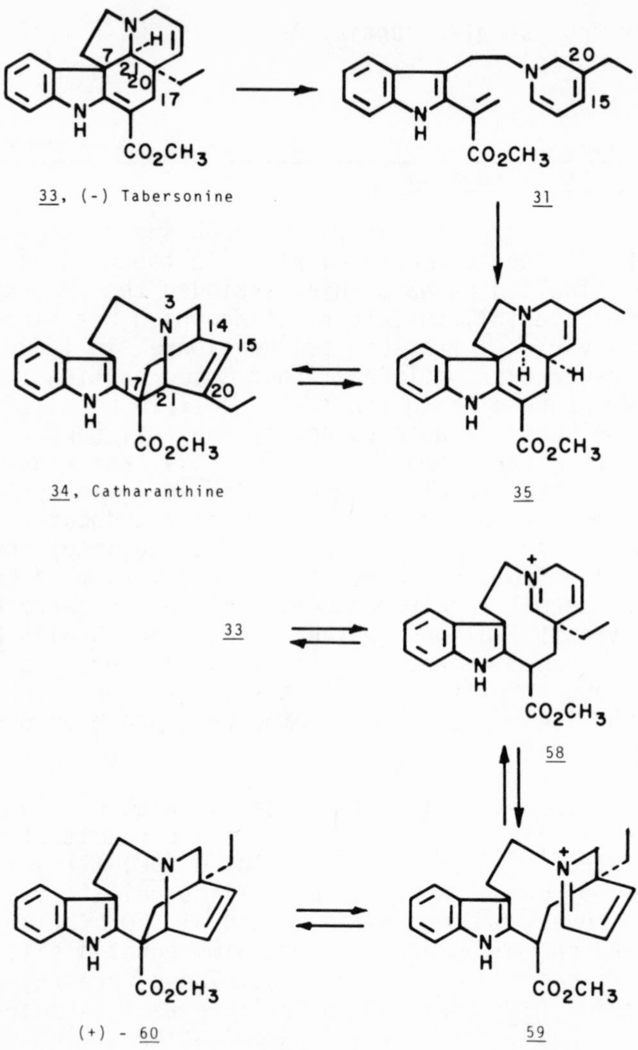

33, (-) Tabersonine

31

34, Catharanthine

35

33

58

(+) - 60

59

Scheme 17

In order to gain further evidence regarding the intermediacy of the secodine esters (31 and 32) a study was made of the thermal reactions of the key members of this series. When xylene solutions of the isomeric alkaloids (-)-tabersonine (33), (+)-catharanthine (34), and (±)-pseudocatharanthine (35) were maintained in sealed tubes for 1.5 hr at the temperatures indicated (Scheme 19), in each case (but with

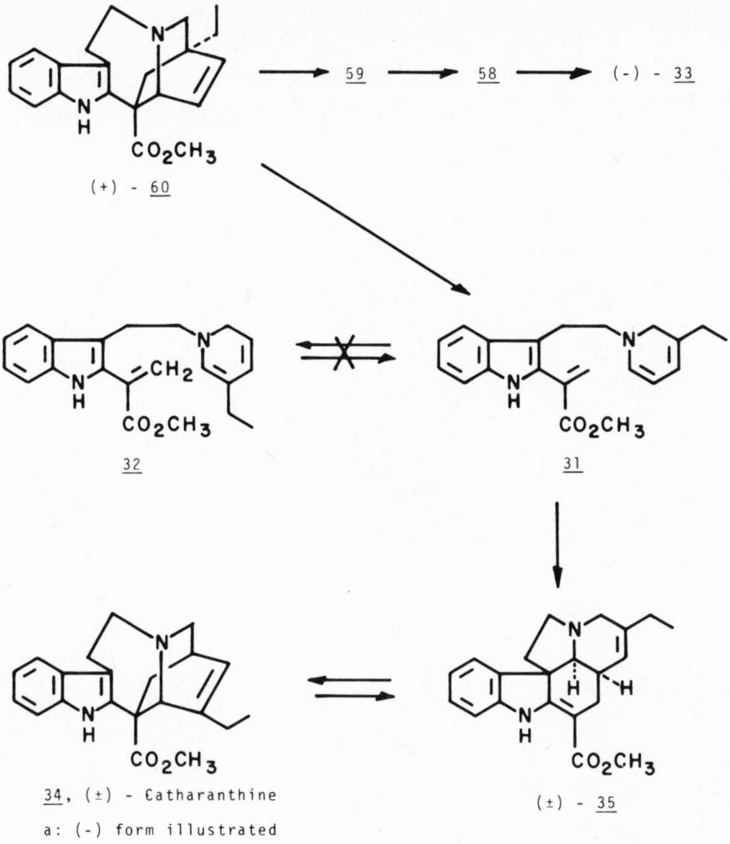

(+) - 60 → 59 → 58 → (-) - 33

32 ⇄ 31 (arrow to 32 crossed out)

34, (±) - Catharanthine

a: (-) form illustrated

(±) - 35

Scheme 18

a different energy requirement) the same products were isolated and characterized as 3-ethylpyridine and 1-methyl-2-hydroxycarbazole (61)[43].

We suggest that the formation of these products takes place by way of a retro-Diels-Alder reaction to afford the fugitive ester[39] followed by an intramolecular rearrangement and hydrogen transfer from the dihydropyridine to the acrylic ester function yielding the hemiketal (62) with loss of 3-ethylpyridine as indicated. Elimination of methanol and further rearrangements then give the carbazole (61). In support of the latter process, 1-methyl-2-methoxycarbazole

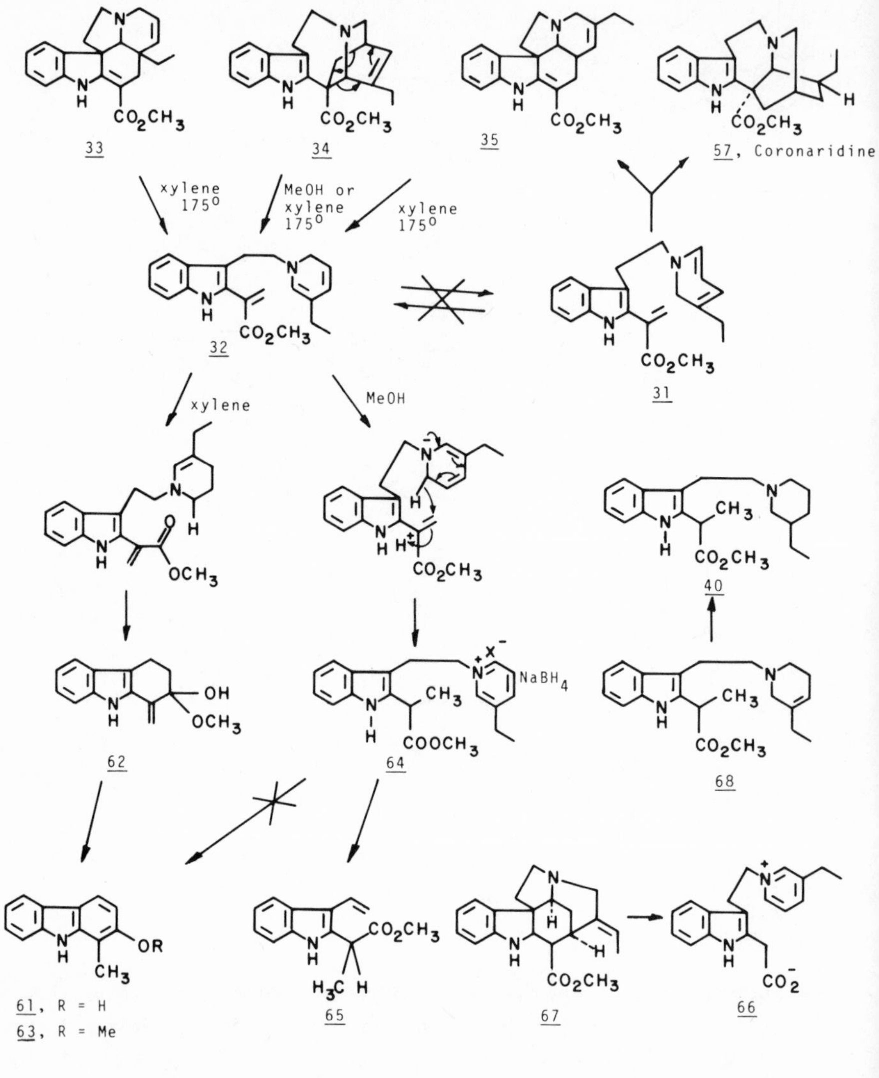

Scheme 19

(63) could be detected and characterized as a minor product of the reaction.

More direct evidence for the formation of 32 was obtained by the capture in 50% yield of the racemic salt 64 when catharanthine was heated in methanol at 140° for 2 hr[43].

In contrast to the intramolecular formation of the carbazole from the dihydropyridineacrylic ester in the aprotic solvent xylene, the availability of solvent protons in the latter case appears to divert the collapse of this intermediate in methanol via an ionic mechanism to the pyridinium salt. When the reaction is carried out in CH_3OD solution, the nmr spectrum of the salt no longer shows a signal at τ 6.12 ($CD(CH_3)CO_2$-CH_3) and the doublet τ 8.62 ($CH(CH_3)CO_2CH_3$) is replaced by a singlet (3 H) in accord with the mechanism 32 → 64 as shown in Scheme 19. This salt is stable in methanol at 175°, but on pyrolysis at this temperature affords the carbazole (61) presumably via elimination of ethylpyridine and cyclization of the resulting vinyl ester (65). The generation in methanol solution[39] could also be rationalized by an ionic mechanism which recalls the formation of the betaine (66) from akuammicine (67)[44]. Since the species 32 and 64 could be reached *in vivo* from stemmadenine, tabersonine and catharanthine, it will be of interest to test these three alkaloids as biochemical precursors for secodines and secamines and also to consider the system 32 → 64 as a labile but isolable biosynthetic entity in the *Aspidosperma* and *Iboga* metabolic grid[43].

Laboratory analogy for secodine formation from salt 64 was obtained by reduction of 64 with sodium borohydride which gave (±)-dihydrosecodine (68) as a crystalline racemic base (35% yield), identical in all spectroscopic properties with the amorphous natural alkaloid obtained from *Rhaza stricta* (A.I. Scott *et al.*, unpublished results). Further (catalytic) reduction of 68 afforded tetrahydrosecodine (40) which also occurs in *R. stricta*. This mode of formation of dihydrosecodine may be contrasted as a biosynthetic model with the extremely facile formation of (±)-tetrahydrosecodine (40) from stemmadenine acetate in 75% yield where it was suggested that rupture of the C-15,C-17 bond was facilitated by platinum-catalyzed isomerization of the 19,20-double bond to the endo position thereby allowing collapse to the secodine system, as discussed earlier.

Finally we note that the earlier difficulties in rationalizing the lack of biochemical conversion of (+)-catharanthine (34) to its 15,20-dihydro derivative coronaridine (57) can now be understood, since in *Catharanthus roseus* it has been found that these *Iboga* alkaloids bear antipodal absolute stereochemistry at C-14,C-17 and C-21[45]. The co-occurrence of such antipodal species in the same plant may once again signify the onset of different synthetases acting on the same intermediate. Thus (-)-coronaridine is detectable at very early stages of germination of *C. roseus* whereas (+)-catharanthine is found only after prolonged germination and seedling growth[13]. The achiral ester dehydrosecodine A (31) or its isomer B (32) could serve as an intermediate which can undergo cyclization (Scheme 19) to either coronaridine (57) or (+)-catharanthine (34) and, predictably, the *pseudo* series represented by 35, discovery of which from a natural plant source has just been reported.*

Andranginine, Pseudocatharanthine and Vindolinine

During the *in vitro* conversion of precondylocarpine (30) and related pentacyclic alkaloids, it was observed that thermolysis of 30) in ethyl acetate solution afforded (in 28% yield) a new racemic pentacyclic compound (69), M.W. 334, corresponding to dehydrogenation of the 336 series The reaction was interpreted as shown in Scheme 20, where the stability of the vinyl dihydropyridine species is believed to have mediated in a relatively high yielding reaction to produce the new system (69) corresponding to the ionic addition depicted. In the course of structural investigation of a Madagascar species undertaken some months after this experiment was completed it was discovered that the compound andranginine (MW 334, $[\alpha]_{300-600} 0^{o}$) was in fact identical in every respect with this novel, partially synthetic alkaloid[46]. The full power of biogenetic type synthesis was thus revealed in two independent studies which converged in a most satisfying manner. The implications of this and related discoveries of racemic alkaloids is discussed in the final section of this review.

Although pseudocatharanthine (35) has never been described as a natural product, its synthesis from catharanthine, tabersonine, and preakuammicine are suggestive that the *Aspidosperma*-like structure of yet another mode of cyclization of the acrylic ester will appear as a natural series.* This postulate may well have extremely

---

*Dr. P. Potier (Gif-sur-Yvette) has informed us that the system (35) has now been isolated from a natural source by Professor J. LeMen.

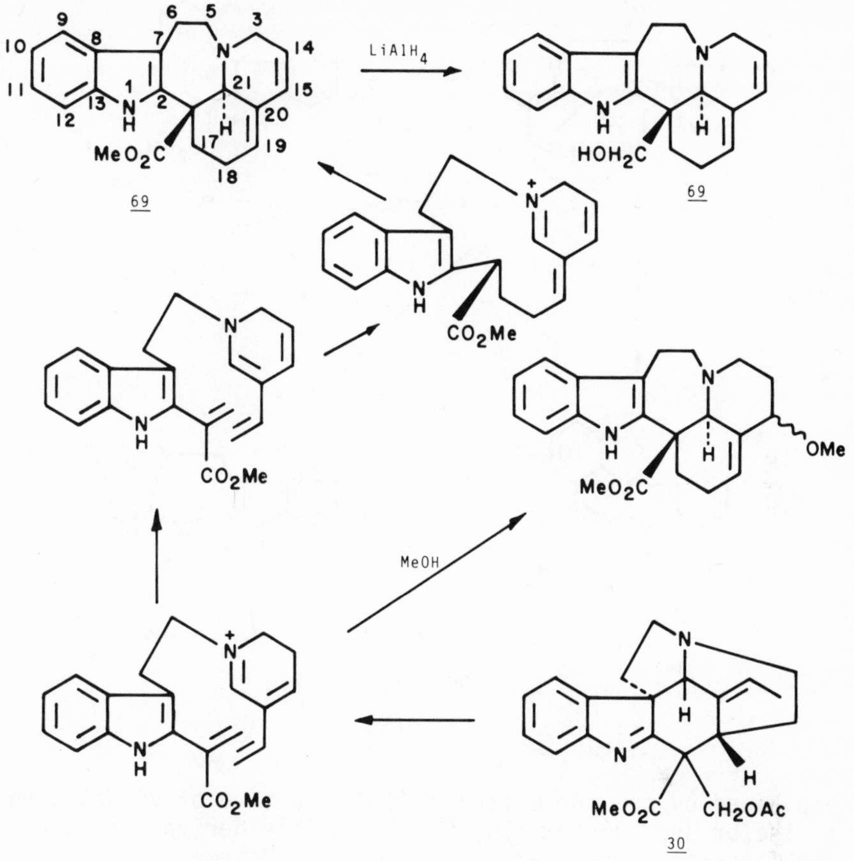

LiAlH$_4$

69

69

CO_2Me

CO_2Me

MeOH

CO_2Me

30

Scheme 20

important consequences for the biosynthesis of the "double" alkaloids discussed below.

In surveying the modes of cyclization of the secodine system, there exists a rather puzzling feature of the structure of a class of hexacyclic alkaloids exemplified by vindolinine (70) inasmuch as the stereochemistry required by the ethano-bridge (C-19, C-5) confers a *trans-syn* configuration on the fusion of rings D and E, a feature which is absent from all known *Aspidosperma* alkaloids. Elsewhere[47], we comment on this fact which has now been satisfactorily

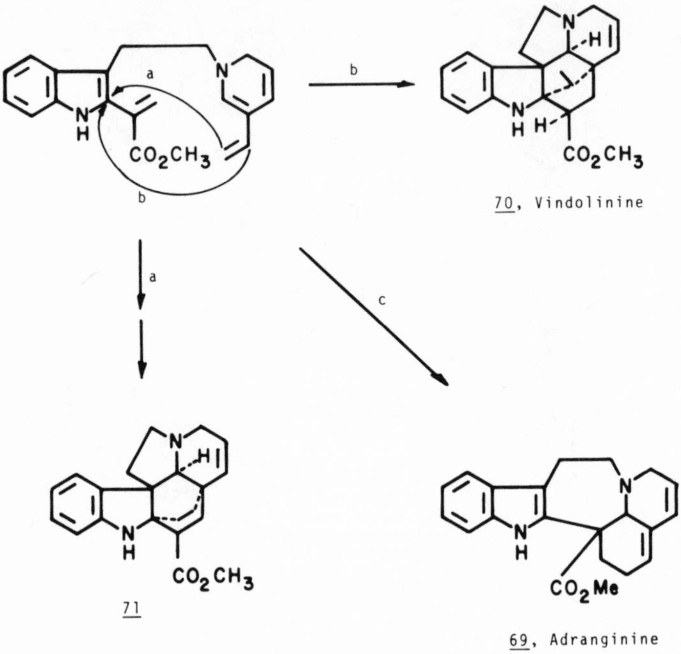

Scheme 21

explained by the correction of the structure of vindolinine to the bridged system (70)[48]. A possible derivation based on biogenetic condensations is shown in Scheme 21, which incidentally allows prediction of the relative stereochemistry of the asymmetric centers of vindolinine and also traces the connection with the hexacyclic series (71).

STEREOCHEMICAL AND STRUCTURAL RELATIONSHIPS WITHIN ALKALOID FAMILIES

Strychnos series

A perennial problem encountered in the theory of *Strychnos* biosynthesis has been the occurrence of enantiomeric and diastereoisomeric pairs of alkaloids exemplified by (+)-and (-)-akuammicine (67) and (-)-lochneridine and (+)-20 *epi*lochneridine. The crux of the stereochemical

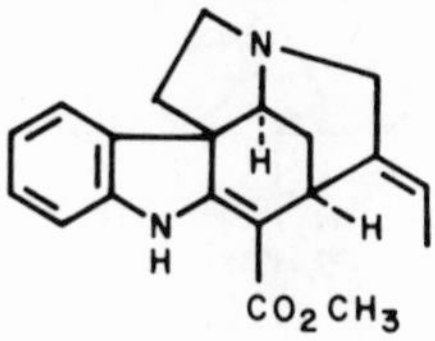

67, (-) - Akuammicine

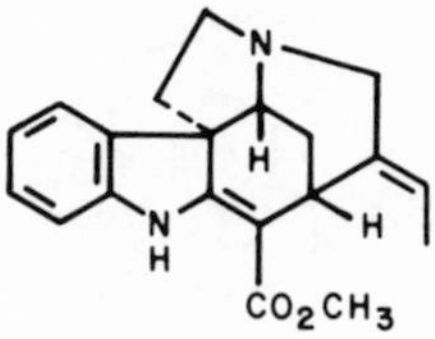

67, (+) - Akuammicine

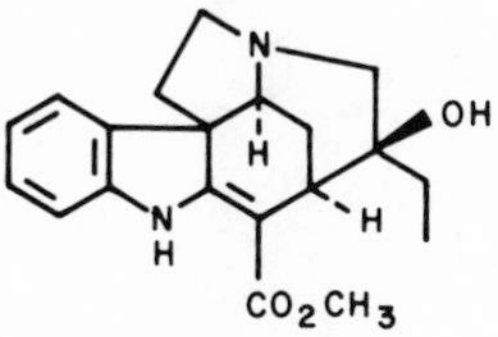

(-) - Lochneridine

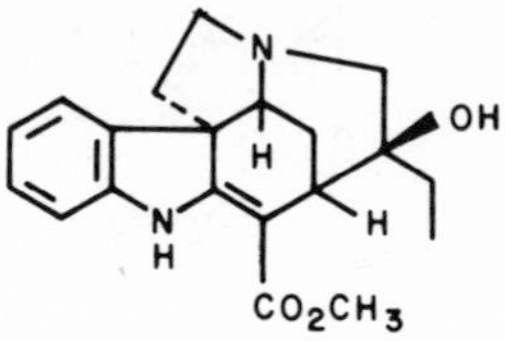

(+) - Lochneridine

enigma in this series concerns the apparent inversion at C-15 which, in the absence of any satisfactory rationale for the racemization or epimerization of this center, has led to the logical suggestion that C-15 retain its absolute stereochemical identity (αH or βH) from an earlier precursor of the *Corynanthe* series, or indeed that the inverted βH configuration comes all the way from a secoiridoid or "unnatural" configuration at this site. However, a recent series of experiments[49] portrayed in Scheme 22 has shown that epimerization of C-3, C-7 and C-15 is accomplished by a simple equilibration experiment in methanol solution at 95° (50 hr): this result now rationalizes the appearance of the so-called antipodal *Strychnos* family. The implications of these findings are discussed in the concluding section of this review. At this point we note that the intermediate *seco* system (71) may serve as a useful and easily accessible relay for total synthesis in this field. These experiments conducted with 19,20α-dihydroakuammicine (R=H) also constitute a model for the lochneridine series (R=OH).

Secodines and related compounds

The discovery, beginning in 1968, of a whole new range

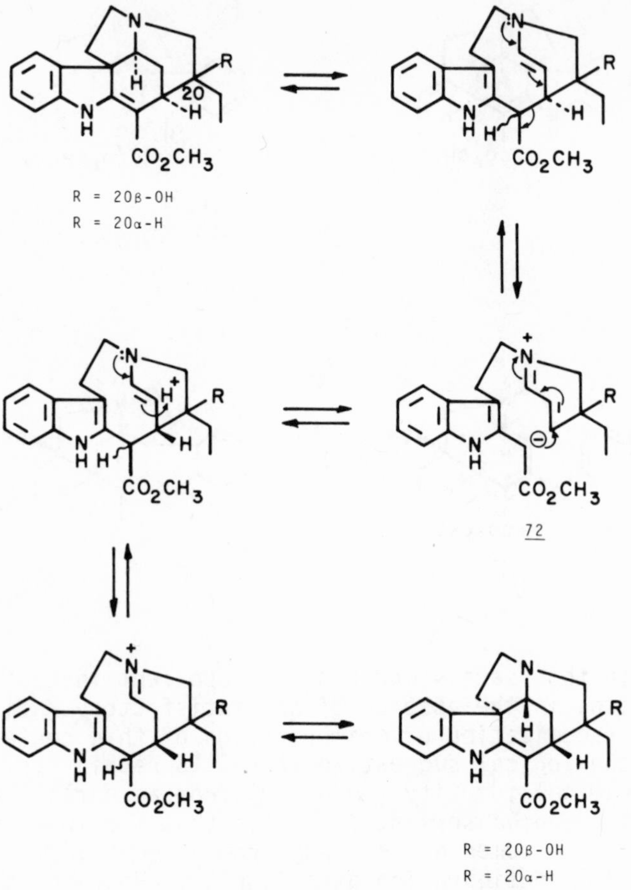

R = 20β-OH
R = 20α-H

72

R = 20β-OH
R = 20α-H

Scheme 22

of "*chano*" alkaloids based on the secodine array[50] was welcome testimony to the "336 concept" engendered by our preliminary experiments carried out the previous year. A selection of these structures and their chemical interrelationship is shown in Scheme 23. Secodine (53) is involved in several important biochemical transformations leading to the various classes (*Aspidosperma*, *Iboga*, and Apparicine series)[51]. The chemical manipulation of the fully activated dihydropyridine system in Scheme 23 has so far been studied indirectly[36-39] but a recent stabilization experiment[52] on some simpler models augurs well for more intensive and successful studies on the capture of 32 and its relatives.

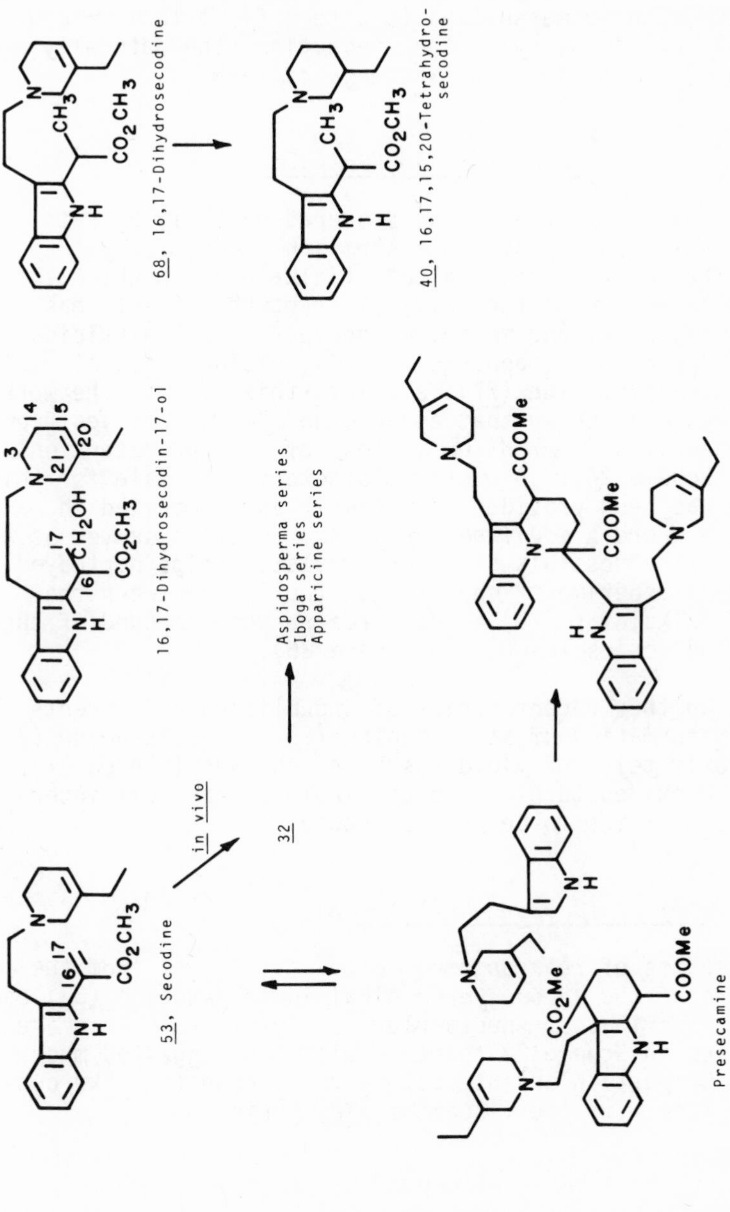

53, Secodine

16,17-Dihydrosecodin-17-ol

68, 16,17-Dihydrosecodine

in vivo

32

Aspidosperma series
Iboga series
Apparicine series

40, 16,17,15,20-Tetrahydro-secodine

Presecamine

Scheme 23. Chano alkaloids and their relationships.

An extremely efficient method of arriving at the tetrahydro secodine alkaloids (as 40) was portrayed in Scheme 19 where the *in vitro* chemistry of tabersonine, and especially catharanthine, can be marshalled to afford first the pyridinium salt (64) and then, by further reduction, the naturally occurring 40 which is found in *Rhazya stricta*.

N-oxides as substrates for "Nor"-alkaloids

An intriguing reaction discovered recently by Potier and his colleagues provides a striking model for yet another puzzling feature of a set of alkaloids in which the ethanamino bridge of the original tryptophan moiety has been shortened by one methylene group[53]. Such alkaloids as vallesamine (73), apparicine (74), uleine (75), olivacine (76), and ellipticine (77) fall into this class. The work of Kutney[54] has shown that apparicine (74) is derived from tryptophan in *A. pyricollum* by loss of C-2 and retention of C-3 (Scheme 24). A most satisfactory rationale for this process has been provided by Potier[55] as summarized in Scheme 25 where a modified Polonovsky reaction serves as a model for the possible biogenetic relationships portrayed. An elegant experiment[56] has forged the link between the vobasine alkaloids (29) and the rearranged compound of the ervatamine series ($R{=}CHCH_3$) (Scheme 26).

In another recent series of unpublished experiments, the transformation of stemmadenine (29) to vallesamine (73) in approximately 20% yield has been achieved (Scheme 27). Whether N-oxides turn out to be viable biochemical intermediates remains to be proved, however.

Reactions of Aspidosperma Alkaloids

A series of rearrangement products derived from the structure of the *Aspidosperma* alkaloid tabersonine (33) can only be described as spectacular[57]. Several of these are reproduced in Scheme 28 together with the suggested mechanism. Of particular interest is the biogenetic-like connection made with the vincamine (78) series.

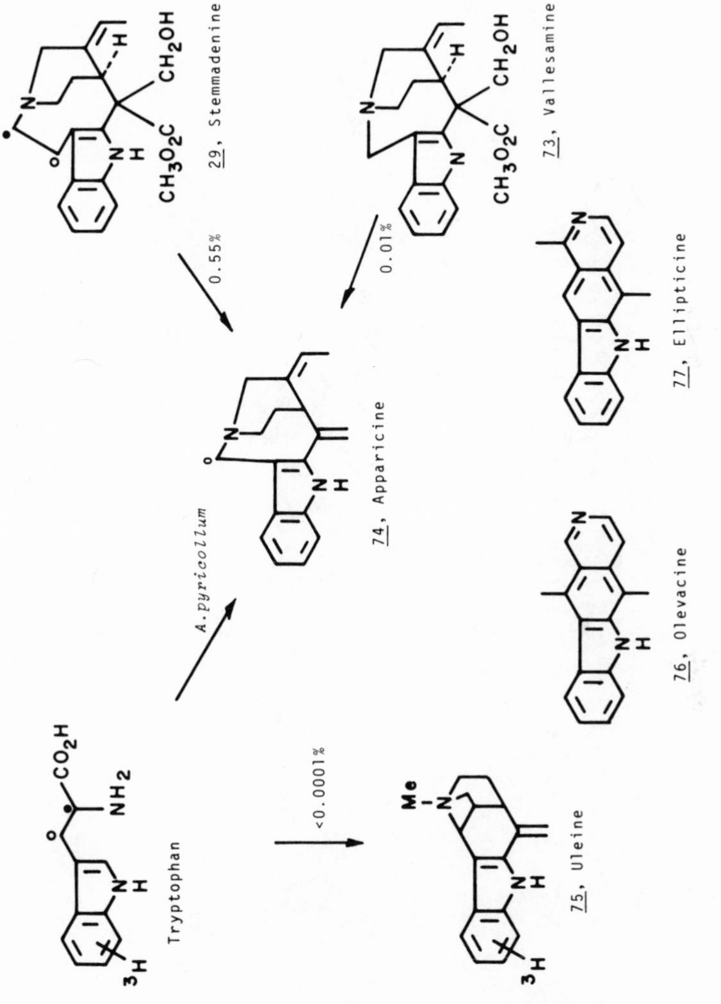

Tryptophan

^3H CO_2H NH_2

A. pyricollum

0.55%

29, Stemmadenine

CH_3O_2C CH_2OH

74, Apparicine

<0.0001%

0.01%

73, Vallesamine

CH_3O_2C CH_2OH

75, Uleine

76, Olevacine

77, Ellipticine

Scheme 24

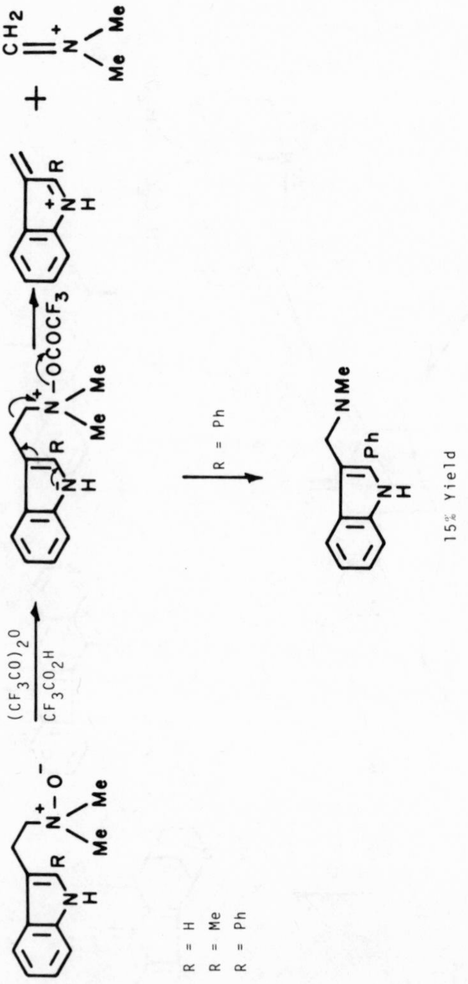

N
H
R
N
O
Me
Me
$(CF_3CO)_2O$
CF_3CO_2H
N
H
R
N
OCOCF$_3$
Me
Me
N
H
R
+
CH$_2$
N
Me
Me
R = H
R = Me
R = Ph
R = Ph
N
H
Ph
NMe
15% Yield

Scheme 25

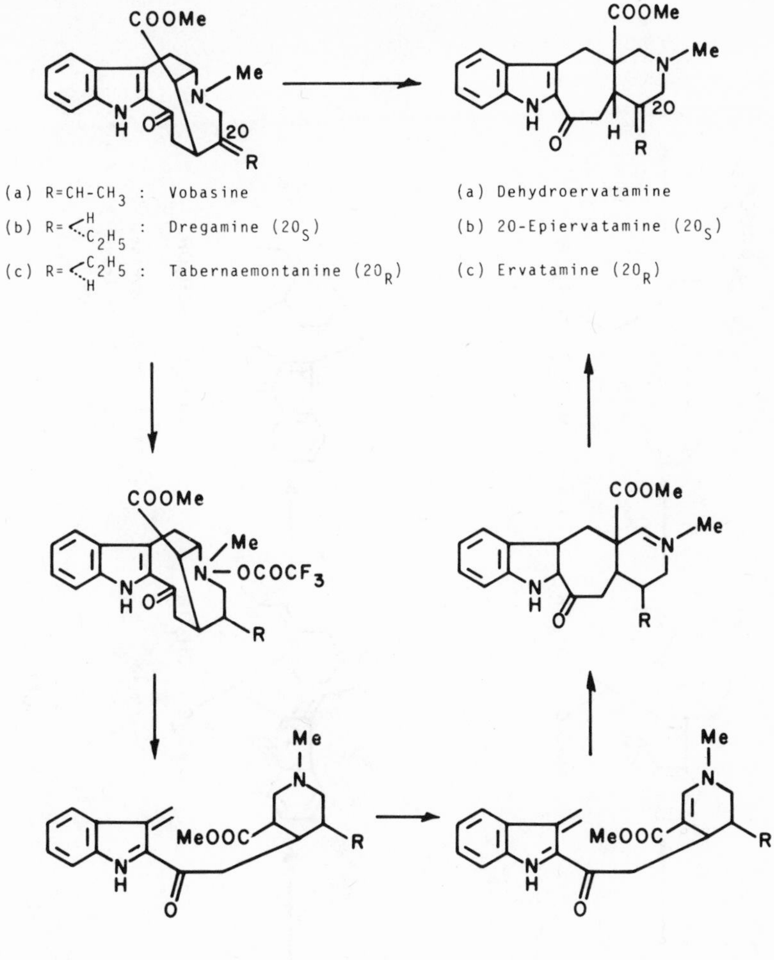

(a) R=CH-CH_3 : Vobasine
(b) R= : Dregamine (20_S)
(c) R= : Tabernaemontanine (20_R)

(a) Dehydroervatamine
(b) 20-Epiervatamine (20_S)
(c) Ervatamine (20_R)

Scheme 26

"Double" or Dimeric Indolic Alkaloids

The chemistry of these complex substances which have gained prominence due to the anti-tumor properties of several members of the *Aspidosperma-Iboga* double alkaloids, e.g. vinblastine, VLB (79) and vincristine, VLC (80) has been treated in an excellent recent review[58]. From the standpoint of biogenetic-type synthesis it appears certain that many (but by no means all) of these structures could be readily formed as artefacts. Examples of successful

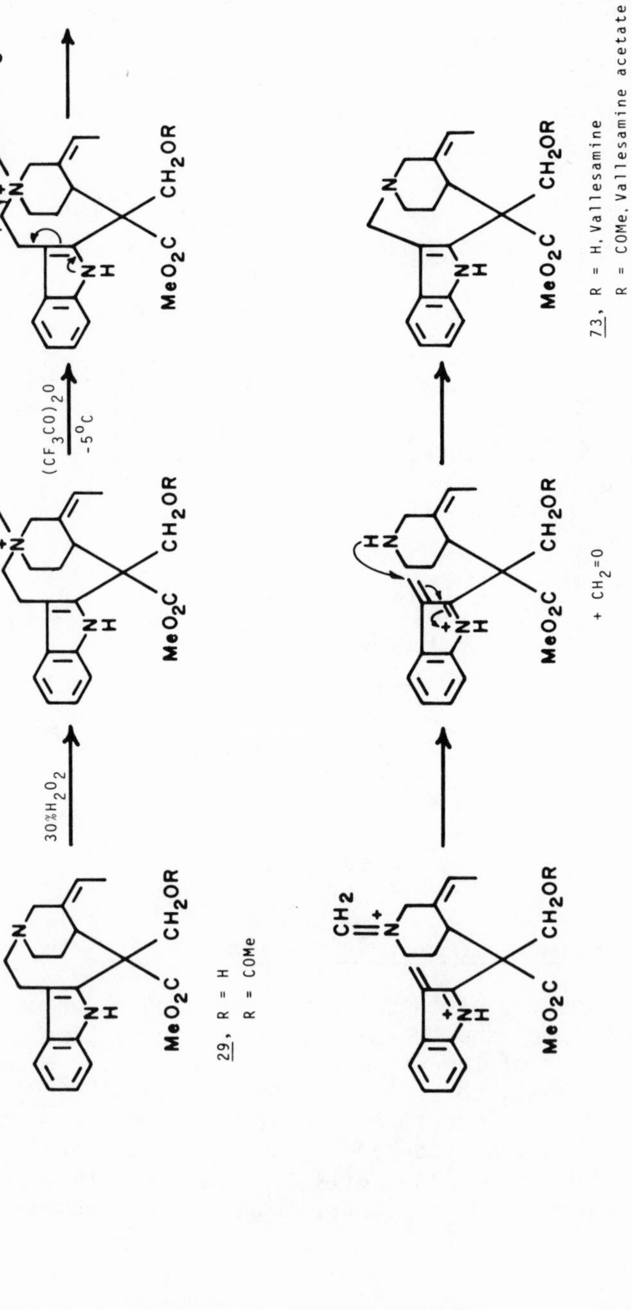

29, R = H
R = COMe

73, R = H, Vallesamine
R = COMe, Vallesamine acetate

Scheme 27

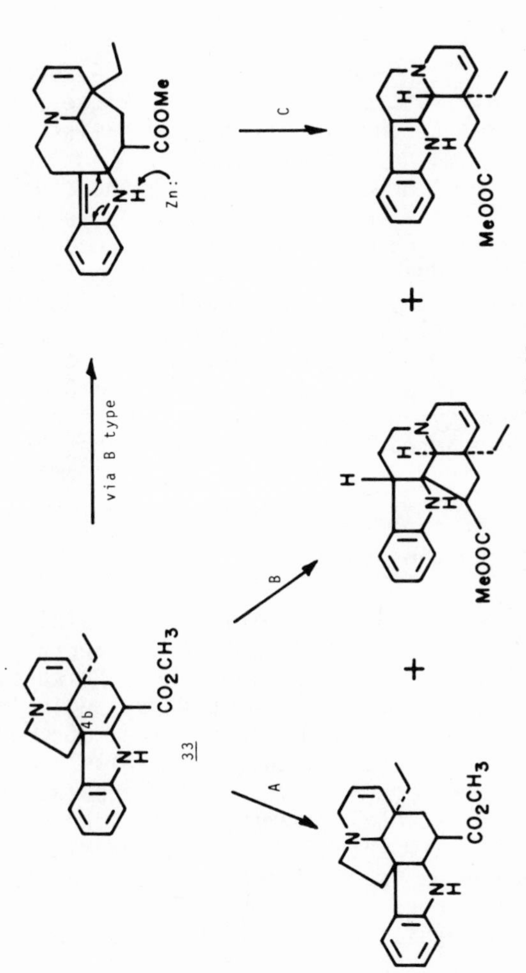

33

4b

via B type

Zn:

COOMe

A

B

C

16 Methoxycarbonyl-14,15-dehydrovallesamidine

78, "1,16-Secovincamine"

Scheme 28. Reaction of tabersonine with $Zn/CuSO_4/AcOH/110^{o}$.

coupling reactions include the preparation of geissospermine (81) which is formed from geissoschizine (1) and geissoschizol (82) under extremely mild conditions, the classical work on the Calebash-curare dimers[58] and a more recent biogenetic-type synthesis[59] of villalstonine (83) as illustrated in Scheme 29.

It seems unlikely that the anti-tumor series exemplified by VLB (79), VLC (80) and leurosidine (84) occurs without enzymic intervention in *C. roseus*, since a coupling reaction which leads to the desired stereochemistry at C-18 in the dimer remains to be described. Considerable progress has been made however in refining the coupling reactions of substituted, modified structures such as 85 and 86 with vindoline (88) to give dimeric products whose stereochemistry is under investigation (Scheme 30, *in vitro*)[60,61,62,63]. Preliminary feeding experiments (Scheme 30, *in vivo*) have not settled the problem in *C. roseus*; the specific incorporation of vindolinine (88) and catharanthine (34) into VLB is approximately 0.05% and 0.005% respectively. An intriguing possibility, which is currently being tested in the author's laboratory, is that the *pseudo* series (35) may be involved.

SUMMARY, CONCLUSIONS AND PROGNOSIS

Within the last decade, and particularly since 1968, the apparent diversity and complexity of the approximately 1000 members of the mevalonate-derived indole alkaloids has been rationalized in terms of a main pathway beginning with vincoside and passing through geissoschizine, stemmadenine⇄ preakuammicine, tabersonine and their more highly oxygenated, rearranged and degraded derivatives. The important fact that all of the observed skeletal rearrangements take place on a preformed alkaloidal template such as geissoschizine and stemmadenine can now be used not only to clear up several remaining mysteries, e.g. the *Corynanthe* → *Strychnos* mechanism, but also to predict within each category new classes of alkaloid based for example on the "336" concept, which represent highly reactive species upon which the rearrangement mechanisms may operate.

We feel that many short and eventually high-yielding syntheses of little-studied and rare alkaloidal structures await the development of methodology based on the pathway

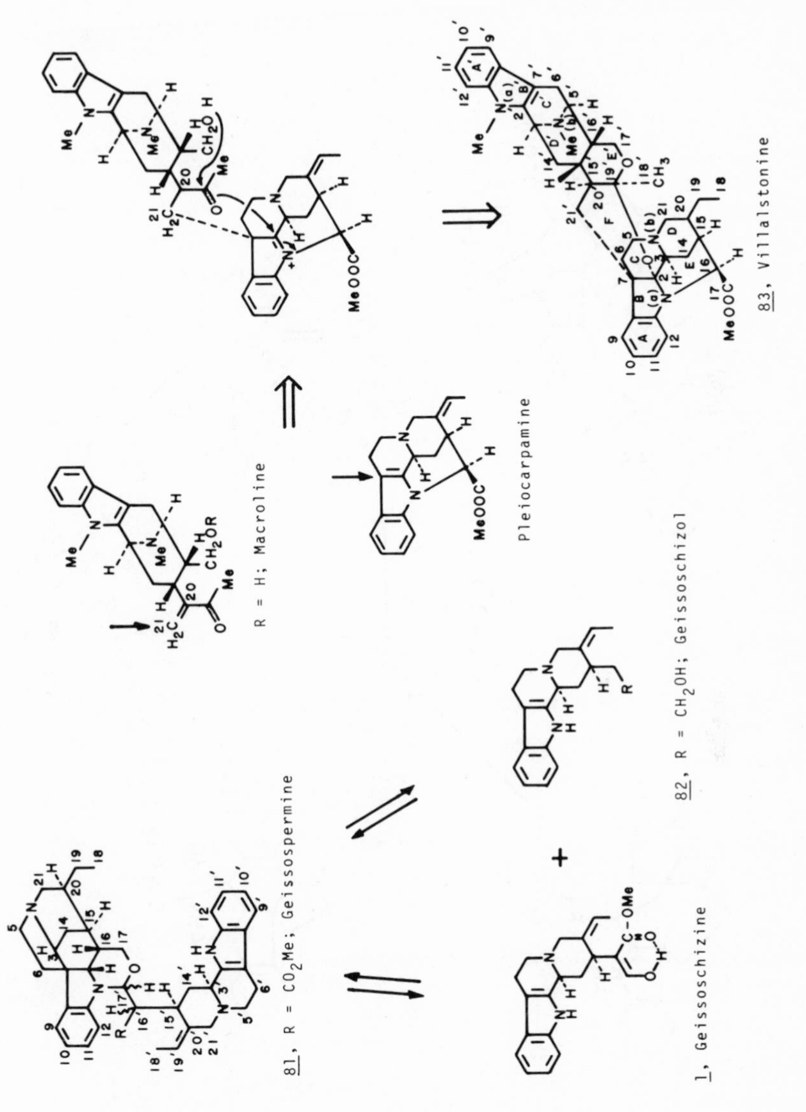
81, R = CO_2Me; Geissospermine
R = H; Macroline
Pleiocarpamine
83, Villalstonine
1, Geissoschizine
82, R = CH_2OH; Geissoschizol

Scheme 29

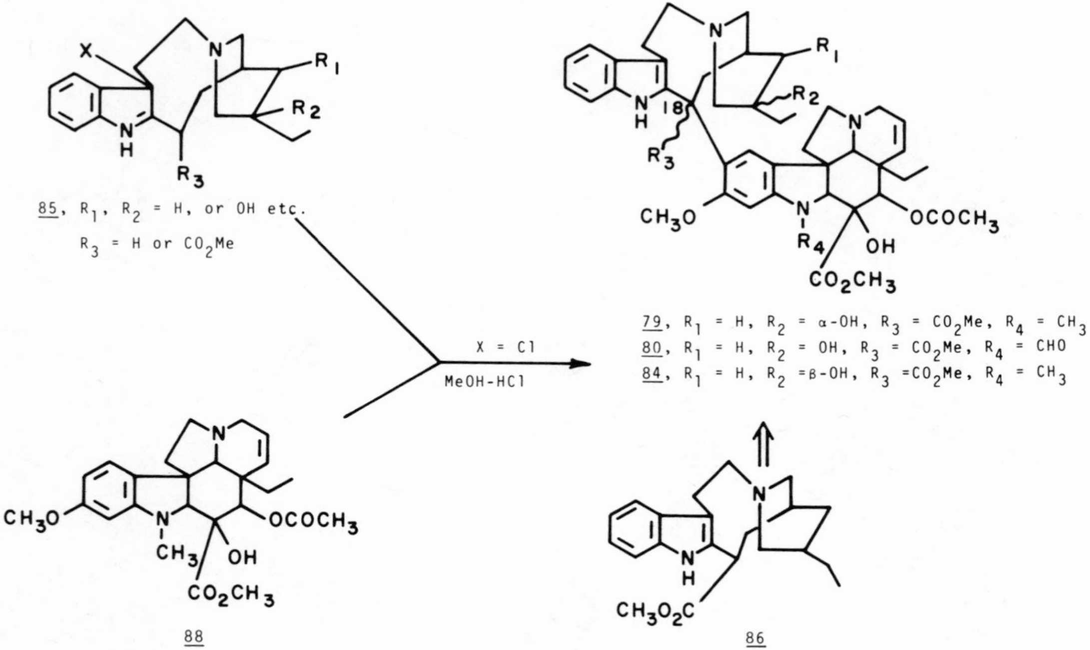

Scheme 30 (*in vitro*)

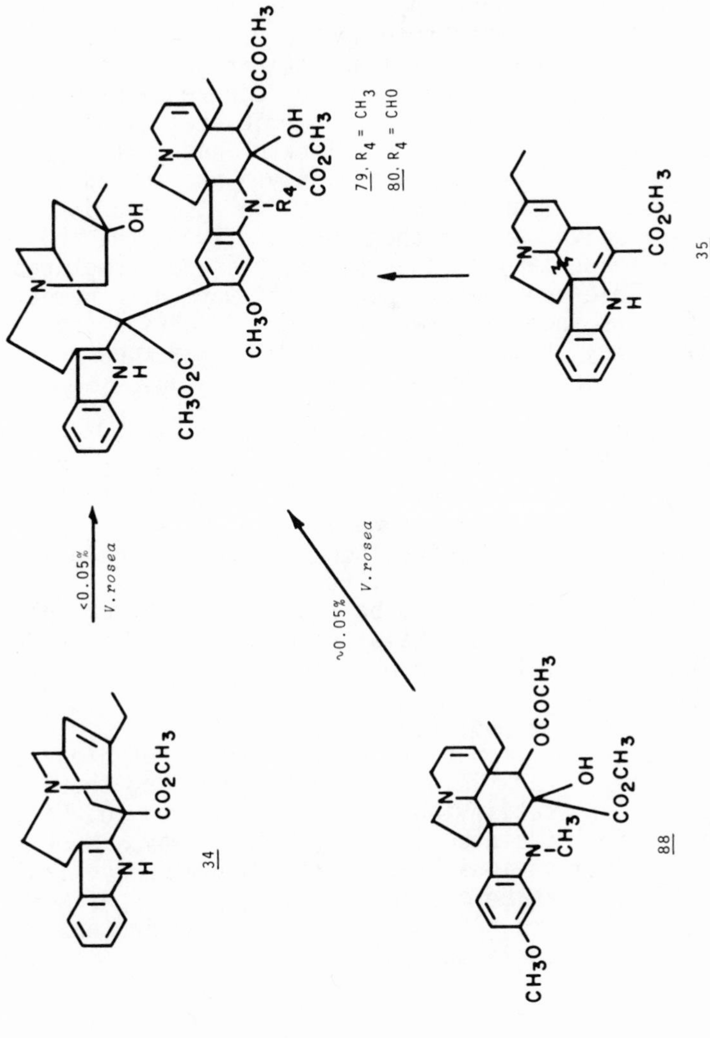

$\underline{34}$ $\xrightarrow[V. rosea]{<0.05\%}$ 79, 80

$\underline{88}$ $\xrightarrow[V. rosea]{\sim 0.05\%}$ 79, 80

$\underline{35}$ ⟶ 79, 80

$\underline{79}$. $R_4 = CH_3$

$\underline{80}$. $R_4 = CHO$

Scheme 30 (*in vivo*)

followed by Nature in arriving at these rather complex and challenging structures. Now that the innate beauty of the biochemical pathway has been revealed (at least in broad detail) there seems no doubt that many improvements can and will be made on existing and still quite primitive methods of inter-familial connection. However, the facility of several of the biogenetic-type syntheses reviewed even in this rather selective survey leads to a highly speculative - even bizarre - suggestion that the occurrence of many racemic, enantiomeric and diastereomeric sets within the Apocynaceae may reflect either a lack of substrate specificity on the part of a rearranging enzyme or, indeed, the complete absence of any such enzyme. In other words the latter alternative implies that, especially in the tropical regions, where so many of the most rearranged structures occur in species subjected to long periods of warmth, a number of transformations including rearrangement, dimerization, and racemization may have taken place without the agency of enzyme control on a highly reactive substrate.

The final suggestion and many others implicit in our survey remain to be tested by rigorous experiment. Such experiments would, we believe, have profound implications for chemotaxonomy and perhaps reduce the apparent multitude of enzymes in plant alkaloid biochemistry to a reasonable number. At the same time the "artificial" nature of many complex alkaloids may have to be recognized, the examples of vallesiachotamine (10), geissospermine (31) and even andranginine (69) falling into this category.

In spite of the many frustrations and experimental difficulties encountered by the investigators of biogenetic-type synthesis in the field of indole alkaloids, there is emerging the most satisfying viewpoint that many of Nature's processes can be simulated in a significant way. Most importantly, work from various laboratories has made available substrates which can be tested in biochemical experiments and which may be used to test model reactions for enzymic and non-enzymic transformations.

Acknowledgements

The work from the author's laboratory described in this review was carried out by a skilled group of postdoctoral fellows - Drs. P.C. Cherry, C.R. Bennett, D. Greenslade,

A.A. Qureshi, P.B. Reichardt, M.B. Slaytor, J.G. Sweeny, C.C. Wei and C.L. Yeh. We thank the National Institutes of Health and National Science Foundation for generous financial support.

REFERENCES

1. Scott, A.I., P.B. Reichardt, M.B. Slaytor & J.G. Sweeny. 1972. Rec. Adv. Phytochem. 6: 117.
2. Hesse, M. 1964, 1968. "Indolalkaloide in Tabellen", Springer, Berlin.
3. The Chemical Society. 1971, 1972, 1973. "The Alkaloids", Specialist Periodical Reports, Vols. 1, 2, 3. The Chemical Society, London.
4. Devon, T.K. & A.I. Scott. 1973. "Handbook of Natural Products", Academic Press, New York.
5. van Tamelen, E.E. 1968. Accts. Chem. Res. 1: 111.
6. Kompis, I., M. Hesse & H. Schmid. 1971. Lloydia, 34: 269.
7. Winterfeldt, E. 1971. Chimia (Switz.) 25: 394.
8. van Tamelen, E.E. 1963. In "Progress in the Chemistry of Natural Products" (L. Zechmeister, ed.) Vol. 20, Springer, New York.
9. Battersby, A.R. 1967. Pure Appl. Chem. 14: 117.
10. Scott, A.I. 1970. Accts. Chem. Res. 3: 151.
11. Battersby, A.R. 1971. In "The Alkaloids" (J.E. Saxton, ed.). Specialist Periodical Reports, Vol. 1, p. 31. The Chemical Society, London.
12. Leete, E. 1972. In "Biosynthesis" (T.A. Geissmann, ed.). Specialist Periodical Reports, Vol. 1, p. 158. The Chemical Society, London.
13. Qureshi, A.A. & A.I. Scott. 1968. Chem. Commun. 948.
14. Battersby, A.R., A.R. Burnett & P.G. Parsons. 1969. J. Chem. Soc. 1193.
15. Kennard, O., P.J. Roberts, N.W. Isaacs, F.H. Allen, W. D.S. Motherwell, K.H. Gibson & A.R. Battersby. 1971. Chem. Commun. 899.
16. De Silva, K.T.D., G.N. Smith & K.E.H. Warren. 1971. Chem. Commun. 905.
17. Blackstock, W.P., R.T. Brown & G.K. Lee. 1971. Chem. Commun. 910.
18. Scott, A.I., P.B. Reichardt, M.B. Slaytor & J.G. Sweeny. 1971. Bioorganic Chem. 1: 157.
19. Brown, R.T. & C.L. Chapple. 1973. Chem. Commun. 886.
20. Harley-Mason, J. & W. Waterfield. 1963. Tetrahedron, 19: 65.

21. Qureshi, A.A. & A.I. Scott. 1968. Chem. Commun. 145.
22. Bindra, S.K. 1971. In "Progress in the Chemistry of Natural Products" (W. Herz, G.W. Kirby & H. Grisebach, eds.) Vol. 29, Springer, New York.
23. Winterfeldt, E., T. Korth, D. Pike & M. Boch. 1972. Angew. Chem. International Edn. 11: 289.
24. Boch, M., T. Korth, J.M. Nelke, D. Pike, H. Radunz & E. Winterfeldt. 1972. Chem.Ber. 105, 2126.
25. van Tamelen, E.E. & L.K. Oliver. 1970. J. Am. Chem. Soc. 92: 2136.
26. Pousset, J.-L., J. Poisson, L. Olivier, J. LeMen & M.-M. Janot. 1965. Compt. Rend. 261: 5538.
27. Oliver, L., J. Levy, J. LeMen, M.-M. Janot, H. Budzikiewica & C. Djerassi. 1965. Bull. Soc. Chim. France, 868.
28. Wenkert, E. 1962. J. Am. Chem. Soc. 84: 98.
29. Wenkert, E. & B. Wickberg. 1965. J. Am. Chem. Soc. 87: 1580.
30. Qureshi, A.A. & A.I. Scott. 1968. Chem. Commun., 947.
31. Brown, R.T., J.S. Hill, G.F. Smith, R.S.T. Stapleford, J. Poisson, M. Muquet & N. Kunesch. 1969. Chem. Commun. 1475.
32. Brown, R.T., J.S. Hill, G.F. Smith & K.S.J. Stapleford. 1971. Tetrahedron, 27: 5217.
33. Poisson, J., M. Muquet & N. Kunesch. 1972. Tetrahedron, 28: 1363.
34. Scott, A.I. 1972. J. Am. Chem. Soc. 94: 8262.
35. Brown, R.T., G.F. Smith, J. Poisson & N. Kunesch. 1973. J. Am. Chem. Soc. 95: 5778.
36. Scott, A.I. & C.C. Wei. 1972. J. Am. Chem. Soc. 94: 8264.
37. Walser, A. & C. Djerassi. 1965. Helv. Chim. Acta, 48: 391.
38. Schumann, D. & H. Schmid. 1963. Helv. Chim. Acta, 46: 196
39. Scott, A.I. & C.C. Wei. 1972. J. Am. Chem. Soc. 94: 8263.
40. Gorman, M., N. Neuss & N.J. Cone. 1965. J. Am. Chem. Soc. 87: 93.
41. Bláha, K. Z. Koblicová & J. Trojanek. 1972. Tetrahedron Lett., 2: 763.
42. Scott, A.I. & C.C. Wei. 1972. J. Am. Chem. Soc. 94: 8266.
43. ______ & P.C. Cherry. 1969. J. Am. Chem. Soc. 91: 5872.
44. Edwards, P.N. & G.P. Smith. 1961. J. Chem. Soc. 1458.
45. Kutney, J.P., K. Fuji, A.M. Treasurywala, J. Fayos, J. Clardy, A. I. Scott & C.C. Wei. J. Am. Chem. Soc. 95: 5407.
46. Kan-Fan, C., G. Massiot, A. Ahond, B.C. Das, H.P. Husson, P. Potier, A.I. Scott & C.C. Wei. 1974. Chem. Commun. 164.

47. Scott, A.I. & A.A. Qureshi, 1974. Bioorganic Chemistry (in press).
48. Ahond, A., M.-M. Janot, N. Langlois, G. Lukacs, P. Potier, P. Rasoanaivo, M. Sangare, N. Neuss, M.J. LeMen, E.W. Hagaman & E. Wenkert. 1974. J. Am. Chem. Soc. 96: 633.
49. Scott, A.I. & C.L. Yeh. 1974. J. Am. Chem. Soc. (in press).
50. Cordell, G.A., G.F. Smith & G.N. Smith. 1970. Chem. Commun. 189.
51. Kutney, J.P. 1972. J. Heterocyclic Chem. 9: supplementary issue and references s-1.
52. Cullen, W.R., J.P. Kutney, V.E. Ridaura, J. Trotter & A. Zanarotti. 1973. J. Am. Chem. Soc. 95: 3058.
53. Ahond, A., A. Cone, C. Kan-Fan, Y. Langlois & P. Potier. 1970. Chem. Commun. 517.
54. Kutney, J.P., V.R. Nelson & D.C. Wigfield. 1969. J. Am. Chem. Soc. 91: 4278.
55. Husson, A., Y. Langlois, C. Riche, H-P. Husson & P. Potier. 1973. Tetrahedron. 29: 3095.
56. Potier, P. & M.-M. Janot. 1973. C.R. Acad. Sci. (Paris), 276C: 1727.
57. Pierson, C., J. Garnier, J. Levy & J. LeMen. 1971. Tetrahedron Lett. 1007.
58. Gorman, A.A., M. Hesse, H. Schmid, P.G. Waser & W.H. Hopft. 1971. In "The Alkaloids" (J.E. Saxton, ed.) Specialist Periodical Reports. Vol. 1. The Chemical Society, London.
59. Burke, D.E., J.M. Cook & P.W. LeQuesne. 1973. J. Am. Chem. Soc. 95: 546.
60. Büchi, G. 1964, 1965. "IUPAC, The Chemistry of Natural Products" Butterworth, London.
61. Neuss, N., M. Gorman, N.J. Cone & L.L. Huckstep. 1968. Tetrahedron Lett. 783.
62. Harley-Mason, J. & A. Rahman. 1967. Chem. Commun. 1048.
63. Kutney, J.P., J. Beck, F. Bylsma & W.J. Cretney. 1968. J. Am. Chem. Soc. 90: 4504.

Chapter Ten

ANTIMICROBIAL AGENTS FROM HIGHER PLANTS

L.A. MITSCHER

Division of Natural Products Chemistry
The Ohio State University, Columbus, Ohio

INTRODUCTION

The modern antibiotic era can be said to have opened on February 12, 1941, with the first clinical trial of penicillin. Within a couple of years man had at last, after eons of adventitious searching, a truly effective and safe agent for the treatment of many systemic bacterial infections. This was shortly followed by the introduction of one after another of the major antibiotic substances which remain the mainstay of clinical therapy of infectuous diseases to this date. Intensive screening of fermentation liquors derived from various microbes, principally fungi and streptomycetes, has resulted in literature descriptions of more than 2,000 individual antibiotics to date, and it has been estimated that more than 20,000 analogs of 6-aminopenicillanic acid alone have been made by chemists working in close collaboration with biochemists and microbiologists. Several of these substances are currently on the market.

Another dimension to this story can be perceived by looking at some statistics relating to the extent to which antibiotics have assumed a dominant role in practical therapeutics. In 1972 alone, 10 billion doses of antibiotics, representing some 2,400,000 kg of these drugs, were certified for use in the United States alone by the Federal Food and Drug Administration. This is sufficient to provide 50 doses per year for every man, woman and child in this country. In 1971, 160 million prescriptions were written for antibiotics making this the most specified single therapeutic category. For some years now an antibiotic, tetracycline, has been the most commonly prescribed generic entity.

From these data, it might be supposed that the treatment of bacterial infections of man would be well under control. However, it is not difficult to document that the need for new antibiotics still exists. Sabin[1] has pointed out that in 1963, infectuous diseases ranked as the number three killer (after cancer and heart disease-stroke) with influenza, pneumonia, tuberculosis, kidney infections, bronchitis and syphilis prominent in the statistics. The relative situation today is virtually unchanged. Considering that mortality represents the smallest portion of those inconvenienced or injured by infection, the magnitude of the challenge remaining is obvious. Perlman[2] points out that only infections due to Gram-positive bacteria, notably staphylococci, streptococci or pneumococci, can be regularly controlled by antibiotics and superinfections by Gram-negative bacteria, and in particular by multiple drug-resistant strains, is an increasing problem. Many strains of *Pseudomonas*, *Proteus*, *Aerobacter*, and *Salmonella* remain largely resistant to existing antibiotics. Microorganisms such as *Providentia*, which were seldom previously encountered, are becoming an increasingly challenging therapeutic problem. Systemically active antifungal agents, antiviral agents and antiprotozoal agents of a new and effective type could come into rapid use because of real difficulties with existing agents. Gonorrhea, although usually treatable by existing agents when used under appropriate conditions, is nevertheless out of control in some areas - especially college campuses.

The human misery of those afflicted with these conditions can only be imagined. To balance the account somewhat the reader need hardly be reminded that very significant strides have been made toward the control of communicable diseases, but the foregoing brief discussion shows that a great deal remains to be accomplished.

Faced with a continuing need for antibiotics and the very large economic stakes for those who are successful in finding them, it is not surprising that intensive efforts are still under way to uncover new agents. It can be stated that we have reached the point of diminishing returns in screening systems confined nearly exclusively to the Streptomycetes and the fungi, even though important agents are still being described from such sources, e.g. tobramycin, from *Streptomyces tenebrarius*. Many laboratories are broadening their search to include more unusual sources and the recent literature describes such agents as ambutyrosin from *Bacillus circulans* and the gentamicin C complex from *Micromonospora purpurea*. One notes, in examining this literature, how frequently these agents resemble in structural terms the antibiotics one finds among streptomycete fermentation liquors.

It is the thesis of this chapter that there are as yet incompletely exploited alternative sources of antimicrobial agents whose structures and modes of action may very likely differ from those from the more well-worked microbial sources. Compounds in extracts of higher plants have the potential of filling this need and numerous investigators currently aware of this are actively pursuing such studies. Because the literature is voluminous and scattered, some arbitrary decisions were made of necessity. The coverage is limited primarily to the years 1967-1974. Phytoalexins, a subject of great intrinsic interest, have been largely neglected, in part because of recent excellent chapters in this series dealing directly and indirectly with this subject. Those agents which appear to have their greatest potential for the control of plant pathogens have not been thoroughly covered. Furthermore, antifungal agents are incompletely reviewed also because of the excellent recent treatment in this series by Stoessl[3].

ANCIENT HISTORY

Man has doubtless been aware, at least dimly for centuries, that antimicrobial agents are present in higher plants, because the folk literature contains frequent reference to such uses stretching back for at least 6,000 years. The variable quality of such preparations, the inadequately controlled conditions of their use, the fact that results obtained in research on antibiotics are strongly influenced by the methodology used for extraction and measuring antibacterial activity, and a host of other complicating factors make it impossible even in retrospect to make much use of such reports. The persistence of such reports over a period of centuries is undoubtedly attributable both to the dire need for such agents and the fact that such treatment occasionally worked. It is significant that no generally recognized valuable therapeutic regimen resulted from such work.

ANTIMICROBIAL AGENTS FROM HIGHER PLANTS IN THE EARLY YEARS OF THE ANTIBIOTIC ERA (1940-1967)

Stimulated by the advent of penicillin, tyrothricin, and streptomycins and by the advent of more reliable microbial testing systems, large scale screening programs for antimicrobial agents from higher plants were undertaken in the 1940's using methodology which enables one to judge the relative worth of the results. This work has been reviewed in some detail[4,5,6,7,8,9]. These studies were, surprisingly, in most cases not followed through at the time to the isolation of pure individual constituents. It is obvious that a large range of potency possibilities exist, ranging all the way from a large content of a weak agent to a low content of a strongly active one. Thus, simple zone sizes are not in themselves compelling indications of the likelihood that a given plant contains a highly active agent. The main lesson to be gained from these studies is that a very large range of vascular plants contain extractable antimicrobial agents and that, if one is willing to include enough species in the study and to consider activity against Gram-positive bacteria, one can expect about 1 in 5 of the plants under study to contain detectable activity. We have taken a number of plants listed in one or another of such screening reports and subjected these to a confirmatory study using somewhat different methodology and have found that the reconfirmation rate runs about 20% in our hands[10]. Nevertheless, this is sufficient to provide us with enough leads to keep us fully occupied for a long time to come. The very numbers of species cited by various authors encourages the belief that clinically active agents may be found given sufficiently determined examination. An element that is usually missing from literature reports, even those published quite recently, is some indication of the relative *in vivo* toxicity of purified agents. The single most prized attribute of an antibiotic to be used to treat systemic infections is its selective toxicity for pathogens. To find that strychnine had *in vitro* potency would not likely encourage its use as an antibiotic. It should also be noted that a pharmacological agent which had some antibiotic properties might be tolerated in clinical use, but almost certainly an antibiotic would have to be devoid of pronounced pharmacodynamic activity unless used in life-threatening or severe infections. Many reports on volatile oils suffer from the disadvantage of poor selectivity as the agents are most often phenols and alcohols which disrupt cell membranes in the host as well as in invasive microorganisms. Thus these materials are disinfectants on the whole and not antibiotics. Typical screening efforts are unlikely to turn up phytoalexins unless the specimen chosen is not recognized as being diseased when collected.

RECENT WORK ON ANTIMICROBIAL AGENTS FROM HIGHER PLANTS (1967-1974).

The surge of interest in antimicrobial agents from higher plants subsided somewhat when the remarkable selectivity and high potency of the Streptomycete-derived agents became apparent in the late 1940's and early 1950's. Another reason for this was the relative ease of accession of multitudes of microbial species, the convenience of producing quantities of the desired agents, and the large capital investment in fermentation facilities which had been made. There has been a recovery of interest in this area now that there are clear signs of overwork in screening the Streptomycetes. In quite recent times, several screening reports have appeared. Murikami and coworkers report screening 300 crude drugs and 200 common wild plants[11]. Shcherbanovskii and Nilov looked for inhibitors of alcohol fermentation among higher plants, finding that saponins were apparently responsible for the activities seen[12]. *Pittosporum glabratum, P. ralphii, P. crassifolium, Yucca aloifolia, Y. filimentosa* and *Primula veris* were found to be active. Mathes looked at the elaboration of antimicrobial agents from plant tissue cultures[13], finding that in most cases antimicrobial agents could be so produced just as well as they were by the whole plant. The cultured cells prepared from *Populus tremuloides*, in particular, had broad spectrum activity. The bacteriostatic activity of several crude drugs was compared with a variety of plant secondary metabolites, such as berberine, shikonin, paeonol, etc[14]. *Phellodendron* cortex, *Lithospermum* root, *Rheum* rhizome and *Angelica* root were especially examined. Makarenko used paper chromatographic procedures to compare agents produced by different plants[15]. *Spirea mongolica* and *S. nipponica* extracts showed limited spectra. Several species of *Discorea* were found to contain steroidal saponins with antifungal (but not antibacterial) properties[16]. *D. takoro* gave dioscin, gracillin and dioscin prosapogenin as active agents. This was followed up by study of the activity of 23 steroidal sapogenins and saponins, which indicated that only the saponins are active[17]. The well known hemolytic activity of saponins probably prevent these agents from playing a role in systemic infections of humans. Khanin and coworkers screened a variety of plant extracts using liquid CO_2 as an extractant. Such common spice plants as cinnamon bark, clove flower buds, wormwood, coriander seed and sweet-scented pepper gave extracts active against yeast as well as bacteria[18]. Examination by tlc of a number of extracts of higher plants showing antifungal properties led to the conclusion, in part because of the regular hemolysis caused by the active plants, that saponins were probably present[19]. The apparent parallel in mechanism between saponins and polyene macrolide antibiotics, both of which kill fungi by binding to the cell membrane and destroying its integrity, is striking and has received some experimenta attention[20]. Soeding and Doerffling also screened a variety of higher plant extracts for anti Gram-positive and anti Gram-negative activity, finding *Acer pseudoplatanus, Helianthus annus* (sunflower) and *Pisum sativum* (pea shoots) to be of some interest[21]. Malcolm and Sofowora[22] found both *Combretum micranthum* root and *Dracaena mannii* stem bark to have broad spectrum activity *in vitro* while *Terminalia avicennioides* was only anti Gram-positive among a group of plants selected from Nigerian folk remedies. Of 300 botanically classified Indian plants, 56 were found to have some activity[23]. Among several species of St. John's wort cultivated in Uzbekistan, five were found to be active against *Staphylococcus aureus in vitro*, with *Hypericum calycinum* being most potent[24]. In an examination of extracts of 27 higher plants from Egypt[25], activity was found only in the dried residues of tomato tops, eggplant tops and guava leaves. Solis tested extracts of 27 legumes against four bacteria[26]. Of those found to be active, vitexin was proposed to be the responsible agent in *Tamarindus indica*. Vichkanova and coworkers tested saponins from 18 different plant families, reporting that 9 of 62 preparations were active - mainly against fungi and protozoa[27]. *In vitro* antimicrobial test data on a variety of garden plants (particularly *Allium* sp.) showed that the juices were active against Gram-negative organisms[28]. Of 13 species of ferns from Gujarat State (India), three (*Adiantum trapiziforme*, *Aleuritopteris farinosa* and *Pteris vittata*) contained antimicrobial constituents[29]. A number of plants show activity against *Bacillus thuringiensis* and entomogenous bacteria; the hydroxylic carboxylic acids from *Pinus rigida* have been most thoroughly studied[30]. In a survey of 24 Minnesotan aquatic plants for antimicrobial activity, *Nymphaea tuberosa*, *Nuphar variegatum* and *Sparganium fluctuans* were found to be of special interest[31,32]. Mitscher and coworkers screened 154 species and found potentially interesting activity present in extracts of *Amorpha fruiticosa*, *Chelidonium majus*, *Chrysanthemum anethifolium*, *Hunemania fumariaefolia*, *Ptelea trifoliata*, *Solanum pseudocapsicum*, and *Thalictrum rugosum*[10]. Another 48 plants showed weak activity. An examination of plants from the Pacific area found *Dioscorea bulbifera*, *Intsia bijuga*, *Thespesia populnea* and *Premna taitensis* to be of greatest interest[33]. These were culled from the 12 antimicrobial extracts of 110 plants tested. As part of a continuing large scale phytochemical survey of plants, Farnsworth and coworkers screened 400 species and some 101 showed at least marginal activity against one species[34,35]. *Dryopteris marginalis*, *Hydrophyllum appendiculatum* and *Piper methysticum* were singled out as the most interesting for further study.

Thus, it is evident that screening studies continue to turn up new leads and that this phase of the work is inherently more rapid than the isolation, identification and evaluation of the active constituents. It is difficult to compare screening studies with one another because of the widely differing methodologies used in the various laboratories, if for no other reason than the use of quite different strains of microorganisms employed. Use of two different strains of the same bacterium can result in dramatically different potencies. Media constituents can, in some cases, also play a dramatic role in determining the apparent potency of a substance. At the very least, it is hoped that journals will no longer publish papers in which the strain designation and origin of the strain are not specified. This would help the various laboratories to correlate their findings better. This problem is not as serious now as it was a decade and more ago. Despite these problems, these screening reports are a fruitful field for future work.

A comparatively large number of papers has been published dealing with antimicrobially active agents present in a single plant. These data are assembled in Table 1 in alphabetical order rather than according to botanical relationships in order to make the data more readily useful to the chemist.

Table 1. Studies Involving Antimicrobial Activity from a Single Plant Species in the Period 1967-1974

Plant	Activity	Reference
Achillea salicifolia	Bacteria, fungi, protozoa	36
Allium kurrat (leaf)	Gram-pos., Gram-neg.	37
Allium sativum (bulb)	Yeast, fungi	38
Ampelopsis brevipedunculata	*E. coli*, *Staph. aureus*	39
Anagallis arvensis	Fungi	40
Anagallis arvensis var. *phoenicea* (stem, leaf)	Fungi	41
Artemisia tridentata	Gram-pos., Gram-neg.	42
Calophyllum inophyllum (root bark)	Gram-pos.	43
Cannabis sativa (tissue culture)	Gram-pos., Gram-neg.	44
Capsicum annuum	Yeast	45
Capsicum annuum	Yeast, bacilli	46
Cassia appendiculata		47

Table 1 - continued

Plant	Activity	Reference
Echium lycopsis	Staph. and Enterococci	48
Echium spinescens	Staph. and Enterococci	48
Echium vulgare	Staph. and Enterococci	48
Enicostemma litorale	Gram-pos.	49
Glaucium flavium	*B. subtilis*	50
Gossypium hirsutum (stalk)	Gram-pos., Gram-neg., fungi.	51
Helichrysum arenarium	Bacteria	52
Hordeum vulgare	Bacteria	53
Hypericum elongatum	Bacteria	54
Hypericum perforatum	Bacteria	55
Hypericum perforatum	Bacteria	54
Hypericum perforatum	Bacteria	56
Hypericum scabrum	Bacteria	54
Ichthyothere cunabi	Gram-pos.	57
Lysimachia nummularia (aerial portion)	Gram-pos., Gram-neg.	58
Melodinus suaveolens	*Staph. aureus*, some Gram-neg.	59
Nigella sativa (seed)	Gram-pos., Gram-neg.	60
Nuphar luteum ss	Bacteria	61
Nymphaea tuberosa (leaf, stem)	Mycobacteria	32
Olea europaea (green olive)	Lactic acid bact.	62
Oryza sativa (grain)	Fungi	63
Passiflora mollisima (fruit rind)	Fungi, yeast, Gram-pos.	64
Phaseolus vulgaris (leaf)	Gram-neg.	65
Phytolacca americana	Bacteria, fungi, protozoa	36
Pinus monticola	Fungi	66
Quercus macrocarpa (acorn)	Gram-pos.	67
Sanicula europaea (leaf, root)	Fungi	68
Sinapis arvensis	Gram-pos., Gram-neg.	69
Solanum carolinense (fruit)	Gram-neg., Mycobact.	70
Tropaeolum major	Bacteria, fungi, protozoa	71

Table 1 - continued

Plant	Activity	Reference
Vaccinium macrocarpon (juice)	Gram-pos., Gram-neg.	71
Vaccinium oxycoccus (juice)	Gram-pos., Gram-neg.	71
Vaccinium vitis-ideae (juice)	Gram-pos., Gram-neg.	71
Vitex agnuscastus (leaf)	Gram-pos.	72
Withania coagulans (whole fruit)	Bacteria, helminths	73

For the average reader whose language skills are limited to the traditional English, French, and German, much of this literature is somewhat difficult to evaluate, as one must depend upon synopsizing services, such as Chemical Abstracts, a essential bits of information are inevitably unavailable from these sources. Nevertheless, it is clear that several individuals thought sufficiently highly of the potential of these plants to devote considerable time toward their exploration.

In at least a few cases, the fractionation work was pursued to the stage where the toxicity of the preparations was established, and, in a few cases, animal protection studies with infected mice were used to demonstrate *in vivo* activity[59-67] in addition to the more commonly studied toxicity[48]. The most intense interest is associate, of course, with those studies which have been pursued farthest with seemingly favorable results.

The antimicrobial activity of volatile oils used for the preservation and flavoring of food has been known since very ancient times. This property has not been found useful in the treatment of systemic infections of humans. The phenols present, and perhaps other constituents too, are antiseptic due to their capacity to interfere with the selective permeability of cell walls on which life depends. Unfortunately, the selectivity of this action in discriminating between human and acterial membranes is insufficient for the treatment of systemic infectious diseases. Nevertheless, work on these materials continues, particularly in the Indian subcontinent and in Russia and its ancillary states. These references are set out in Table 2.

Table 2. Recent Studies on Plants Whose Antibacterial and Antifungal Properties are Attributable to the Presence of Volatile Oils

Plant	Reference
Abies sibirica	74
Acanthospermum hispidum	75
Acorus calamus	76
Aegle marmelos	77
Allium cepa	78
Allium sativum	79
Anethum sowa	77
Artemisia cina	80
Artemisia frigida	81
Artemisia gmelini	81
Artemisia martjanovii	81
Artemisia nova	82
Artemisia tridentata	82
Blumea erintha	75
Blumea laciniata	76
Calendula officinalis	83
Cassia occidentalis	84
Chloroxylon swietenia	77
Colubrina asiatica	76
Cupressus torulosa	85
Curcuma zedoaria	86
Echinopanax elatum	80
Eucalyptus globulus	85
Eugenia bracteata	86
Eugenia jambolana	86
"Grapes"	81
Heracleum sibericum	88
Hyptis suaveolens	77
Hyptis suaveolens	76
Libanotis intermedia	89
Litsea chinensis	76
Matricaria chamomilla	90
Mentha arvensis var. *piperascens*	91
Micromeria serphyllifolia	88
Monarda citriodora	92
Monarda ramoleyi	92
Monarda violacea	92
Ocimum basilicum	93
Ocimum canum	77

Table 2 - continued

Plant	Reference
Ocimum thyrsiflorum	86
Piper cubeba	76
Piper nigrum	75
Polyalthia longifolia	75
Psoralea drupacea	94
Phytolacca americana	80
Pyrethrum parthenifolium	88
Rosmarinus officinalis	86
Satureia taurica	95
Satureia taurica	88
Thymus dzevanovskyi	88
Thymus pseudohumillimus	88
Thymus tauricus	88
Thymus vulgaris	96
Toddalia asiatica	77
Tropaeolum major	80
Valeriana exaltata	97
Valeriana nitida	97
Valeriana stolonifera	97
Zanthoxylum alatum	75

Closely related to these studies are several studies of common citrus and spice plants[98,99,100,101,102,103,104,105,106].

Because volatile oils appear to have little chance of assuming a significant role in the chemotherapy of systemic human infections, most investigators in the field now pay little heed to such reports.

The last category of screening reports that bears some relation to our task is that consisting of the screening of numbers of pure chemical compounds which have been isolated from natural sources, usually for some other purpose, or compounds related to these substances. Rieche has discussed the antimicrobial properties of various mustard oil glycosides[108]. The plants which contain these materials frequently display antimicrobial properties, and it is thought that enzymatic or spontaneous hydrolysis of the sugar moieties followed by rearrangement to alkyl and aryl isothiocyanates is responsible for this activity. Isothiocyanates may well work by addition of protein and nucleic acid NH_2- and SH-groups *in vivo*. The products of these reactions should, in all likelihood, not function biochemically as effectively as the normal biopolymer does. Even if this is not the primary mechanism of action, their reactivity suggests that their use in systemic human therapy would likely be accompanied by a significant toxic liability so that their present anticipated role, if any, would be as disinfectants. Wolters has examined the antimicrobial action of numerous neutral steroid saponins[20,108]. These compounds are primarily antifungal in spectrum and appear to work by complexing with sterols in the cell membrane and, thus, interfere with the cell's normally required semipermeability. Polyene macrolide antibiotics operate by an analogous mechanism. As bacteria do not have significant sterol concentrations in their cell membranes, the relative lack of activity displayed by saponins toward bacteria seems readily rationalized. Mammalian cell walls contain significant amounts of sterol, hence systemic use of these agents against human fungal infections does not seem to be an attractive prospect. In addition to the anticipated membrane damage, glycosides of this type have well-known surface activity which results in hemolysis of red blood corpuscles. This characteristic would also argue against systemic administration of these compounds.

In screening a number of alkaloids isolated from Uzbekistani plants the following were recorded as being active, primarily against Gram-positive microorganisms: *d*-roemerine, roemerine ethochloride, glaucine methiodide, dorydine methiodide, *d*-roemerine methochloride and thalicmine methiodide[109].

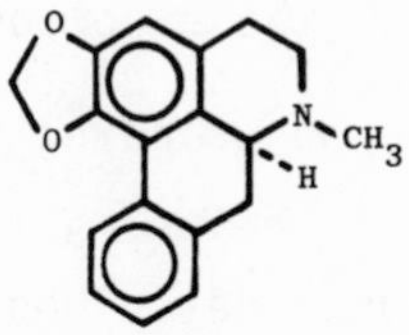

d-Roemerine

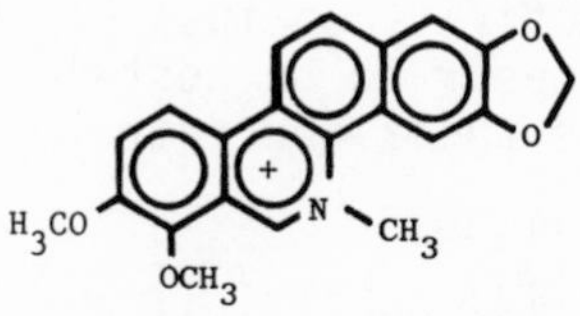

Chelerythrine

Mitscher and coworkers, in a similar study[10], showed that affinisine, chelerythrine, conessine, dihydroberberine chloride, dihydrohimbacine, doryafranine, hydrastine, ibogamine, l-methocanadine chloride, oxyacanthine, palmatine chloride, sanguinarine nitrate and solanidine had weak antibacterial activity against a number of standard strains of microorganisms. Another 15 substances had narrower antimicrobial spectra. Chelerythrine was the only alkaloid in this group to show activity at 100 μg/ml in this screening test. An activity level of 100 μg/ml is considered to be the threshold of meaningful antibiosis in potential clinical terms. Berberine, berberrubine and its alkyl ethers, jatorrhizine and its alkyl ethers and demethylberberine have also been shown to be active antibacterial agents[110]. Berberine has, of course, been known to have antibacterial properties for a long time. Of the steroidal glycoalkaloids, α-tomatin α-soladulcine, α-solanine, and α-chaconine have been shown to be active against yeast and fungi, though not against bacteria[111]. These findings are reminiscent of the saponin work discussed above.

Morris examined 24 terpenes and found pinene, pulegone, menthone, 4-terpineol, citral a, citral b, citronellol and linalool to inhibit *Bacillus thuringiensis*[112]. Several sesquiterpene lactones are active against yeast and fungi, with some showing activity against protozoa[113].

Jurd and his group[114,115,116] and Dadak[117] have been looking at a number of antimicrobial natural phenols and related compounds, largely in the hope of finding safer natural fungicides for agricultural use, as the specific potencies of these compounds appear to be too low for possible systemic use. Obtusastyrine and ethers of umbelliferone (including herniarin) show antifungal activity but littl antibacterial potency. The phenolic OH-groups are quite significant, as might be expected, for antibacterial activit In particular, O-methylation of hydroxycoumarins decreases their activity against bacteria. Many of these compounds would appear to be post-infectional inhibitors, at least in part, as their concentration increases upon wounding the plants.

It is doubtful if any of the compounds described in these screening reports have the potency or inherent selectivity desired for a systemically useful antibiotic, but thei activity is evidence that a wide variety of secondary plant metabolites show antimicrobial activity in a variety of test systems and make it a certainty that isolation of the responsible agent from additional crude plant extracts will result in interesting chemical compounds whose structures will be very different from those isolated from microbial sources.

INDIVIDUAL PLANTS AND THEIR CONSTITUENTS

Numerous investigations have been carried out on individual plants and the responsible antimicrobial agent has been identified in a gratifying number of cases. It was the lack of this sort of definitive follow-through which made the earlier literature on antimicrobial agents from higher plants so frustrating to evaluate. Once pure, the real antimicrobial potential for a given agent can be established both *in vitro* and *in vivo* and a definitive decision made as to the role, if any, these agents might play in the treatment of human infections. Unfortunately, it is only in rare instances that even animal protection data can be found in the literature of these materials. It is also noteworthy that, with the exception of the saponins, work on the molecular mode of action of these agents is almost wholly lacking. For this reason, a biological subdivision of the subject matter is presently impossible and the following section deals with related chemical substances instead.

Aliphatic molecules, including polyacetylenes

The long chain fatty acids have long been known to have antimicrobial properties, particularly in the form of their salts (soaps). Skinner has reviewed the earlier work on this subject[9]. These substances have never found more than topical use in human therapy. Still, some recent studies have concerned newer members of this family. *Sapium japonicum*[118] shows antifungal properties due to the presence of methyl 8-hydroxy-5,6-octadienoate. *Cnicus benedictus* has antimicrobial activity due to the presence of cnicin and some polyacetylenes[119]. Passicol, isolated from cut rinds of banana passion fruit (*Passiflora mollisima*), is probably a polyacetylene[64,120]. It is non-toxic to mice and rabbits. *Bidens cernua* yields a volatile oil whose anti Gram-positive, antiyeast and antifungal activity are due to aromatic hydrocarbons and 1-phenylhepta-1,3,5-triyne. Some long chain unsaturated polyalcohols have also been described[121]. Infected *Carthamus tinctorius* (safflower) owes its antifungal activity to safynol[122], while the avocado pear owes its activity against Gram-positive organisms to 1,2,3-trihydroxyheptadec-16-ene[123]. Plant polyacetylenes have been studied extensivel

$$CH_3-CH=CH-C \equiv C-C \equiv C-C \equiv C-CH=CH-CHOH-CH_2OH$$

safynol

$$CH_2=CH-(CH_2)_{12}-CHOH-CHOH-CH_2OH$$

1,2,3-trihydroxyheptadec-16-ene

by Jones and Bohlmann and their respective groups and many of these should, undoubtedly, be active. A number of these have been isolated from various microorganisms as antibacterial agents. None of these has seen commercial use, partly because of their pronounced instability.

Acids, phenols and related alcohols

The point was made above that many simple phenols, alcohols and acids - especially those occurring in volatile oils - have activity as disinfectants because they destroy membrane integrity. It is not, therefore, surprising that several plants have been shown to owe their biopotency to such constituents. *Nymphaea tuberosa* owes its anti-*Mycobacterium smegmatis* activity to tannic acid, gallic acid and ethyl gallate[32]. The root extract of *Paeonia decora* owes its activity to tannic acid and benzoic and gallic acids[124]. The only antibacterial compounds isolatable from *Eugenia javanica* were complex tannins[125]. Tanninols of *Alnus incana* (speckeled alder) show antibacterial activity *in vitro* and in mice[126], and scopoletin and caffeic acid are partly responsible for the antifungal activity of potato tubers[127]. More complex phenols which may operate by a more subtle mechanism include gossypol (anti-Gram positive, antiyeast) from cottonseed oil[128] and damaged *Gossypium* tissues[129], and uliginosins A and B from *Hypericum uliginosum*[130]. Uliginosin A and analogs have recently been synthesized[131]. Russian workers have been especially interested in the properties of novoimanine (active in mice against *Staph. aureus* infections), isolated from St. Johns wort (*Hypericum perforatum*)[54,132] and perhaps also in closely related species such as *H. elongatum* and *H. scarbrum*[54]. Hops (*Humulus lupulu*s has been known for a very long time to have antibacterial activity - due to humulone and lupulone and a variety of less well known phenols. Teuber has recently shown that the

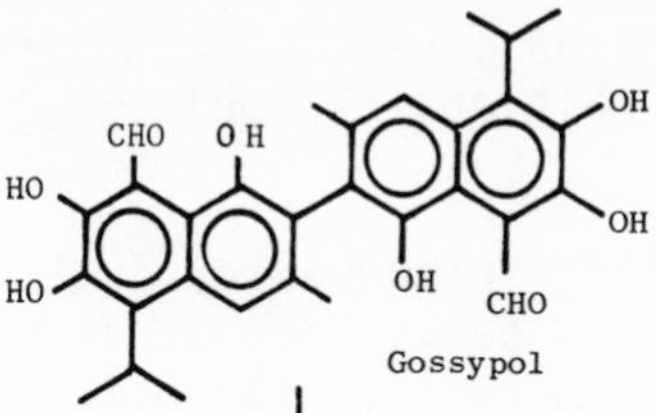

Gossypol

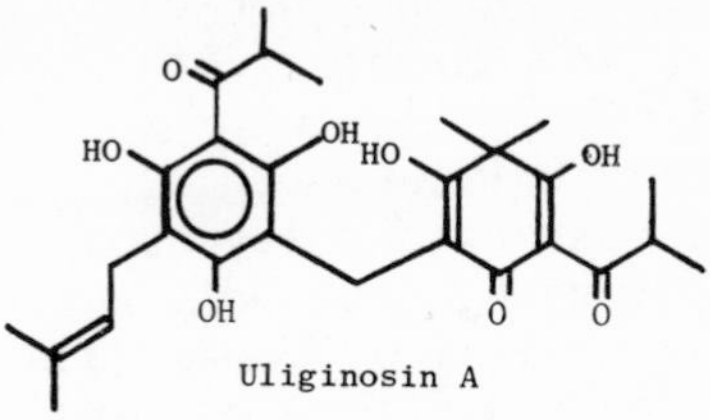

Uliginosin A

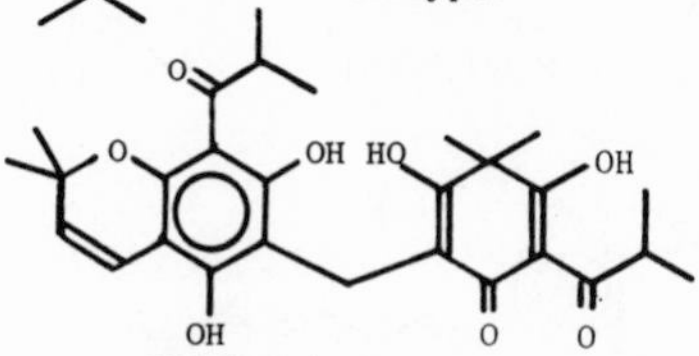

Uliginosin B

stereoisomeric isohumulones are much less potent and their concentration in beer is probably too low to be of significance[133]. While beer has been consumed by humans for centuries and some of the discomfort accompanying the "morning after" syndrome following overindulgence is due to the anti-Gram positive constituents, no practical therapeutic use of these compounds seems likely for the treatment of systemic infections. The antibacterial and antifungal activity of *Thymus dzevanovskyi* and *Satureja taurica* has been attributed to "phenols"[134].

The flavone, isoflavone, flavonol groups are phenolic and some of these compounds have antimicrobial properties. The immature fruits of *Lupinus luteus* produce the antifungal isoflavone, luteone[135]. Poplar resin contains the bacteriostatic agents galangine and pinocembrine[136]. Ramaswamy and coworkers looked at a variety of citrus bioflavonoids and concluded that flavonols were generally more active than

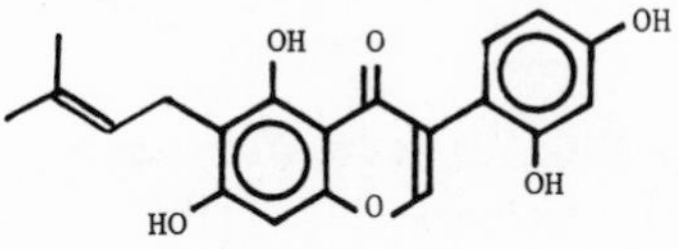

Luteone

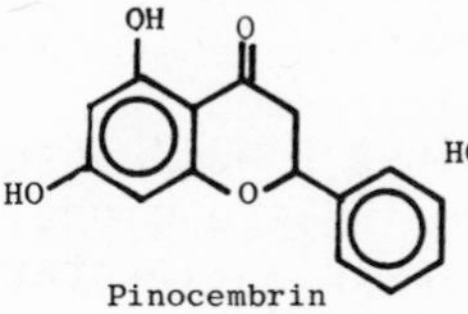

Pinocembrin

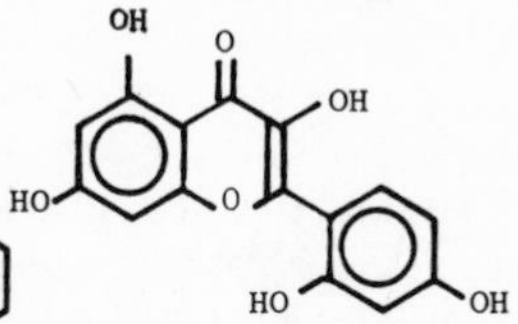

Morin

flavones against bacteria[137]. Morin was especially active. Naringenin and hesperetin were the most active flavones, and the glycosides were inactive. Interestingly, the dihydro analogs were more active against bacteria. It should be noted that these compounds have notable chelating activity and this could conceivably be more significant than their phenolic character with regard to their antimicrobial activity.

Most of the styrenes are phenolic in character. Obtusastyrene, from *Dalbergia obtusa*, is bacteriocidal at 25 μg/ml against *B. cereus* and *Staph. aureus*[138]. The oil from the seeds of *Psoralea corylifolia* contains bakuchiol which is quite potent against *Staph. aureus*[139].

Obtusastyrene Bakuchiol

Quinones

Quinones result from the oxidation of phenols and many quinoid products have been isolated from higher plant extracts and shown to have antimicrobial properties. Quinones are well known to interfere with cellular redox reactions in mammalian as well as microbial systems so their therapeutic index is often disappointing in clinical trials. The anti Gram-positive and antiyeast activity of *Ceratostigma plumbaginoides, Diospyros virginiana* and *D. caucasia* is attributed to plumbagin [134-140], which is also responsible for the anti Gram-positive activity of *Plumbago auriculata* and *P. zeylanica*, used in folk medicine to treat urinary tract infections[141]. Plumbagin is also responsible for the *in*

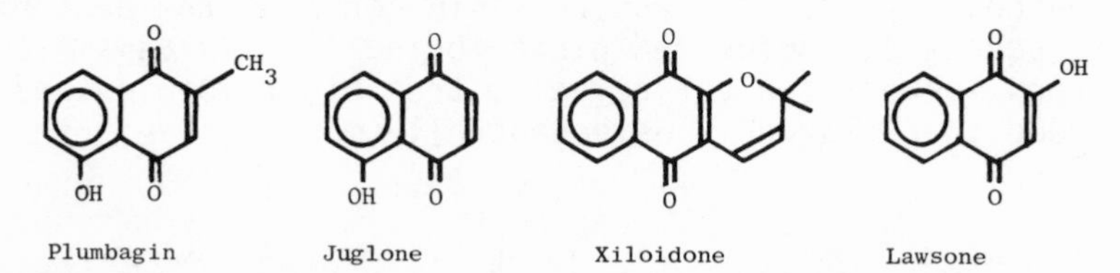

Plumbagin Juglone Xiloidone Lawsone

vitro activity against *Bacillus* spp. and *Candida albicans* of *Plumbago scandens*[142]. Juglone, of course, has long been known to have mild antibacterial and antifungal properties[143-144]. Goncalves de Lima and coworkers converted lapachol, a well known constituent of *Tabebuea* spp., to xiloidone with pyridine and showed the product to be antibiotic[145]. The antiyeast and antibacterial properties of *Lawsonia inermis* to have been attributed to lawsone, those of *Pyrola secunda* to chimaphilin, of *Echium rubrum* to

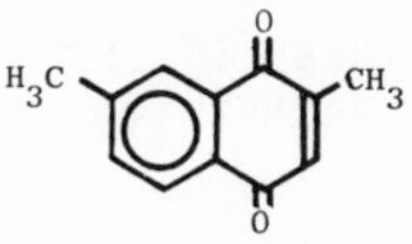

Chimaphilin

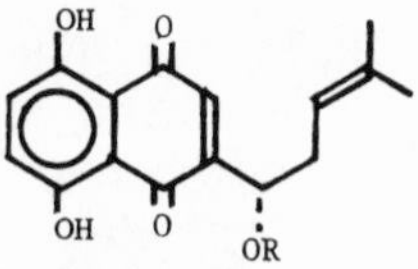

R=H=Shikonin
R=$COCH_3$=Shikonin Acetate

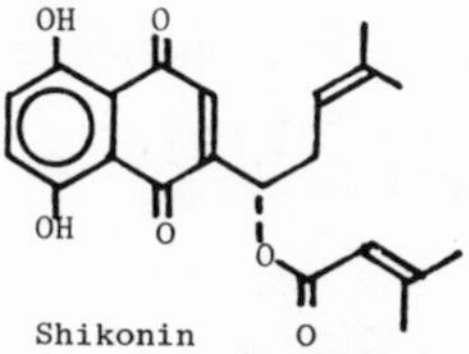

Shikonin β,β-dimethylacrylate

shikonin and of *Diospyros virginiana* to "naphthoquinones"[146]. None of these was as active as juglone. *Arnebia nobilis* contains several naphthoquinones, one of which is shikonin (alkanin)[147]. The others are close analogs (shikonin β,β-dimethylacrylate, shikonin acetate, and dihydrohydroxyshikonin tiglate [most active]).

Based on their generally low activity and generally low therapeutic index, none of these compounds appears to have clinical potential for systemic infections.

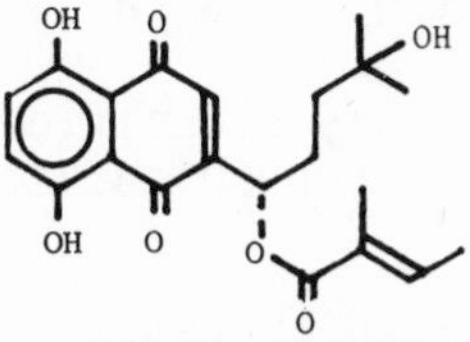

Dihydrohydroxyshikonin tiglate

Glycosides and their derived aglycones

A large number of glycosides and their aglycones have antimicrobial activity. This subject has been ably reviewed recently by Tschesche[148]. These compounds can be divided roughly into two groups: Those active with the sugars attached, and those active as the aglycones. The former group comprises mainly the saponins[108,149]. These substances have an affinity for sterols in cell membranes and apparently exert their action by disrupting membrane function. As complexing agents, some considerable specificityis seen from one agent to another (for example, digitonin is quite active while digitoxin is much weaker[20,108]). Removal of the sugars, in these cases, results in loss of activity. They are predominantly active against fungi, whose cell walls contain sterols, rather than bacteria, whose cell walls do not. Recent studies involving compounds of this class include those on the antiyeast but not antibacterial saponins of *Pittosporum* spp., *Yucca* spp., *Primula veris* and *P. vulgaris*[134,140]; on avenacin from oats (*Avena sativa*)[150]; on steroidal alkaloidal glycosides[111]; and on eupteleosides A and B, antifungal and antiyeast triterpenoid glycosides from *Euptelea polyandra*[151].

The second class is present in plants as inactive glycosides and would not be detected in a typical antibiotic screening program. Activation requires hydrolysis. Sometimes this occurs within the plant by liberation of the appropriate enzyme following injury, and sometimes via nonspecific glycosidases present in an invasive microorganism. Fungi, in particular, are prone to commit suicide in this fashion! Thus many of these compounds can be regarded as phytoalexins and postinfectional inhibitors. Screening programs will, in fact, often pick up such compounds. Their use, however, at present seems largely to be restricted to crop protection although great potential exists here for the development of selective agents for use in human systemic infections based upon presumed differences in enzymatic capacity between the host and the invasive parasite.

Glycosides are generally classified on the basis of their aglycones, and it is convenient to continue this for our purposes.

Mustard oil glycosides are widely distributed in nature and their antimicrobial potential becomes expressed after enzymatic or acid hydrolysis followed by a Lossen rearrangement to a variety of isothiocyanates. It is thought that these are active due to reaction with NH_2- and SH-groups on biopolymers[107]. This kind of reaction would probably be nondiscriminate and lead to host toxicity in preformed isothiocyanates whereas the glycosides could conceivably act more selectively if the needed glycosidase were present in the microorganism and not the host. A recent study on seeds of *Capparis decidua*, which yield the inactive glycoside glucocaparin, has shown that the latter is enzymatically converted to the bacteriocide, methyl isothiocyanate[152]. In *Tropaeolus majus* benzylisothiocyanate is released which is active against bacteria and fungi[134].

Various *Arctostaphylos* species have been used many years as urinary antiseptics. Apparently the activity is due to liberation of hydroquinone from the glycoside, arbutin[153,154]. The activity is too weak to have excited continuous medical interest.

Unsaturated terpenes

A number of α-methylenelactone sesquiterpenes have shown strong cytotoxic activity and, therefore, been of interest as antitumor substances (see Chapter 7). These compounds are thought to act at least in part by Michael-type addition of NH_2- and SH-groups of essential biopolymers. It seems inevitable, therefore, that some compounds of this type should turn up as antimicrobial agents from higher plants. Recently, for example, cnicin, from *Cnicus benedictus*, was reported to have weak antibacterial properties[155]. Vichkanova and coworkers studied several sesquiterpene lactones and found 7 of 14 to be active[113]. Several simpler compounds of this general chemical type have recently been described, for example, protoanemonin from *Ranunculus oxyspermus*[156]. Protoanemonin had been shown years before to be the active agent from *Anemone pulsatilla*. The diastereoisomeric glucosides nartheside A and B from *Narthecium ossifragum* give a racemic aglycone active against *B. subtilis*[157]. Much interest has been shown in the resistance to fungal attack of tulips[149,158,159,160] probably due to phytoalexins. It is doubtful at the moment whether such highly reactive compounds could find a place in human therapy for non-life threatening conditions.

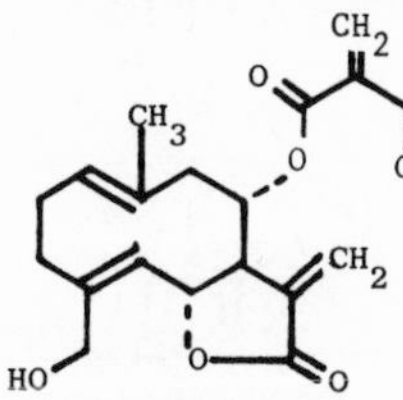

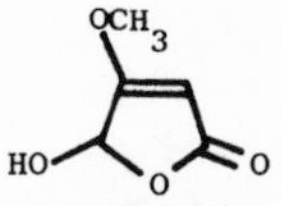

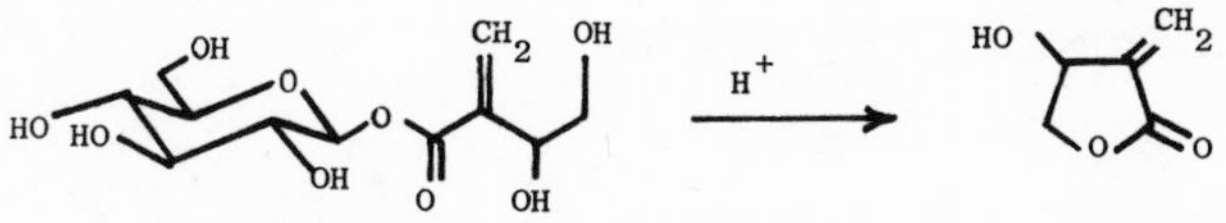

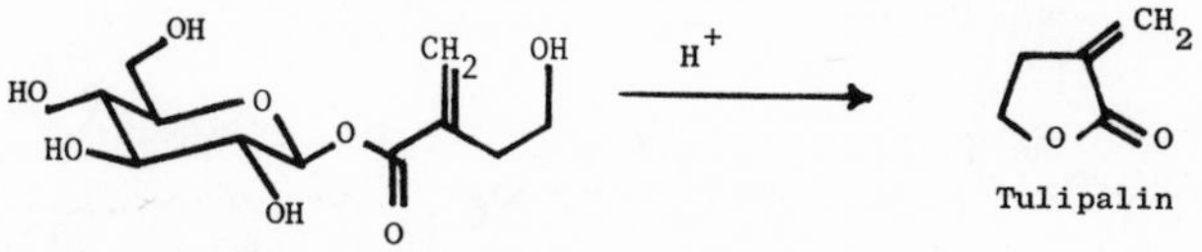

Cnicin

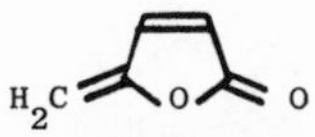

Protoanemonin

Narthecium Aglucone

Tuliposide B

Tuliposide A

Tulipalin

Alkaloids

The alkaloids have long held a special fascination for the natural products chemist and the fact that several see use in the clinic, although not as antibiotics, makes it highly interesting that numerous alkaloids show antimicrobial activity. For example, people keep rediscovering the antimicrobial activity of the ubiquitous alkaloid, berberine. Thus, Kowalewski and coworkers obtained it, as one might expect, from *Berberis vulgaris* and showed it to be active against 70% of 196 strains of *Staph. aureus* and not to be cross resistant with antibiotics presently used in the clinic with the exception of streptomycin[161]. Berberine and several analogs (berberubine and its alkyl ethers, jatorrhizine and its alkyl ethers, and "demethylberberine") are active against bacteria and fungi, with berberine more potent, but narrower in spectrum than the others[110]. The activity of roemerine[109] has already been described. Rather more unusual types are hordatine A and B, antifungal agents from

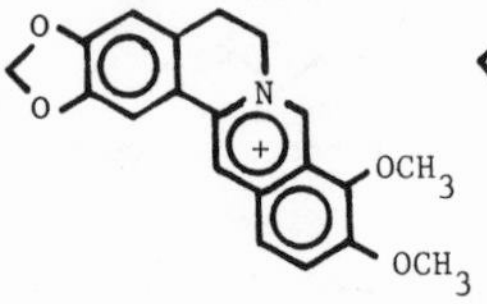

Berberine 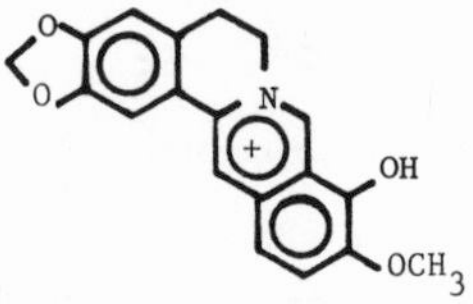 Berberubine 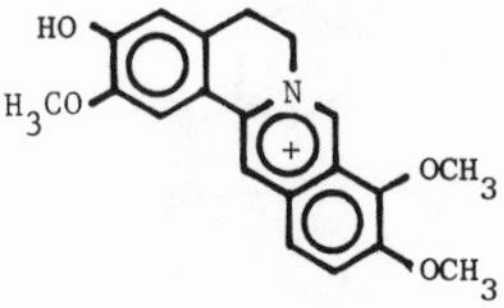 Jatorrhizine

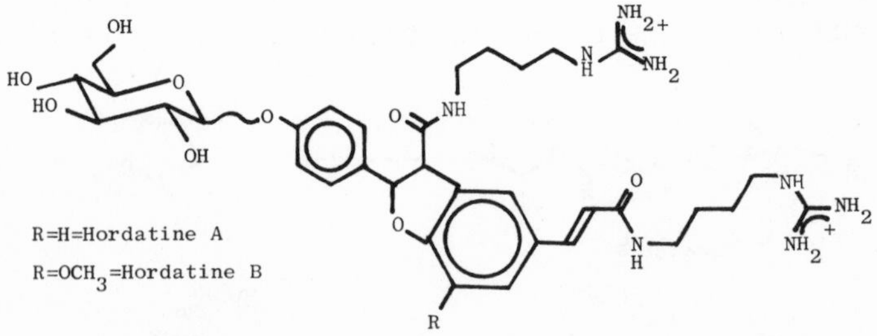

R=H=Hordatine A

R=OCH_3=Hordatine B

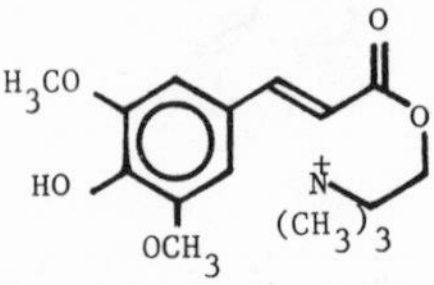

Sinapine 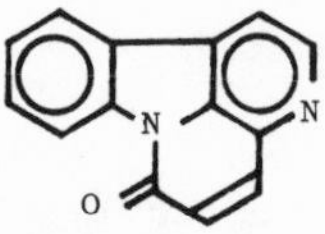 Canthin-6-one 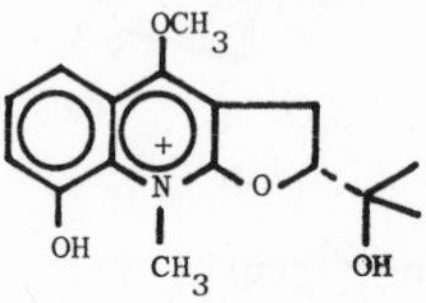 Pteleatinium Salt

barley seedlings[162], and sinapine[163], an antibiotic quaternary base from *Sinapis alba* seeds.

Following a fairly standardized protocol[10], the broad spectrum activity of *Zanthoxylum elephantiasis* was shown to be due exclusively to canthin-6-one[164]. This alkaloid had been isolated previously from the same source, but its antimicrobial activity had not been suspected. Using the same approach, the new antifungal and antimycobacterial alkaloid pteleatinium chloride was isolated from *Ptelea trifoliata*[165]. An investigation of the roots of *Thalictrum rugosum* showed antimycobacterial activity for berberine, obamegine, thalidasine, thalrugosine, thalrugosidine and thaliglucinone (alkaloid D)[166].

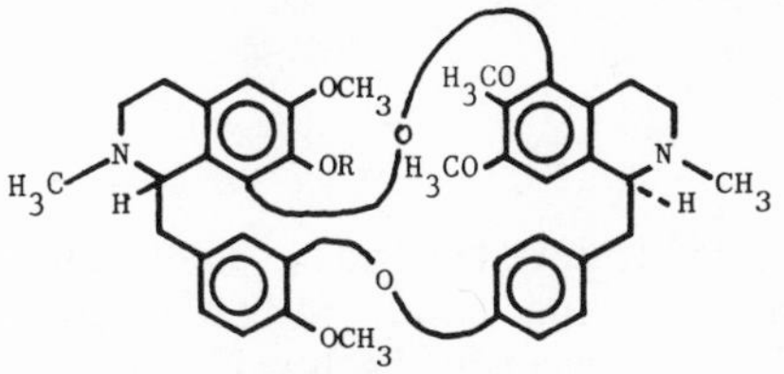

R=CH_3=Thalidasine
R=H=Thalrugosidine

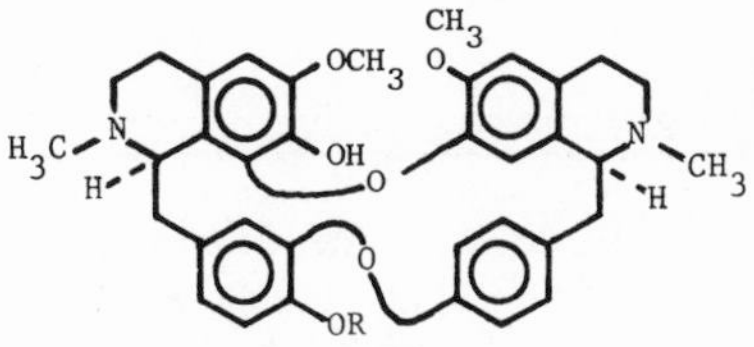

R=H=Obamegine
R=CH_3=Thalrugosine

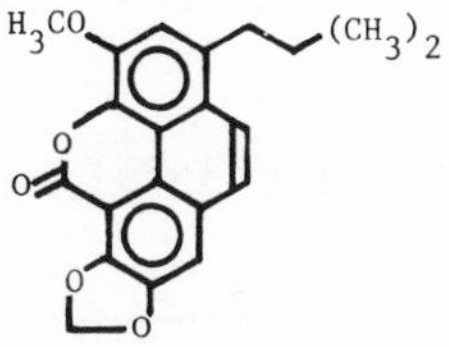

Thaliglucinone

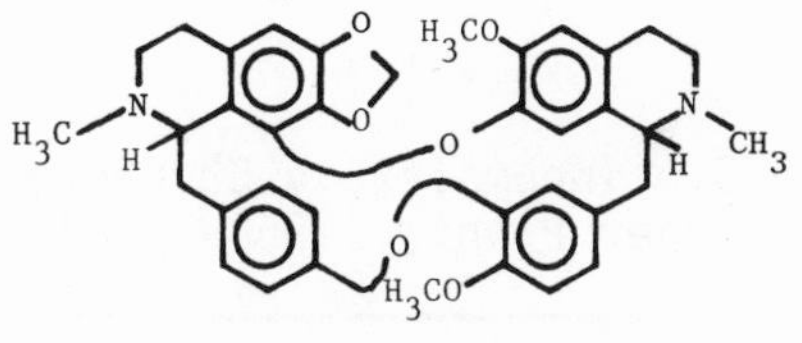

Cepharanthine

This is consistent with a report that (+)-cepharanthine is highly effective against human tuberculosis and leprosy[167]. A later investigation of *Thalictrum polygamum* showed the activity of that plant to be due to thalicarpine, berberubine and thaliglucinone[168]. Also thalrugosamine was isolated in very small quantity from the mother-liquors of an earlier study of *Thalictrum rugosum* and was shown to have weak activity against *M. smegmatis in vitro*[169].

The discovery of so many new alkaloids with potentially interesting activity - even from species which had been fairly thoroughly explored before - is a promising development. Certainly there is no question that new substances

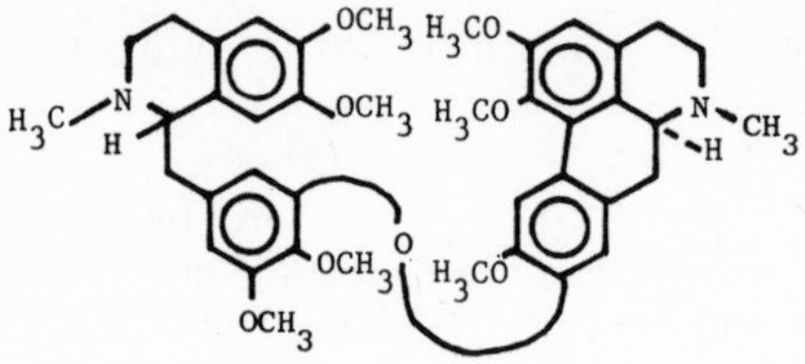

Thalicarpine 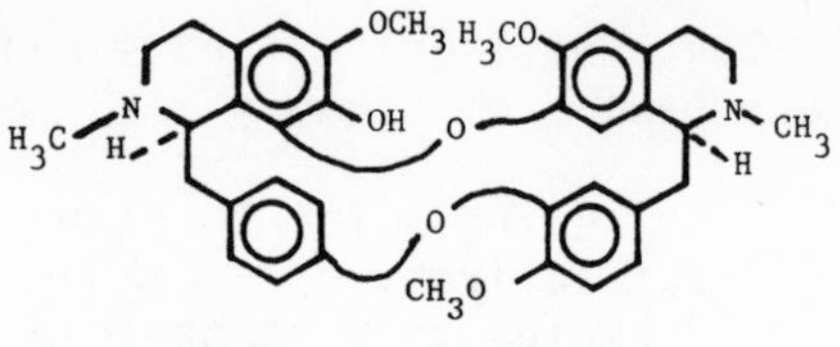 Thalrugosamine

will emerge from studies following up a physiological activity; this approach is much more likely to yield useful compounds than random chemical exploration, where all too often the products reported are those present in the largest amount or those easiest to purify and identify.

Miscellaneous substances

There are several papers dealing with individual compounds isolated from various plants and about which only a limited amount of structural information is available. These are set out in Table 3.

Table 3. Incompletely Characterized Antimicrobial Agents from Higher Plants (1968-1974)

Plant	Agent	Reference
Capsicum annum	Capsaicin	170
Cyathus helenae	Cyathin	171
Emblica officinalis	Phyllemblin	172
Erysimium perofskianum	Glucoerysolin	173
Garcinia morella	α_1-and γ-guttiferins	174
Garcinia morella	Neomorellin and X-guttiferin	175
Garcinia morella	Morellin	176
Maytenus sp.	Maytenin	177
Prionostemma aspera	Pristimerin	178
Unlisted	Lutenurin	179

SUMMARY

From this brief account it is apparent not only that antimicrobial agents are distributed widely among the higher plants but that study of these substances is being pursued vigorously in laboratories throughout the world. Those agents identified to date are almost invariably substances which natural products chemists would identify as typical secondary metabolites, and their structures are, with only rare exceptions, widely different from those of the antibiotics coming from the lower plants (bacteria, Streptomycetes, fungi etc.). While no compound from a higher plant has come into significant clinical use, most of the studies have not been pursued far enough to make such a decision on rational ground. The wide structural diversity and the very number of agents potentially available makes it highly likely that such an agent could emerge from this group and earn a significant place in human therapy. Agricultural use of such agents seems inherently more likely. Many of the compounds reported to date have weak biopotency, possess narrow spectra or present too high a risk to the patient and, so, will not see clinical use. Of course, the majority of the Streptomycete antibiotics could also be described in these terms. If there is a new penicillin or tetracycline replacement lurking in one of the papers cited in this review, its properties are not, at present, obvious.

ACKNOWLEDGMENT

This study was supported in part by United States Public Health Service Grant No. AI 09846A5.

REFERENCES

1. Sabin, A. 1969. Control of infectuous diseases. Antimicrob. Agts. Chemother. 1.
2. Perlman, D. 1970. Antibiotics. In "Medicinal Chemistry" (A. Burger, ed.) 3rd. Ed., p. 305. Wiley (Interscience), New York.
3. Stoessl, A. 1970. Antifungal compounds produced by higher plants. In "Recent Advances in Phytochemistry" (C. Steelink & V.C. Runeckles, eds.) Vol. 3. p. 143. Appleton-Century-Crofts, New York.

4. Burkholder, P.R. & G.M Sharma. 1969. Antimicrobial agents from the sea. Lloydia, 32: 466.
5. Cavallito, C.J. 1951. Antibiotics from plants. In "Medicinal Chemistry". (C.M. Suter, ed.) Vol. 1, p. 222. Wiley, New York.
6. Irving, G.W., Jr. 1949. Antibiotics from higher plants. In "Antibiotics". (G.W. Irving, Jr. & H.T. Herrick, eds.) p.187. Chemical Publishing Company, New York.
7. Korzybski, T., Z. Kowszyk-Gindifer & W. Kurylowicz. 1967. Antibiotics from higher plants. In "Antibiotics" Vol. 2, p. 1438. Pergamon Press, New York.
8. Nickell, L.G. 1959. Antimicrobial activity of vascular plants. Economic Botany, 13: 281.
9. Skinner, F.A. 1965. Antibiotics, In "Modern Methods of Plant Analysis." (K. Paech & M.V. Tracey, eds.) Vol. 3, p. 626. Springer, Berlin.
10. Mitscher, L.A., R.P. Leu, M.S. Bathala, W.-N. Wu, J.L. Beal & R. White. 1972. Antimicrobial agents from higher plants. I. Introduction, rationale and methodology. Lloydia, 35: 157.
11. Murakami, N.T. Hamada, K. Kondo & K. Ando. 1966. Screening tests of plant extracts for antitumor and antibiotic action. Kumamoto Pharm. Bull. 7: 19. *Via* Chem. Abs. 66: 84401e (1967).
12. Shcherbanovskii, L.R. & G.I. Nilov. 1967. The inhibitors of alcohol fermentation among higher plants. Rast. Resur. 3: 451. *Via* Chem. Abs. 68: 28439c (1968).
13. Mathes, M.C. 1967. The secretion of antimicrobial materials by various isolated plant tissues. Lloydia, 30: 177.
14. Inagaki, I., M. Yamazaki, A. Takahashi, K. Ooye, S. Shibata & S. Takeda. 1967. Bacteriostatic effects of the constituents of several crude drugs. Effects of cortex phellodendri, radix lithospermi, cortex radicis moutan, rhizoma rhei and radix angelicae. Nagoya Shiritsu Daigaku Yakugakubu Kenkyu Nempo, 27. *Via* Chem. Abs. 69: 41955h (1967).
15. Makarenko, N.G. 1967. Detection of individual phytocidal agents by paper chromatography. Mikroorganizmy Zelenye Rast. 131. *Via* Chem. Abs. 69: 8916y (1968).
16. Imai, S., T. Murata, S. Fujioka, E. Murata & M. Goto. 1967. Crude drugs and oriental crude drug preparations by bioassay. XXI. Search for biologically active plant ingredients by means of antimicrobial tests. 3. Antifungal principles of *Dioscorea* species.

Takeda Kenkyusho Nempo, 26: 66. *Via* Chem. Abs. 68: 66709y (1968).

17. _____, S. Fujioka, E. Murata, M. Goto, T. Kawasaki & T. Yamauchi. 1967. Bioassay of crude drugs and oriental crude drug preparations. XXII. Search for biologically active plant ingredients by means of antimicrobial tests. 4. Antifungal activity of dioscin and related compounds. Takeda Kenkyusho Nempo, 26: *Via* Chem. Abs. 68: 76007 g (1968).

18. Khanin, M.L., A.I. Korotyaev, A.F. Prokopchuk, T.V. Perova & O.F. Vyazemskii. 1968. Antibiotic properties of extracts obtained from medicinal plants by extraction with liquid carbon dioxide. Khim. Farm. Zh., 40. *Via* Chem. Abs. 69: 944k (1968).

19. Ferenczy, L., K. Horvath & J. Zsolt. 1966. Thin-layer chromatography of antifungal compounds from higher plants. Herba Hung. 5: 88 *Via* Chem. Abs. 68: 46992g (1968).

20. Wolters, B. 1966. Antimicrobial activity of plant steroids and triterpenes. Planta Med., 14: 392.

21. Soeding, H. & K. Doerffling. 1968. Paper chromatographic analysis of bactericidal substances in plant extracts and possible relation of these to plant growth inhibitors. Arch. Mikrobiol. 60: 182.

22. Malcolm, J.A. & E.A. Sofowora. 1969. Antimicrobial activity of selected Nigerian folk remedies and their constituent plants. Lloydia, 32: 512.

23. Bhakuni, D.S., M.L. Dhar, M. Dhar, B.N. Dhawan & B.N. Mehrotra. 1969. Screening of Indian plants for biological activity. Ind. J. Exp. Biol. 7: 250. *Via* Chem. Abs. 72: 51790b (1970).

24. Shakirova, K.K., A.D. Garagulya & R.L. Khazanovich. 1970. Antimicrobial properties of some species of St. John's wort cultivated in Uzbekistan. Mikrobiol. Zh. (Kiev), 32: 494. *Via* Chem. Abs. 74: 34570d (1971).

25. Zayed, M.N., M.S.M. Saber, Y. Abd-el-Malek & M. Monib. 1971. Antibacterial substances in dry residues of certain higher Egyptian plants. Zentralbl. Bakteriol. Parasitenk. Infektionskr. Hyg., Abt. 2, 126: 615. *Via* Chem. Abs. 76: 108655a (1972).

26. Solis, C.S. 1969. Antibacterial and antibiotic properties of the leguminosae. Acta Maniliana, Ser. A. 52. *Via* Chem. Abs. 74: 20326s (1971).

27. Vichkanova, S.A., V.V. Adgina, L.V. Makarova & M.A. Rubinchik. 1971. Antimicrobial preparations of saponins. Tr. Vses. Nauch.-Issled. Inst. Lek. Rest.

14: 191. *Via* Chem Abs. 78: 155106y (1973).
28. Abdou, I.A., A.A. Abou-Zeid, M.R. El-Sherbeeny & Z.H. Abou-El-Gheat. 1972. Antimicrobial activities of *Allium sativum, Allium cepa, Raphanus sativus, Capsicum frutescens, Eruca sativa, Allium kurrat* on bacteria. Qual. Plant. Mater. Veg. 22: 29. *Via* Chem. Abs. 78: 80226b (1973).
29. Kshirsagar, M.K. & A.R. Mehta. 1972. Survey of ferns in Gujarat State (India) for presence of antibacterial substances of ferns. Planta Med. 22: 386.
30. Maksymiuk, B. 1970. Occurrence and nature of antibacterial substances in plants affecting *Bacillus thuringiensis* and other entomogenous bacteria. J. Invertebr. Pathol. 15: 356.
31. Su, K.L., E.J. Staba & Y. Abul-Hajj. 1972. Aquatic plants from Minnesota. III. Antimicrobial effects. U.S. Nat. Tech. Inform. Serv. PB Rep. No. 2095-30. *Via* Chem. Abs. 77: 111445h (1972).
32. _____, Y. Abul-Hajj & E.J. Staba. 1973. Antimicrobial effects of aquatic plants from Minnesota. Lloydia, 36: 80.
33. Norton, T.R., M.L. Bristol, G.W. Read, O.A. Bushnell, M. Kashiwagi, C.M. Okinaga & C.S. Oda. 1973. Pharmacological evaluation of medicinal plants from Western Samoa. J. Pharm. Sci. 62: 1077.
34. Farnsworth, N.R., L.K. Henry, G.H. Svoboda, R.N. Blomster, H.H.S. Fong, M.W. Quimby & M.J. Yates. 1968. Biological and phytochemical evaluation of plants. II. Test results from an additional two hundred accessions. Lloydia, 31: 237.
35. Fong, H.H.S., N.R. Farnsworth, L.K. Henry, G.H. Svoboda & M.J. Yates. 1972. Biological and phytochemical evaluation of plants. X. Test results from a third two-hundred accessions. Lloydia, 35: 35.
36. Vichkanova, S.A., L.V. Makarova, M.A. Rubinchik & V.V. Adgina. 1970. Antimicrobial characteristics of fatty esters. Tr. Vses. Nauch.-Issled. Inst. Lek. Rast. 14: 221. *Via* Chem. Abs. 78: 155109b (1973).
37. Sharraf, A., I. Abdon, M. Hassan, M. Yossif & S.A.R. Negm. 1969. Pharmacochemical studies on *Allium kurrat*. Qual. Plant. Mater. Veg. 17: 313. *Via* Chem. Abs. 72: 119928k (1970).
38. Kabelik, J. 1970. Antimicrobial properties of garlic. Pharmazie, 25: 266.
39. Yen, K.-Y., W.-H. Lee & J.-F. Lee. 1967. Pharmacologically effective components of local plants in Taiwan.

II. Antibiotic activity and pharmacological action of *Ampelopsis brevipedunculata* var. *hancei*. T'ai Wan Yao Hsueh Tsa Chih, 19: 15. *Via* Chem. Abs. 73: 13031t (1970).
40. Khanna, S.G.S., Y.L. Nene, C.K. Banerjee & P.N. Thapliyal. 19-7. A note on the isolation and chemical characterization of antifungal agents from extracts of *Anagallis arvensis*. Indian Phytopathol. 20: 64. *Via* Chem. Abs. 68: 57373g (1968).
41. Staron, T., C. Allard, S. Heitz & D. Billet. 1969. Purification and antifungal properties of a triterpene glycoside isolated from the stems and leaves of *Anagallis arvensis* var. *phoenicea*. Phytiat.-Phytopharm., 18: 161. *Via* Chem. Abs. 75: 59769j (1971).
42. Ramirez, C. 1969. Antibacterial action of nonvolatile substances extracted from *Artemisia tridentata* subspecies *tridentata*. Can. J. Microbiol. 15: 1341.
43. Radhakrishnan, P. & P.A. Kurup. 1970. Antibacterial principle of the root bark of *Calophyllum inophyllum*: isolation and antibacterial activity. Indian J. Exp. Biol. 8: 39. *Via* Chem. Abs. 72: 108242f (1970).
44. Veliky, I.A. & K. Genest. 1972. Growth and metabolites of *Cannabis sativa* cell suspension cultures. Lloydia, 35: 450.
45. Gal, I.E. 1966. Composition of capsicidin. Z. Lebensm.-Unters. Forsch. 132: 82. *Via* Chem. Abs. 66: 43849p (1967).
46. _____. 1968. Antibacterial activity of the spice, paprika. Testing of capsicidin and capsaicin activity. Z. Lebensm.-Unters. Forsch., 138: 86. *Via* Chem. Abs. 69: 104148v (1968).
47. Gonçalves de Lima, O., I.L. D'Albuquerque, M.A.P. Borba, G.M. Maciel & J.F. de Mello. 1966. Antibiotic substances from higher plants. XXIII. Initial observations on the antibiotic activity of *Cassia appendiculata* extracts. Rev. Inst. Antibiot., Univ. Recife, 6: 3. *Via* Chem. Abs. 68: 62631n (1968).
48. Peter, M.H., M. Peter & G. Racz. 1970. Antibiotic effect of *Echium* extracts. Stud. Cercet. Biol., Ser. Bot. 22: 71. *Via* Chem. Abs. 73: 95755u (1970).
49. Natarajan, P.N. & S. Prasad. 1972. Chemical investigations of *Enicostemma litorale*. Planta Med. 22: 42.
50. Altinkurt, O. 1969. *Glaucium flavum* (yellow hornpoppy) and *Glaucium rubrum* (red hornpoppy); use in pharmaceuticals, insecticides, and antibiotics. Turk.

Hij. Tecr. Biyol. Derg. 29: 113. *Via* Chem. Abs. 73: 11384m (1970).

51. Muslimov, Z.M., M. Kh. Avazkhodzhaev, & A. Kh. Khankhodzhaev. 1972. Antibiotic properties of an ether extract of wood from cotton stalks. Uzb. Biol. Zh. 16: 14. *Via* Chem. Abs. 77: 83898n (1972).
52. Bel'tyukova, K.G. 1968. Plant antibiotic, arenarin. Mikrobiol. Zh. (Kiev), 30: 390. *Via* Chem. Abs. 70: 35049v (1969).
53. Ezhov, N.V. & I.S. Novotel'nov. 1967. Hordecin, its properties and probable mechanism of action. Prikl. Biokhim. Mikrobiol. 3: 178. *Via* Chem. Abs. 67: 20143k (1967).
54. Babaeva, N.G. 1963. Antibacterial effect of various forms of *Hypericum perforatum*. Fiziol. Patol. Pishch. Sist. 251. *Via* Chem. Abs. 28861n (1970).
55. Gurevich, A.I., V.N. Dobrynin, M.N. Kolosov, S.A. Popravko, I.D. Ryabova, B.K. Chernov, N.A. Derbentseva, B.E. Aizenman, & A.D. Garagulya. 1971. Hyperforin, an antibiotic from *Hypericum perforatum*. Antibiotiki (Moscow), 16: 510.
56. Omel'chuk-Myakushko, T. & A. Ya. Fastovskaya. 1968. Dynamics of the accumulation of antimicrobic substances in *Hypericum perforatum*. Rast. Resur. 4: 346. *Via* Chem. Abs. 70: 9367c (1970).
57. Gonçalves de Lima, O., I.L. D'Albuquerque, M.A.P. Borba & G.M. Maciel. 1966. Antibiotic substances from higher plants. XXIV. Isolation and preliminary study of a diterpene component from *Ichthyothere cunabi*. Rev. Inst. Antibiot., Univ., Recife, 6: 17. *Via* Chem. Abs. 68: 47019g (1968).
58. Racz, G., I. Fuzi & L. Domokos. 1965. Antibiotic activity of the extracts from the aerial portion of the Romanian plant *Lysimachia nummularia*. Rev. Med. 11: *Via* Chem. Abs. 67: 14802h (1967).
59. Au, K.S. & D.E. Gray. 1969. New antibiotic. Biochem. Pharmacol. 18: 2673.
60. Toppozada, H.H., H.A. Mazloum & M. El-Dakhakhny. 1965. The antibacterial properties of *Nigella sativa* seeds. Active principle with some clinical applications. J. Egypt. Med. Asso., Spec. Number, 48: 187. *Via* Chem. Abs. 66: 92390r (1967).
61. LaLonde, R.T. 1970. Aquatic plant chemistry. Its application to water and pollution control. U.S. Clearinghouse Fed. Sci. Tech. Inform., P.B. Rep. 192810. *Via* Chem. Abs. 74: 39208b (1971).

62. Fleming, H.P. & J.L. Etchells. 1967. Occurrence of an inhibitor of lactic acid bacteria in green olives. Appl. Microbiol. 15: 1178.
63. Yochizawa, K., M. Itoh, K. Aida & K. Otsuka. 1970. Growth-inhibition against *Aspergillus oryzae* No. 40 by the compound III.3 in rice grain. Agr. Biol. Chem. 34: 1262.
64. Nicolls, J.M., J. Birner & P. Forsell. 1973. Passicol, an antibacterial and antifungal agent produced by *Passiflora* plant species: qualitative and quantitative range of activity. Antimicrob. Agts. Chemother. 3: 110.
65. Rudolf, K. & O. Cinar. 1971. Bacteriostatic compounds from bean leaves. Acta Phytopathol. 6: 105.
66. Hoff, R.J. 1970. Inhibitory compounds of *Pinus Monticola* resistant and susceptible to *Cronartium ribicola*. Can. J. Bot. 48: 371.
67. Dooley, T.P. & R.E. Gibson. 1966. Isolation of an antimicrobial substance from acorn extract. Antimicrob. Agts. Chemother. 480.
68. Trzenschik, U., R. Przyborowski, K. Hiller & B. Linzer. 1967. Components of some Saniculoideae. VII. Antimicrobial properties of *Sanicula* saponins. Pharmazie, 22: 715.
69. Zawahry, M.R. & N.S. El Din. 1965. Isolation of new crystalline active principles and fractions from certain Crucifereae plants cultivated in Egypt. Sci. Pharm. Proc. 25th. 1: 337. *Via* Chem. Abs. 69: 93666j (1968).
70. Pitts, O.M., H.S. Thompson & J.H. Hoch. 1969. Antibacterial activity of *Solanum carolinense*. J. Pharm. Sci. 58: 379.
71. Matzner, F. 1971. Quality and components of cultured American cranberry (*Vaccinium macrocarpon*), European cranberry (*V. oxycoccus*) and wild cranberry (*V. vitisideae*). Ind. Obst-Gemueseverwert. 56: 27. *Via* Chem. Abs. 75: 4222v (1971).
72. DeCapite, L. 1967. Histology, anatomy and antibiotic properties of *Vitex agnuscastus*. Ann. Fac. Agr. Univ. Studi Perugia, 22: 109. *Via* Chem. Abs. 71: 73980v (1969).
73. Gaind, K.N. & R.D. Budhiraja. 1967. Antibacterial and anthelmintic activity of *Withania coagulans*. Indian J. Pharm. 29: 185.
74. Stephanov, E.V. & M.A. Komarova. 1972. Composition and antimicrobial activity of volatile substances of a

preparation from *Abies sibirica* needles. Izv. Sib. Otd. Akad. Nauk S.S.S.R., Ser. Biol. Nauk. 38. *Via* Chem. Abs. 78: 616r (1973).

75. Jain, S.R. & A. Kar. 1971. The antibacterial activity of some essential oils and their combinations. Planta Medica, 20: 118.
76. Kar, A. & S.R. Jain. 1971. Antibacterial evaluation of some indigenous medicinal volatile oils. Qual. Plant. Mater. Veg. 20: 231. *Via* Chem. Abs. 75: 95772p (1971).
77. Rao, B.G.V.N. & S. Adinarayana. 1970. Antimicrobial action of some essential oils. Riechst., Aromen, Koerperpflegem. 20: 215. *Via* Chem. Abs. 73: 128145a (1970).
78. Korableva, N.P. & L.M. Potapova. 1966. Role of essential oils in resistance of onion to microorganisms. Biokhim. Osn. Zashch. Rast., Akad. Nauk SSSR, Inst. Biokhim. 79. *Via* Chem. Abs. 66: 92478a (1967).
79. Guven, K.C., A. Aktulga, & E.T. Cetin. 1972. Thin-layer chromatography and antibacterial activity of garlic oil. Eczacilik Bul. 14: 51. *Via* Chem. Abs. 78: 80159g (1973).
80. Vichkanova, S.A., V.V. Adgina & S.B. Izosimova. 1969. Essential oils as a source of new antifungal preparations. Fitontsidy, Mater. Soveshch. 6th, 123. *Via* Chem. Abs. 78: 67456g (1973).
81. Solov'eva, M., T.P. Berezovskaya, R.V. Usynina, E. Serykh, R. Uralova, V.V. Dudko & V.I. Velikanova. 1970. Chemical studies of Siberian *Artemisia* species and the possibility of their use in medicine. Postep Dziedzinie Leku Rosl., Pr. Ref. Dosw. Wygloszone Symp. 24. *Via* Chem. Abs. 78: 69217d (1973).
82. Nagy, J.G. & R.P. Tengerdy. 1967. Antibacterial action of essential oils of *Artemisia tridentata* and *Artemisia nova* on aerobic bacteria. Appl. Microbiol. 15: 819.
83. Gracza, L. & K. Szasz. 1968. Examination of active agents of marigold petals (*Calendula officinalis*). Acta Pharm. Hung. 38: 118. *Via* Chem. Abs. 69: 49756q (1968).
84. Gaind, K.N., R.D. Budhiraja & R.N. Kaul. 1966. Antibiotic activity of *Cassia occidentalis*. Indian J. Pharm. 28: 248.
85. Prakash, S., G.K. Sinha & R.C. Pathak. 1972. Antibacterial and antifungal properties of some essential oils extracted from medicinal plants of the Kumaon

region. Indian Oil Soap J. 37: 230. *Via* Chem. Abs. 79: 727y (1973).
86. Rao, B.G.V.N. & S.S. Nigam. 1970. *In vitro* antimicrobial efficiency of essential oils. Ind. J. Med. Res. 58: 627.
87. Perov, N.N. & A.N. Yatsyna. 1968. Antimicrobial properties of volatile compounds from grapes. Prikl. Biokhim. Mikrobiol. 4: 730. *Via* Chem. Abs. 70: 55178q (1969).
88. Chirkina, N.N. & T.P. Khort. 1968. Antibiotic activity of essential oils of some wild plants in the Crimea. Rast. Resur. 4: 186. *Via* Chem. Abs. 69: 99280a (1968).
89. Solodovnichenko, N.M. 1970. Chemical study of the essential oil of *Libanotis intermedia*. Khim. Prir. Soedin. 6: 768. *Via* Chem. Abs. 74: 102955k (1971).
90. Aggag, M.E. & R.T. Yousef. 1972. Antimicrobial activity of chamomile oil. Planta Med. 22: 140.
91. Sanyal, A. & K.C. Varma. 1969. *In vitro* antibacterial and antifungal activity of *Mentha arvensis* var. *piperascens* oil obtained from different sources. Indian J. Microbiol. 9: 23.
92. Kachoyan, V.I. & A.V. Abramyan. 1969. Effect of some essential oils on microorganisms. Biol. Zh. Arm. 22: 106. *Via* Chem. Abs. 72: 40099g (1970).
93. Toleva, P., M. Beshkov, E. Karova & G. Bozhkov. 1970. Physicochemical characteristics and antifungal activity of basil oil and its fractions. Nauch. Tr. Vissh. Inst. Khraint. Vkusova Prom., Plovdiv, 17: 49. *Via* Chem. Abs. 79: 35034x (1973).
94. Bondarenko, A.S., B.E. Aizenman, V. Prikhod'ko, A.A. Meshcheryakov, T.I. Skarbogat'ko, & E.L. Mishenkova. 1972. Antibiotic properties of the essential oil *Psoralea drupacea*. Mikrobiol. Zh. (Kiev), 34: 612. *Via* Chem. Abs. 79: 27846y (1973).
95. Khort, T.P. & T.I. Ruleva. 1969. *Satureia taurica*, a new wild, essential oil plant of Crimea. Rast. Resur. 5: 116. *Via* Chem. Abs. 70: 112400g (1969).
96. Allegrini, J., M. DeBuochberg, M. Simeon, H. Maillols & A. Boillot. 1973. Essential oil emulsions. Manufacture and application in microbiology. Trav. Soc. Pharm. Montpellier, 33: 73. *Via* Chem. Abs. 80: 19364v (1974).
97. Kornievskii, Yu. I., A.S. Rybal'chenko & M.V. Steblyuk. 1970. Antimicrobial properties of essential oils of *Valeriana stolonifera*, *nitida* and *exaltata*.

Zh. Mikrobiol., Epidemiol. Immunobiol. 47: 137. *Via* Chem. Abs. 74: 95967k (1971).

98. Subba, M.S., T.C. Soumithri & R.S. Rao. 1967. Antimicrobial action of citrus oils. J. Food Sci. 32: 225.
99. Joubert, L. & M. Gattefosse. 1968. Bactericidal properties of essential oils in veterinary preventive treatment and therapy. Mezhdunar. Kongr. Efirnym Maslam, (Mater.) 4th, 1, 99. *Via* Chem. Abs. 78: 119653r (1973.
100. Birggal, E. 1969. Essential oils as bactericidal and bacteriostatic agents. Medicinal plants in phytobalneology. Kosmet.-Parfum-Drogen Rundsch. 16: 109. *Via* Chem. Abs. 72: 59005v (1970).
101. Skvortsov, S.S. & V.A. Khan-Fimina. 1969. Effect of volatile antimicrobial substances of plants on micro- and macroorganisms. Fitontsidy, Mater. Soveshch. 6th, 207. *Via* Chem. Abs. 78: 38665u (1973).
102. Dabbah, R., V.M. Edwards & W.A. Moats. 1970. Antimicrobial activity of some citrus fruit oils on selected food-borne bacteria. Appl. Microbiol. 19: 27.
103. Dovgich, N.A. 1971. Antimicrobial effect of essential oils. Mikrobiol. Zh. (Kiev), 33: 253. *Via* Chem. Abs. 75: 72948m (1971).
104. Rao, B.G.V.N. 1971. Antimicrobial effect of some essential oils. IV. Effect of organic compounds. Riechst., Aromen, Koerperpflegem. 21: 10, 12, 14, 16. *Via* Chem. Abs. 74: 136975g (1971).
105. Ramadan, F.M., H.T. El-Zanfaly, A.M. Alian & F.A. El-Wakeil. 1972. Antibacterial effects of some essential oils. II. Semisolid agar phase. Chem. Mikrobiol. Technol. Lebensm. 1: 96. *Via* Chem. Abs. 77: 122533m (1972).
106. ______, R.T. El-Zanfaly, F.A. El-Wakeil & A.M. Alian. 1972. Antibacterial effects of some essential oils. I. Use of agar diffusion method. Chem. Mikrobiol. Technol. Lebensm. 2: 51. *Via* Chem. Abs. 77: 122532k (1972).
107. Rieche, A. 1967. Mustard oil formers. Sulfur Inst. J. 3: 16. *Via* Chem. Abs. 67: 96777r (1967).
108. Wolters, B. 1968. Antibiotic effect of saponins. IV. Antibiotic effect of neutral steroid glycosides with and without saponin characteristics. Planta Med., 16: 114.
109. Akhmedov, N.A. & S.F. Fakhrutdinov. 1967. Antibacterial

action of some alkaloids. Farmakol. Alkaloidov Glikozidov, 245. *Via* Chem. Abs. 70: 1162m (1969).

110. Radosevic, A. & D. Barkovic. 1969. Microbiological testing of berberine and some semisynthetic berberine homologs for antibacterial and antimycotic action. Acta Pharm. Jugoslav. 19: 75. *Via* Chem. Abs. 72: 75988y (1970).
111. Tukalo, E.A., E.S. Leplya, & B.J. Ivanchenko. 1972. Antibiotic action of glycoside alkaloid preparations. Biol. Nauki. 15: 99. *Via* Chem. Abs. 78: 93033m (1973).
112. Morris, O.N. 1972. Inhibitory effects of foliage extracts of some trees on commercial *Bacillus thuringiensis*. Can. Entomol. 104: 1357.
113. Vichkanova, S.A., M.A. Rubinchik & V.V. Adgina. 1971. Antimicrobial activity of sesquiterpene lactones from Compositae. Tr. Vses. Nauch.-Issled. Inst. Lek. Rast. 14: 230. *Via* Chem. Abs. 78: 155110v (1973).
114. Jurd, L., A.D. King, Jr., K.L. Mihara & W.L. Stanley. 1971. Antimicrobial properties of natural phenols and related compounds. I. Obtusastyrene. Appl. Microbiol. 21: 507.
115. _____, A.D. King, Jr. & K. Mihara. 1971. Antimicrobial properties of umbelliferone derivatives. Phytochemistry, 10: 2965.
116. _____, J. Corse, A.D. King, Jr., H. Bayne & K. Mihara. 1971. Antimicrobial properties of 6,7-dihydroxy-, 7,8-dihydroxy-, 6-hydroxy- and 8-hydroxycoumarins. Phytochemistry, 10: 2971.
117. Dadak, V. 1967. Antibiotic effectiveness of natural coumarins. VIII. Effect of ostruthin (6-geranyl-7-hydroxycoumarin) on *Saccharomyces cerevisiae*. Pharmazie, 22: 216.
118. Ohigashi, H., K. Kawazu, H. Egawa & T. Mitsui. 1972. Antifungal constituent of *Sapium japonicum*. Agr. Biol. Chem., 36: 1399.
119. Vanhaelen-Fastre, R. 1968. *Cnicus benedictus*. Separation of antimicrobial constituents. Plant. Med. Phytother. 2: 294. *Via* Chem. Abs. 70: 75114u (1969).
120. Birner, J. & J.M. Nicolls. 1973. Passicol, an antibacterial and antifungal agent produced by *Passiflora* plant species: Preparation and physiochemical characteristics. Antimicrob. Agts. Chemother. 3: 105.
121. Bondarenko, A.S., L.A. Bakina, E.M. Kleiner, V.I. Sheichenko, M.A. Gil'zin, A.S. Khokhlov, N.M.

Poddubnaya & T.I. Skorobogat'ko. 1968. Biological properties and chemical nature of the antibiotic from *Bidens cernua*. Antibiotiki, 13: 157. *Via* Chem. Abs. 68: 85073c (1968).
122. Allen, E.H. & C.A. Thomas. 1971. Trans-trans-3,11-tridecadiene-5,7,9-triyne-1,2-diol, an antifungal agent from diseased safflower (*Carthamnus tinctorius*) Phytochemistry 10: 1579.
123. Neeman, L., A. Lifshitz & Y. Kashman. 1970. New antibacterial agent isolated from the avocado pear. Appl. Microbiol. 19: 470.
124. Ulubelen, A., E.T. Cetin, S. Isildatici & S. Ozturk. 1968. Phytochemical investigation of *Paeonia decora*. Lloydia, 31: 249.
125. Mitscher, L.A., W.-N. Wu & J.L. Beal. 1973. The isolation and structural characterization of 5-0-methyl-desmethoxymatteucinol from *Eugenia javanica*. Lloydia, 36: 422.
126. Mityagina, Z.M., N.R. Pshenichnova, & Yu. P. Starikova. 1967. Tannal-type preparations from the speckled alder and their antibacterial properties. Tr. Peru. Farm. Inst. 51. *Via* Chem. Abs. 71: 19790v (1969).
127. Ozeretskovskaya, O.L., N.I. Vasyakova & M.A. Davydova. 1968. Antibiotic substances from potato tubers. Pribl. Biokhim. Mikrobiol. 4: 698. *Via* Chem. Abs. 70: 65208p (1970).
128. Margalith, P. 1967. Inhibitory effect of gossypol on microorganisms. Appl. Microbiol. 15: 952.
129. Bell, A.A. 1967. Formation of gossypol in infected or chemically irritated tissues of *Gossypium* (cotton) species. Phytopathology, 57: 759.
130. Parker, W.L. & F. Johnson. 1968. The structure determination of antibiotic compounds from *Hypericum uliginosum*. I. J. Amer. Chem. Soc. 90: 4716.
131. Meikle, T. & R. Stevens. 1972. Synthesis of the antibiotics from higher plants: pteleatinium chloride, a new quaternary quinoline alkaloid from *Ptelea trifoliata* with antitubercular and antiyeast activity. Chem. Commun. 1040.
132. Negrash, A.K. & P. Ya. Pochinok. 1969. Comparative study of chemotherapeutic and pharmacologic properties of antimicrobial preparations from common St. John's wort. Fitontsidy, Mater. Soveshch. 6th, 198. *Via* Chem. Abs. 78: 66908n (1973).
133. Teuber, M. 1970. Low antibiotic potency of isohumulone. Appl. Microbiol. 19: 871.

134. Shcherbanovskii, L.R. & G.I. Nilov. 1969. Plant antibiotics suppressing wine yeast and lactic acid and acetic acid bacteria. Fitonsidy, Mater. Soveshch. 6th, 109. *Via* Chem. Abs. 78: 53207q (1973).
135. Fikui, H., H. Egawa & K. Koshimizu. 1973. New antifungal isoflavone from immature fruits of *Lupinus luteus*. Agr. Biol. Chem. 37: 417.
136. Villanueva, V.R., M. Babier, M. Gonnet & P. Lavie. 1970. Propolis flavonoids. Isolation of a new bacteriostatic substance pinocembrin (5,7-dihydroxyflavanone). Ann Inst. Pasteur, Paris, 118: 84.
137. Ramaswamy, A.S., S. Jayaraman, M. Sirsi & K.H. Rao. 1972. Antibacterial action of some naturally occurring citrus bioflavonoids. Indian J. Exp. Biol., 10: 72.
138. King, A.D., Jr., H.G. Bayne, L. Jurd C. Case. 1972. Antimicrobial properties of natural phenols and related compounds: obtusastyrene and dihydro-obtusastyrene. Antimicrob. Agts. Chemother. 1: 263.
139. Mehta, G., U.R. Nayak & S. Dev. 1973. Oil from the seeds of *Psoralea corylifolia*. Tetrahedron, 29: 1119.
140. Shcherbanovskii, L.R. & G.I. Nilov. 1969. *Ceratostigma plumbaginoides*; producer of an antibiotic which suppresses the development of yeasts and lactic and acetic acid bacteria. Rast. Resur. 8: 112. *Via* Chem. Abs. 72: 87177w (1970).
141. Van der Vijver, L.M. & A.P. Loetter. 1971. Constituents in the roots of *Plumbago auriculala* and *Plumbago zeylanica* responsible for antibacterial activity. Planta Med. 20: 8.
142. Gonçalves de Lima, O., I. Leoncio d'Albuquerque, G.M. Maciel & C.N. Maria. 1968. Antimicrobial substances of higher plants. XXVII. Isolation of plumbagin from *Plumbago scandens*. Rev. Inst. Antibiot., Univ. Recife, 8: 95. *Via* Chem. Abs. 74: 39197x (1971).
143. Ikekawa, T., E.L. Wang, M. Hamada, T. Takeuchi & H. Umezawa. 1967. Isolation and identification of the antifungal active substance in walnuts. Chem. Pharm. Bull. 15: 242.
144. Ahmad, S., M.A. Wahid & A.Q.S. Bukhari. 1973. Fungistatic action of Juglans. Antimicrob. Agts. Chemother. 3: 436.
145. Gonçalves de Lima, O., I.L. D'Albuquerque, M.A.P. Borda & J.F. de Mello. 1966. Antibiotic substances from higher plants. XXV. Isolation of xiloidone (dehydro-

lapachone) by the conversion of lapachol in the presence of pyridine. Rev. Inst. Antibiot., Univ. Recife, 6: 23. *Via* Chem. Abs. 68: 95580e (1968).

146. Shcherbanovskii, L.R., G.I. Nilov, Z.D. Rabinovich & V.A. Gorina. 1972. Plant naphthoquinones as inhibitors of yeasts as well as lactic acid and acetic acid bacteria. Rast. Resur. 8: 112. *Via* Chem. Abs. 76: 136281u (1972).
147. Shukla, Y.N., J.S. Tandon, D.S.Bhakuni & M.M. Dhar. 1969. Constituents of the antibiotic fraction of *Arnebia nobilis*. Experientia, 25: 357.
148. Tschesche, R. 1971. Advances in the chemistry of antibiotic substances from higher plants, In "Pharmacognosy and Phytochemistry". (H. Wagner & L. Horhammer, eds.) p. 274. Springer, New York.
149. _____, F.J. Kaemmerer, G. Wulff & F. Schoenbeck. 1968. Antibacterially active substances of *Tulipa gesneriana*. Tetrahedron Lett. 701.
150. _____, H.H.C. Jha & G. Wulff. 1973. Triterpenes XXIX. The structure of the avenacines. Tetrahedron, 29: 629.
151. Goto, M., S. Imai, T. Murata, T. Noguchi & S. Fujioka. 1970. Antimicrobial glycosides of *Euptelea polyandra*. 1. Isolation, constitutions and antimicrobial activities of eupteleoside A and B. Yakugaku Zasshi, 90: 736. *Via* Chem. Abs. 73: 84655k (1970).
152. Juneja, T.R., K.N. Gaind & A.S. Panesar. 1970. *Capparis decidua*. Study of isothiocyanate glucoside. Res. Bull. Panjab Univ., Sci. 21: 519. *Via* Chem. Abs. 77: 72529s (1972).
153. Constantine, G.H., Jr., P. Catalfomo, K. Sheth & L.A. Sciuchetti. 1966. Phytochemical study of *Arctostaphylos columbiana* and *Arctostaphylos patula*. J. Pharm. Sci. 55: 1378.
154. Frohne, D. 1969. Urinary disinfectant activity of bearberry leaf extracts. Planta Med. 18: 1.
155. Vanhaelen-Fastre, R. 1972. Antibiotic and cytotoxic activities of cnicin, isolated from *Cnicus benedictus*. J. Pharm. Belg. 27: 683.
156. Tocan, V. & O. Baron. 1969. Antibiotic activity of protoanemonin from *Ranunculus oxyspermus*. Boll. Chim. Farm. 108: 789. *Via* Chem. Abs. 73: 84762t (1970).
157. Tshchesche, R. & H.J. Hoppe. 1971. Glycosides with lactone-forming aglycones. V. Nartheside A and B, two diastereoisomeric lactone-glycosides and their

antibiotic aglycone from bog asphodel (*Narthecium ossifragum*). Chem. Ber. 104: 3573.
158. Bergman, B.H.H. & J.C.M. Beijersbergen. 1968. Fungitoxic substance extracted from tulips and its possible role as a protectant against disease. Neth. J. Plant Pathol. 74: 157. *Via* Chem. Abs. 72: 97395e (1970).
159. _____, J.C.M. Beijersbergen, J.C. Overeem & A. Kaars Sijesteijn. 1967. Isolation and identification of α-methylenebutyrolactone, a fungitoxic substance from tulips. Rec. Trav. Chim. Pays-Bas. 86: 709.
160. Schroeder, C. 1972. Host-parasite relation of tulips with *Botrytis* species. I. Stability and activity of the tuliposides. Z. Pfanzenkr. Pflanzenschutz. 79: 1.
161. Kowalewski, Z., W. Kedzia & I. Mirska. 1972. Effect of berberine sulfate on *Staphylococci*. Arch. Immunol. Ther. Exp. 20: 353.
162. Stoessl, A. & C.H. Unwin. 1970. Antifungal factors in barley. V. Antifungal activity of the hordatines. Can. J. Bot. 48: 465.
163. Steffens, H. 1972. Alkaloid-containing antibiotic extracts from *Sinapis alba*. Ger. Offen. 2,046,756. *Via* Chem. Abs. 77: 52311m (1972).
164. Mitscher, L.A., H.D.H. Showalter, M.T. Shipchandler, R.P. Leu & J.L. Beal. 1972. Antimicrobial agents from higher plants. IV. *Zanthoxylum elephantiasis*. Isolation and identification of canthin-6-one. Lloydia, 35: 177.
165. _____, M.S. Bathala & J.L. Beal. 1971. Antibiotics from higher plants: pteleatinium chloride, a new quaternary quinoline alkaloid from *Ptelea trifoliata* with antitubercular and antiyeast activity. Chem. Commun. 1040.
166. _____, W.-N. Wu, R.W. Doskotch & J.L. Beal. 1972. Antimicrobial agents from higher plants. II. Alkaloids from *Thalictrum rugosum*. Lloydia, 35: 167.
167. Lewis, A. & R.G. Shepherd. 1970. In "Medicinal Chemistry". (A. Burger, Ed.) 3rd ed., p. 409. Wiley (Interscience), New York.
168. Gharbo, S.A., J.L. Beal, R.W. Doskotch and L.A. Mitscher. 1973. Alkaloids of Thalictrum. XIV. Isolation of alkaloids having antimicrobial activity from *Thalictrum polygamum*. Lloydia, 36: 349.
169. Mitscher, L.A., W.-N. Wu, & J.L. Beal. 1972. Antibiotics from higher plants. *Thalictrum rugosum*. Thalrugosa-

mine, a new *bis*benzylisoquinoline alkaloid active vs. *Mycobacterium smegmatis*. Experientia, 28: 500.
170. Gal, I.E. 1969. Antibacterial effect of capsaicin. Elelmiszervizsgalati Kozlem. 15: 80. *Via* Chem. Abs. 71: 120796b (1969).
171. Johri, B.N. 1973. Evaluation of the effect of various substances on the action of cyathin by paper-strip technique. Hindustan Antibiot. Bull., 15: 104. *Via* Chem. Abs. 80: 10705z (1975).
172. Khanna, P. & T.N. Nag. 1973. Isolation, Identification and screening of phyllemblin from *Emblica officinalis* tissue culture. Indian J. Pharm. 35: 23. *Via* Chem. Abs. 78: 133346z (1973).
173. Borkowski, B., Z. Kowalewski, W. Wozniak & B. Murawska. 1965. Glucoerysolin. Pol. 50: 185. *Via* Chem. Abs. 66: 5749v (1967).
174. Santhanam, K. & P.L.N. Rao. 1969. Antibiotic principles of *Garcinia morella*. XIII. Antimicrobial activity and toxicity of α_1- and γ-guttiferins and their derivatives Indian J. Exp. Biol. 7: 34.
175. Sani, B.P. & P.L.N. Rao. 1969. Antibiotic principles of *Garcinia morella*. XIV. Neomorellin and X-guttiferin. Indian J. Chem. 7: 680.
176. Verma, S.C.L. & P.L.N. Rao. 1967. Antibiotic principles of *Cannabis sativa* cell suspension cultures. Lloydia, 35: 450.
177. Gonçalves de Lima, O., I. Leoncio d'Albuquerque, J.S. Coelho, D.G. Martins, A.L. Lacerda & G.M. Maciel. 1969. Antimicrobial substances in higher plants. XXXI. Antimicrobial and antitumoral activity of maytenin isolated from the cortical root zone of *Maytenus* species. Rev. Inst. Antibiot., Univ. Fed. Pernambuco, Recife, 9: 17. *Via* Chem. Abs. 75: 95370f (1971).
178. ____, I. Leoncio d'Albuquerque, J.S. de B. Coelho, G.M. Maciel, M. da S.B. Calvalcanti, D.G. Martins & A.L. Lacerda. 1969. Antimicrobial substances in higher plants. XXX. Antimicrobial and antineoplastic activity of pristimerin isolated from *Prionostemma aspera*, from the humid bushes of the Pernambuco region. Rev. Inst. Antibiot., Univ. Fed. Pernambuco, Recife, 9: 3. *Via* Chem. Abs. 75: 95369n (1971).
179. Rubinchik, M.A. & S.A. Vichkanova. 1969. Antitrichomonad activity of lutenurin, an antibiotic of plant derivation. Antibiotiki (Moscow), 14: 926.

Chapter Eleven

STRUCTURE OF THE INSECT ANTIFEEDANT AZADIRACHTIN

KOJI NAKANISHI

Department of Chemistry
Columbia University
New York, New York

INTRODUCTION

Preliminary studies using the leaf extracts of *Azadirachta indica* A. Juss (*Melia azadirachta* L., *Melia indica*, neem or nim tree) collected in Mombasa, Kenya, showed that when the crude extract was applied to coffee leaves, its potency in causing morphological changes in *Antestiopsis* (coffee bug) was stronger than that of an equivalent weight of β-ecdysone (insect molting hormone). It was during attempts to isolate the active compound that we encountered azadirachtin.

Azadirachtin was first isolated by Butterworth and Morgan from the seeds of *M. azadirachta* and the closely related species *M. azedarach* using a feeding inhibition test for *Schistocerca gregaria* (desert locust).[1,2] The limiting concentration to cause 100% inhibition of feeding is 40 μg/l, or, when impregnated on to filter paper,

1 ng/cm^2 [3]. Azadirachtin has also been shown to be an antifeedant* and systemic growth disruptor against insects other than the desert locust[4,5]. According to feeding assays carried out by Drs. G. Ludwick and G. Staal, Zoecon Corp., on *Heliothis virescens* (tobacco bud-worm) larvae, and feeding injection assays carried out by Prof. H. Röller, Texas A&M University, on *Pieris brassica* (cabbage moth) adults, *Galleria mellonella* larvae and *Teticuli termis* (termite) larvae, azadirachtin is a feeding inhibitor; however, it appears not to be the compound possessing stronger hormonal activity than β-ecdysone.

As described above, azadirachtin is a potent insect antifeedant. In addition, the following properties are positive attributes when it is considered from the viewpoint of practical application as an antifeedant[6,7,8,9]. Namely: although it has a complex structure, it is readily available, i.e., 300 g of the plant seeds afford 800 mg of pure azadirachtin, and the tree is quite common in Africa and India; it is presumably nontoxic to birds since it is their favorite fruit, and to humans because the twigs have long been used as a chewing stick to prevent tooth infection; it is chemically unstable. Obviously further studies are necessary to explore its practical usage.

TRITERPENOIDS AND LIMONOIDS FROM THE NEEM TREE

The neem tree has been the subject of numerous studies because it has long been used in India and Africa for various insect repellent[10,11,12,13,14,15] and medicinal[16,17] purposes. Lavie and co-workers have shown that meliantriol[18], a triucallane (20S) triterpenoid (see Fig. 1) which was isolated from the fruit, showed distinct antifeeding activity against the desert locust, *Schistocerca gregaria* Forsk. They have also isolated other triterpenoids having the same carbon skeleton,[19] and have converted the euphane skeleton into an apo-euphane skeleton (8β-Me) by reaction with perbenzoic acid (7,8-α-epoxide formation) followed by $SnCl_4$

---

*An antifeedant functions through contact and therefore insects may approach the target but do not feed on it. On the other hand, a repellent travels through air and prevents the approach of insects.

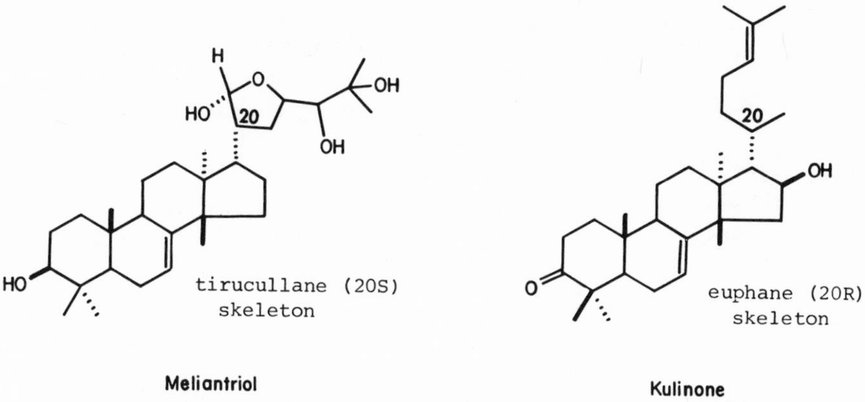

tirucullane (20S) skeleton

Meliantriol

euphane (20R) skeleton

Kulinone

Figure 1. Structures of meliantriol and kulinone.

cleavage (7α-OH/8β-Me/14-ene), thus showing the biogenetic relation between the melianes (e.g., meliantriol) and meliacins (Limonoids, see Fig. 2).[20] More recently, Chiang and Chang[21] have determined the structures of four euphane (20R) triterpenoids, exemplified by kulinone (see Fig. 1), which were isolated from the bark.

In addition to the triterpenoids described above, numerous limonoids have been isolated from the leaves, fruit, wood, bark and seed oil of this tree. Limonoids are tetranortriterpenoids (C_{26}) which lack C-24 to C-27 of the apo-euphane skeleton, and contain a terminal furan ring (the first member in this group to be fully characterized was limonin[22], the bitter principle of lemon).

The seed oil has given nimbin,[23] salannin[24] (see Fig. 2), meldenin,[25] azadirone,[26] azadiradione,[26] and epoxyazadiradione.[26] Nimbolin B,[27] which was isolated together with nimbolin A[27] (see Fig. 2) from the trunk wood, is biogenetically related to the other C-seco compounds, nimbin and salannin. The leaves have given another C-seco limonoid, nimbolide[28] while the total amorphous bitter has given nimbidinin,[29] a 12-keto compound.

Azadirachtin is another limonoid having a complex structure with 16 oxygen atoms. Morgan and co-workers[30]

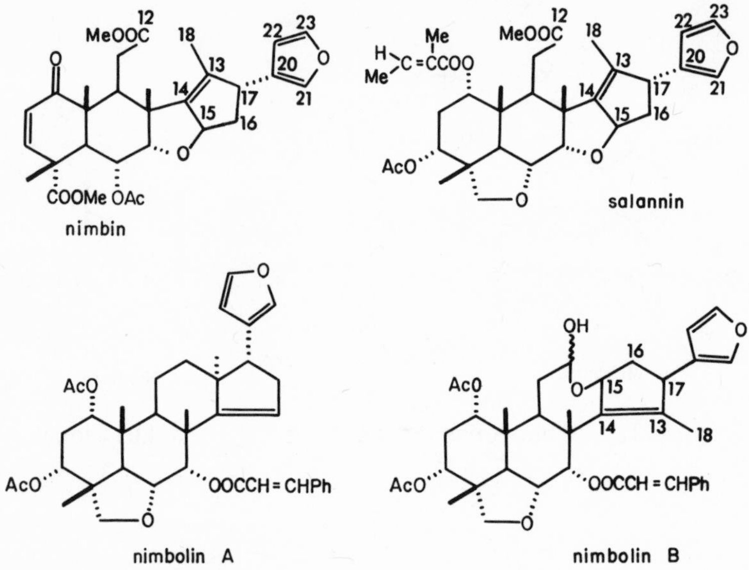

Figure 2. Structures of nimbin, salannin, nimbolin A, and nimbolin B.

assigned to azadirachtin a molecular formula of $C_{35}H_{44}O_{16}$, and also the partial structures and functional groups shown in Figure 3. The difference between these results and the structure derived in the following[31] is the point of attachment of the ring A acetate and tiglate groups, and the fact that the second acetate is actually a quaternary methyl group having its pmr signal at the exceptionally low field of 2.06 ppm (found by cmr). All other findings are in agreement with the derived structure.

ISOLATION

The neem berries (1 Kg) collected in Mombasa, Kenya were air-dried and the seeds (300 g) were separated from the flesh by waiting for the flesh to become rotten. (In our hands, we were unable to isolate azadirachtin by mascerating the whole fresh berries, possibly because of enzymatic degradation.) The seeds were mascerated in a

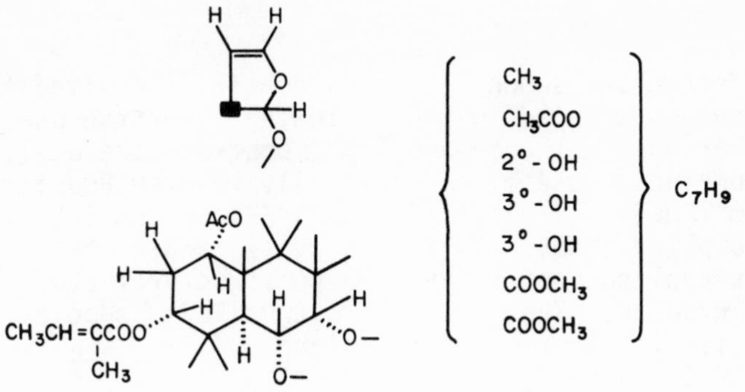

Figure 3. Partial structure of azadirachtin ($C_{35}H_{44}O_{16}$).

blender with 1 liter of ethanol, the slurry was filtered, and the residue was re-extracted with another liter of ethanol. The combined extracts were concentrated in vacuo at 40^{0}, and the residual oil was partitioned between 300 ml each of ethyl acetate and water to extract protein material. The ethyl acetate layer was concentrated to give 19.2 g of oil, which was partitioned between 300 ml each of 5% aqueous methanol and hexane. Concentration of the methanol layer gave 11.6 g of neem oil.

Chromatography of the oil on a column of silica gel (600 g) prepared in ether, using 3% ether in acetone as eluant, first gave fractions containing salannin and then azadirachtin. The azadirachtin fraction was rechromatographed on 200 g of silica gel and eluted with a 1:1 mixture of chloroform and ethyl acetate; preparative thin-layer chromatography, developed twice consecutively in ether and then once in 1:1 chloroform/ethyl acetate gave 800 mg of azadirachtin as an amorphous solid. The salannin fraction was also submitted to preparative tlc, the plates being developed twice with 40:1 chloroform/methanol; silica gel chromotography of the major tlc band with 50:1 chloroform/methanol afforded 450 mg of salannin.

STRUCTURAL STUDIES

The structure shown for azadirachtin is derived from the evidence presented in the following. We have used this rather abundant compound to demonstrate the utility of the combined usage of cmr partially relaxed Fourier transform (PRFT)[32,33,34,35,36,37,38,39] and continuous wave decoupling (CWD). As far as we are aware, this is the first application of PRFT/CWD to structural studies of natural products. The structure with a 12/13 seco carbo-skeleton is closely related to salannin.

Azadirachtin is unstable in solution, and merely leaving a $CDCl_3$ solution in an nmr tube for several days or warming to 40° for 30 min in the nmr tube (in an attempt to sharpen some signals) resulted in decomposition. Several attempts at hydrolysis, modification, or derivatization did not yield crystalline products.

The uv band is due to the tiglate and methoxycarbonyl groups (Fig. 4). Infrared studies by varying the concentration in chloroform disclosed the presence of an intramolecularly hydrogen-bonded hydroxyl (3465 cm^{-1}) as well as a free (3580 cm^{-1}) intermolecularly hydrogen-bonded (3380 cm^{-1}) hydroxyl group.

Extensive pmr and cmr studies were carried out on azadirachtin and its 14-acetate. The proton systems were studied by making measurements at 100 MHz and 220 MHz in $CDCl_3$, benzene-d_6, pyridine-d_5, and DMSO-d_6, or mixtures of these solvents, by addition of $Eu(fod)_3$, and by consideration of cmr splitting patterns. Although evidence for the proton systems was based on a combination of various measurements, only $CDCl_3$ data are used in the following discussion for the sake of simplicity (Fig. 5).

It was readily shown that azadirachtin contains a tiglate, an acetate,* two quaternary methyl, two methoxy-

*The cmr, Figure 8, showed that there are only four acyl carbonyls and hence one of the singlets flanking 2 ppm should be assigned to a quaternary methyl. An expanded pmr showed that the 1.92 ppm signal was sharp whereas the 2.06 ppm was broad; hence they were assigned to acetate and methyl respectively.

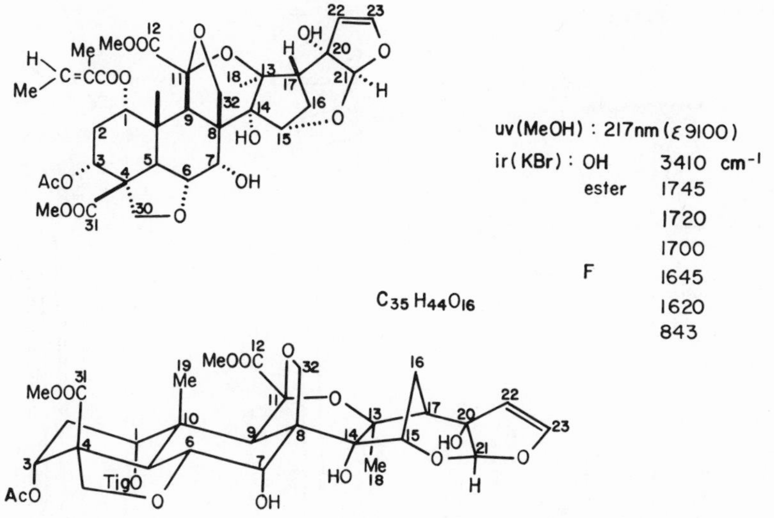

uv(MeOH) : 217nm (ε 9100)

ir(KBr) : OH 3410 cm^{-1}

ester 1745

1720

1700

F 1645

1620

843

$C_{35}H_{44}O_{16}$

Figure 4. Structure and spectral characteristics of azadirachtin.

carbonyls and two isolated methines (9-H and 21-H) (Fig. 6).

In addition it contains the following five proton systems (Figs. 5,6):

i) CH_2 group: a pair of doublets at 3.65/4.16 ppm, AX type, J 9.5 Hz; C-32 CH_2.

ii) CH_2 group: a pair of doublets at 3.75/4.05 ppm, AX type, J 10 Hz; C-30 CH_2.

iii) C-1/C-2/C-3: The half-band width of triplets at 4.76 and 5.49 ppm showed these protons to be equatorial, i.e., axial acyl groups. The same pattern is seen in the pmr of salannin.

iv) C-5/C-6/C-7: A similar pattern is present in the pmr of salannin. Irradiation of a 3.04 ppm signal (hydroxyl, attached to C-7) sharpened the 7-H signal at 4.62 ppm; tickling the 7-H peak collapsed the 4.58 signal (6-H, doublet of doublets, 3 and 12 Hz) into a doublet

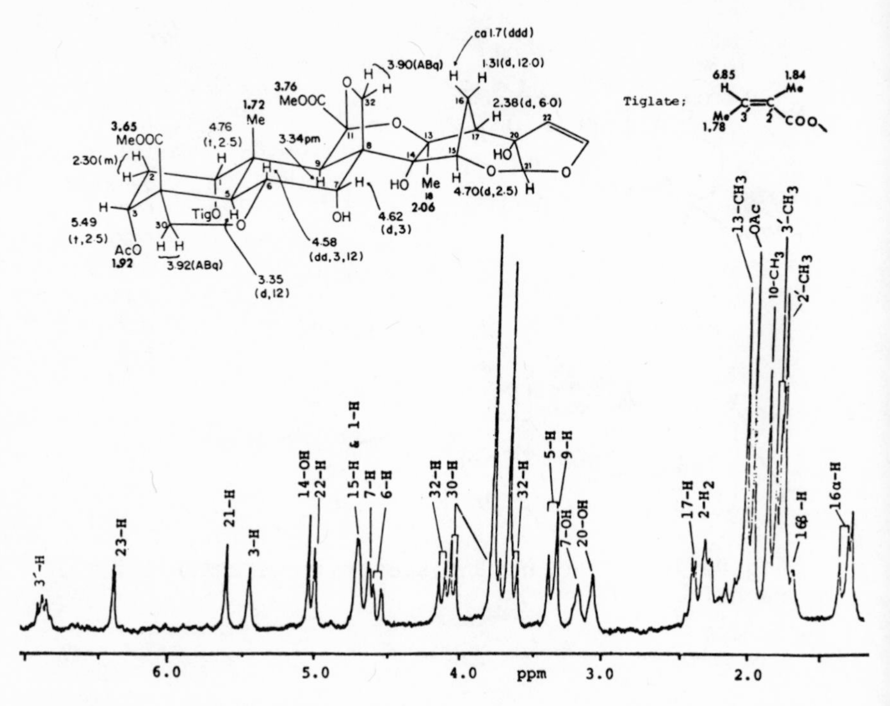

Figure 5. 220 MHz pmr spectrum ($CDCl_3$) of azadirachtin.

(12 Hz), whereas irradiation at 3.35 ppm (5-H) converted the 6-H signal to a doublet (3 Hz).

v) C-15/C-16/C-17: See following for comments.

Three hydroxyl protons were present in the pmr spectrum. Namely, it showed two *tert*-OH signals at 3.04 ppm (20-OH, DMSO-d_6, s, 5.29 ppm) and 5.07 ppm (14-OH, DMSO-d_6, s, 6.33 ppm; low-field due to H-bonding) and one *sec*-OH signal at 3.04 ppm (7-OH, DMSO-d_6, d, 4.89 ppm; this OH is coupled to 7-H as described above).

C-15/C-16/C-17: This four-proton system (Fig. 7) was the most difficult and became clearer only after consideration of the cmr and decoupling of pmr with and without addition of $Eu(fod)_3$. Thus, single frequency irradiation (or narrow band irradiation) at the pmr 1.5 ppm, or midway

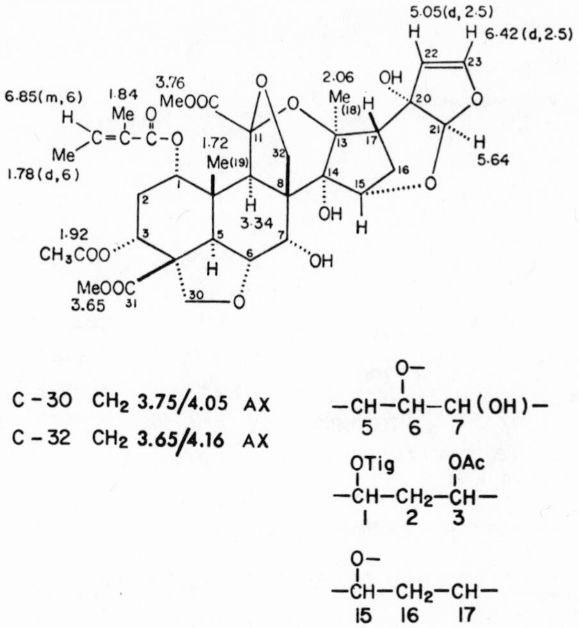

C – 30 CH_2 3.75/4.05 AX

C – 32 CH_2 3.65/4.16 AX

Figure 6. Partial assignments of azadirachtin structure.

between the 16α-H and 16β-H signals (Fig. 5) converted the 25.06 ppm cmr signal into a singlet. Since this 25.06 ppm signal was shown to be due to a methylene carbon by off-resonance decoupling or narrow band decoupling (as depicted in Fig. 8, it appears as a positive triplet upon narrow band irradiation of the pmr at 3.5 ppm) it is clear that the two pmr signals flanking 1.5 ppm, at 1.31 ppm and ca. 1.7 ppm, are the methylene protons attached to the 29.37 ppm carbon. Irradiation of the partially hidden 1.70 ppm 16β-H peak collapsed the 4.70 ppm 15-H, 2.38 ppm 17-H, and 1.3. ppm 16α-H signals into singlets; double irradiation at 4.70 and 2.38 ppm converted the 1.70 ppm 16β-H signal into a 12 Hz doublet.

The nature of the 35 carbons was elucidated by a combination of cmr techniques, i.e., proton noise decoupling (PND, Fig. 8), continuous wave decoupling (CWD, or off-resonance decoupling), partially relaxed Fourier transform

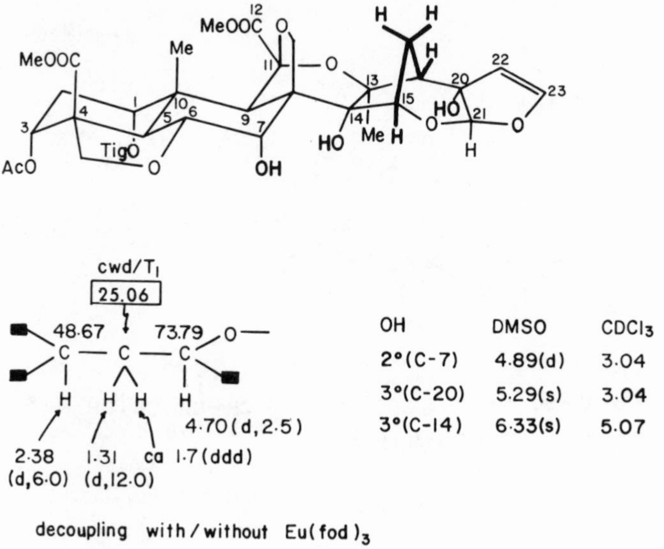

OH	DMSO	$CDCl_3$
2°(C-7)	4.89(d)	3.04
3°(C-20)	5.29(s)	3.04
3°(C-14)	6.33(s)	5.07

Figure 7. Assignment of protons at C-15, C-16, and C-17.

(PRFT), and combined PRFT/CWD. The T_1 relaxation times of ^{13}C contained in alicyclic molecules usually increase in the sequence of CH_2, CH, CH_3 and C with no protons. Thus in the "inversion recovery" or $(180^o - \tau - 90 - T)_n$ sequence, if a suitable interval time τ is chosen, it is possible to separate the cmr peaks into negative and positive peaks. This results in simplification of the congested region in the cmr spectra. In Figure 8 we see that at τ 0.5 sec, only the following peaks which are all due to carbonyl or quaternary carbons appear as inversed negative peaks: the four acyl carbonyl carbons, tiglate C-2' at 128.6, C-11 (ketal C) at 104.1, C-20 at 83.55, C-13 at 69.95, C-14 at 68.53, C-10 at 52.52, C-8 at 50.19, and C-4 at 45.41 ppm.

For measurements of the PRFT/CWD spectrum shown in Figure 8, a narrow band irradiation was applied to the pmr 3.5 region, or close to the 32-H (3.90 ppm) and 5-H (3.35 ppm) signals. Thus, in addition to being separated into positive and negative peaks, the cmr peaks all show residual coupling, if any. Therefore, this not only simplifies the congested region, but also greatly facilitates

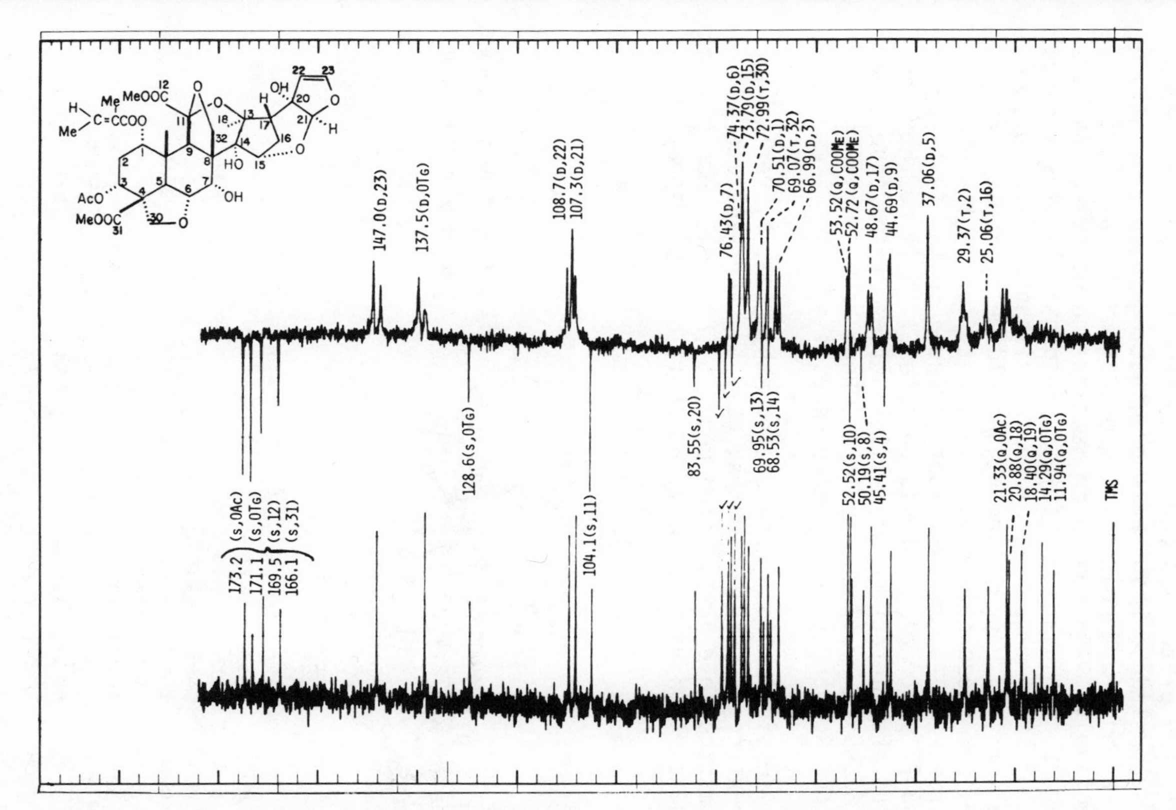

Figure 8. Upper trace: PRFT/CWD ($CDCl_3$) spectrum (8,192 scans). Lower trace: PND ($CDCl_3$) spectrum (32,768 scans).

detection of peak multiplicities.

As the proton 3.5 ppm region (see Fig. 5) was narrow-band decoupled, carbon peaks originating in carbons attached to 32-H and 5-H appear as singlets at 69.07 and 37.06 ppm, respectively. The remaining methine and methylene carbons appear as doublets and triplets, respectively, the residual coupling being proportional to the distance of the proton signals from 3.5 ppm. The high-field methyl peaks are nulled but these can be readily identified in the PND spectrum. The simplifications achieved in the 44-54 ppm and 66-77 ppm regions should be noted. It is not hard to imagine that detection of multiplicities of the five peaks from 66.99 to 70.51 ppm would not have been easy in a straight-forward off-resonance decoupling.

Acetylation with pure acetic anhydride, by refluxing under nitrogen for 10 minutes, gave the *tert*-acetate at C-14 as the sole acetylation product (Fig. 9). Acetylation at the *tert*-OH is presumably a result of acyl migration

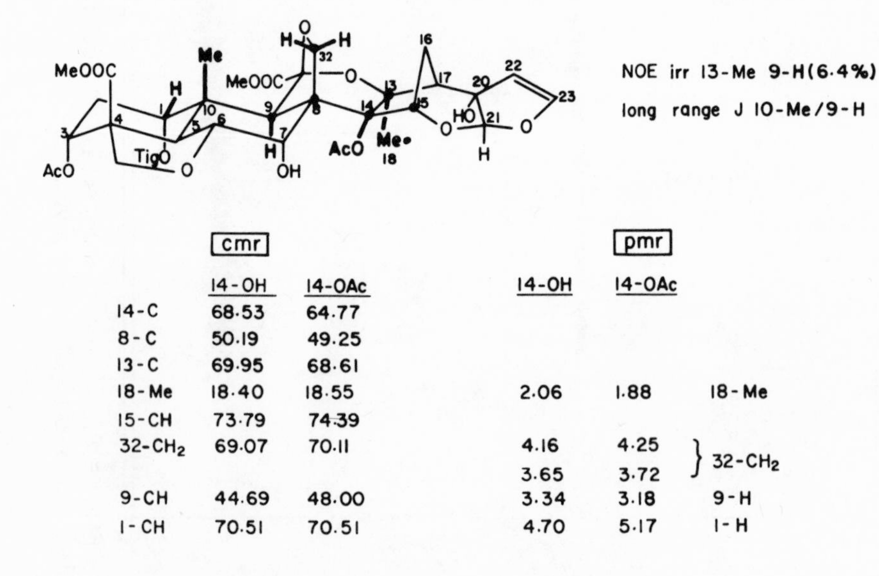

	cmr 14-OH	cmr 14-OAc	pmr 14-OH	pmr 14-OAc	
14-C	68.53	64.77			
8-C	50.19	49.25			
13-C	69.95	68.61			
18-Me	18.40	18.55	2.06	1.88	18-Me
15-CH	73.79	74.39			
32-CH_2	69.07	70.11	4.16	4.25	32-CH_2
			3.65	3.72	
9-CH	44.69	48.00	3.34	3.18	9-H
1-CH	70.51	70.51	4.70	5.17	1-H

Figure 9. Acetyl-azadirachtin.

from 7-OH, the further 7-acetylation being blocked by steric hindrance. Nine cmr peaks underwent significant shifts upon 14-acetylation, and these are listed in Figure 9 (14-C, 8-C, 13-C, 18-Me, 15-CH, 32-CH_2, 9-CH). These carbons, which are marked by black circles in the structure shown, should be placed in close proximity, i.e., α or β, to the tertiary acetoxylated C-14. Variations in the direction of shifts, both of the cmr peaks and pmr peaks, are probably due to different conformations of the six-membered ring comprising the C-11/C-13 ether bridge as well as the loss of H-bonding of 14-OH upon acetylation.

An expected observation was the shift of the 1-H pmr signal at 4.70 ppm in azadirachtin to 5.17 ppm in the acetate. Since the C-1 cmr signal remained constant at 70.51 ppm, the shift must be due to an anisotropic effect, caused by different orientations of the tiglate or the 11-COOMe groups.

The configuration of the C-15/C-16/C-17 bridge should be β since, in the numerous limonoids and euphane/triucallanes isolated from various plant sources, the 17-side chain configuration is always α; the 15-oxygen bond is then necessarily α in order to form the bridged system. Configuration of the dihydrofuran ring should be β, with the 20-OH *cis* to 13-Me in view of the low-field methyl shift (2.06 ppm); if it were α, the 13-Me would be at a higher field due to shielding by the 22-ene.

Finally, a long-range coupling was observed between 10-Me and 9-H, and irradiation of the 13α-Me signal resulted in a 6.4% NOE at the 9-H 3.34 ppm signal.

In spite of the complexity arising from ketal and ether linkages, the structure thus derived is biogenetically closely related to salannin, nimbolin B and nimbin (Fig. 10).

ACKNOWLEDGMENT

The studies were initiated by Dr. D.L. Elder during his stay (1970-1973) as a Research Scientist at the International Centre of Insect Physiology and Ecology (ICIPE), Nairobi, Kenya, and were continued by Dr. P.R. Zanno, Mr. I. Miura and Dr. Elder at Columbia University. We are grateful for support by NIH grant AI-10187. We are also

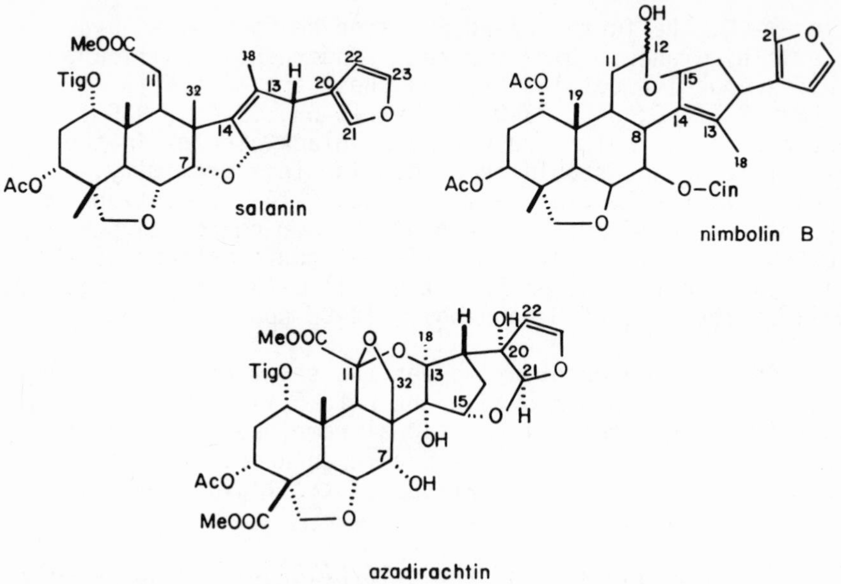

Figure 10. Structures of salannin, nimbolin B and azadirachtin.

grateful to Drs. W. Hauserman and W.F. Wood, ICIPE, for the collection of seeds. The work on Antestiopsis (coffee bug) was carried out in collaboration with K. Leuschner, Kianibu Research Station Foundation, Kenya.

REFERENCES

1. Butterworth, J.H. & E.D. Morgan. 1968. Chem. Commun. 23.
2. Morgan, E.D. & M.D. Thornton. 1973. Phytochemistry, 12: 391.
3. Butterworth, J.H. & E.D. Morgan. 1971. J. Insect Physiol. 17: 969.
4. Gill, J.S. & C.T. Lewis. 1971. Nature, 232: 402.
5. Ruscose, C.N.E. 1972. Nature New Biol. 236: 159.
6. Fraenkel, G. 1959. Science, 129: 1466.
7. Thorsteinson, A.J. 1960. Ann. Rev. Entomol. 5: 193.

8. Beck, S.D. 1965. Ann. Rev. Entomol. 10: 207.
9. Munakata, K. 1970. In "Control of Insect Behavior by Natural Products" (D.L. Wood, R.M. Silverstein & M. Nakajima, eds.) p. 179. Academic Press, New York.
10. Volkonsky, M. 1937. Arch. Inst. Pasteur, Alger. 15: 427.
11. ____. 1937. C.R. Soc. Biol. Paris, 127: 417.
12. Chauvin, R. 1946. C.R. Soc. Biol. Paris, 222: 412.
13. Krishnamurti, B. & D.S. Rao. 1950. Bull. Agric. Coll. Res. Inst., Mysore, 14: 1.
14. Shinha, N.P. & K.C. Gulati. 1968. Bull. Reg. Res. Lab., Jammu, India, 1: 76.
15. McMillan, W.W., M.C. Bowman, R.L. Burton, K.J. Starks & B.R. Wiseman. 1969. J. Econ. Ent. 62: 708.
16. Shopra, R.M. 1958. "Indigenous Drugs of India" 2nd ed., p. 360. Dhur, Calcutta.
17. Watt, J.M. & M.G. Breyer-Brankwijk. 1962. "The Medicinal and Poisonous Plants of Southern and Eastern Africa" 2nd ed., p. 745. E. & S. Livingstone, Edinburgh & London.
18. Lavie, D., M.K. Jain & S.R. Schpan-Gabrielith. 1967. Chem. Commun. 910.
19. ____, M.K. Jain & I. Kirson. 1967. J. Chem. Soc. (C), 1347.
20. ____, & E.C. Levy. 1971. Tetrahedron, 27: 3941.
21. Chiang, C.-K. & F.C. Chang. 1973. Tetrahedron. 29: 1911.
22. Arigoni, D., D.H.R. Barton, E.J. Corey, O. Jeger, L. Caglioti, S. Dev, R.G. Ferrini, E.R. Glazier, A. Melera, S.K. Bradham, K. Schaffner, S. Sternhell, J.F. Templeton & S. Tobinaga. 1960. Experientia, 16: 41.
23. Harris, M., R. Henderson, R. McCrindle, K.H. Overton & D.W. Turner. 1968. Tetrahedron, 24: 1517.
24. Henderson, R., R. McCrindle, A. Melera & K.H. Overton. 1968. Tetrahedron, 24: 1525.
25. Connolly, J.D., K.C. Handa & R. McCrindle. 1968. Tetrahedron Lett. 437.
26. Lavie, D. & K.J. Mahendra. 1967. Chem. Commun. 278.
27. Ekong, D.E.U., C.O. Fakunle, A.K. Fasina & J.I. Okogun. 1969. Chem. Commun. 1166.
28. ____. 1967. Chem. Commun. 808.
29. Mitra, C.R., H.S. Garg & G.N. Pandey. 1970. Tetrahedron Lett. 32: 2761.
30. Butterworth, J.H., E.D. Morgan & G.R. Percy. 1972. J. Chem. Soc., Perkin I: 2445.
31. Zanno, P.R., I. Miura, K. Nakanishi & D.L. Elder. 1975.

J. Am. Chem. Soc. (in press).
32. Vold, R.L., J.S. Waugh, M.P. Klein & D.E. Phelps. 1968. J. Chem. Phys. 48: 3831.
33. Freeman, R. & H.D.W. Hill. 1970. J. Chem. Phys. 53: 4103.
34. Allerhand, A., D. Doddrell, V. Gushko, D.W. Cochran, E. Wenkert, P. J. Lawson & F.R.N. Gurd. 1971. J. Am. Chem. Soc. 93: 544.
35. Doddrell, D. & A. Allerhand. 1971. Proc. Nat. Acad. Sci., U.S. 68: 1083.
36. Levy, G.C., J.D. Cargioli & F.A.L. Anet. 1973. J. Am. Chem. Soc. 95: 1572.
37. Wehrli, F.W. 1973. J. Chem. Soc. (C), 379.
38. Nakanishi, K., V.P. Gullo, I. Muira, T. Govindachari & N. Viswanathan. 1973. J. Am. Chem. Soc. 95: 6473.
39. _____, R. Crouch, I. Muira, X. Dominguez, A. Zamudio & R. Villarreal. 1974. J. Am. Chem. Soc. 96: 4348.
40. Levy, G.C. & G.L. Nelson. 1972. "Carbon-13 Nuclear Magnetic Resonance for Organic Chemists" p. 182. Wiley-Interscience, New York.

INDEX